W0234658

Worldwide Trends in Green Chemistry Education

Worldwide Trends in Green Chemistry Education

Edited by

Vânia Gomes Zuin
Federal University of São Carlos, São Paulo, Brazil
Email: vaniaz@ufscar.br

Liliana Mammino
University of Venda, Thohoyandou, South Africa
Email: sasdestria@yahoo.com

THE QUEEN'S AWARDS
FOR ENTERPRISE:
INTERNATIONAL TRADE
2013

Print ISBN: 978-1-84973-949-8
PDF eISBN: 978-1-78262-194-2

A catalogue record for this book is available from the British Library

© The Royal Society of Chemistry 2015

All rights reserved

Apart from fair dealing for the purposes of research for non-commercial purposes or for private study, criticism or review, as permitted under the Copyright, Designs and Patents Act 1988 and the Copyright and Related Rights Regulations 2003, this publication may not be reproduced, stored or transmitted, in any form or by any means, without the prior permission in writing of The Royal Society of Chemistry or the copyright owner, or in the case of reproduction in accordance with the terms of licences issued by the Copyright Licensing Agency in the UK, or in accordance with the terms of the licences issued by the appropriate Reproduction Rights Organization outside the UK. Enquiries concerning reproduction outside the terms stated here should be sent to The Royal Society of Chemistry at the address printed on this page.

The RSC is not responsible for individual opinions expressed in this work.

The authors have sought to locate owners of all reproduced material not in their own possession and trust that no copyrights have been inadvertently infringed.

Published by The Royal Society of Chemistry,
Thomas Graham House, Science Park, Milton Road,
Cambridge CB4 0WF, UK

Registered Charity Number 207890

Visit our website at www.rsc.org/books

Printed and bound by CPI Group (UK) Ltd, Croydon, CR0 4YY

Foreword

Green Chemistry Education: Worldwide Trends Amidst Changing Times

The time is right to draw the attention of chemists, educators, and others to the global status of green chemistry education. Timely, because of the mismatch between the everyday practice of chemistry teachers at the secondary and post-secondary level and high profile interrelated global initiatives that are guiding scientific and public sustainability discourse. Timely also, because of the opportunity presented to transform that educational practice, to take green and sustainable chemistry out of the aside boxes in textbooks and the margins of curriculum, and infuse it through the body of knowledge included in student learning outcomes and assessments.

While relatively little change is evident over the past several decades in curricular emphases in chemistry, interdisciplinary science is pressing forward with two important initiatives that should push scientific understandings of sustainability onto the agenda of formal and informal science educators. The first initiative rewrites our understanding of the times we live in on our planet, by moving the clock ahead on the geological time scale. An International Union of Geological Sciences blue-ribbon working group of the Sub-commission on Quaternary Stratigraphy is expected to report by 2016 on whether sufficient scientific evidence is present to formally determine that we have moved from the relatively stable interglacial Holocene Epoch to the Anthropocene Epoch [Greek 'anthropo-' (human), and '-cene' (new)], on the geological time scale. Many expect the determination to be that we are in the Anthropocene already, an epoch on the geological time scale that is defined by the human imprint. A leading candidate for the beginning of this epoch is the industrial revolution, when we observe the beginning of steep and steady rises in numerous chemical parameters related to our planetary life support

Worldwide Trends in Green Chemistry Education
Edited by Vânia Gomes Zuin and Liliana Mammino
© The Royal Society of Chemistry 2015
Published by the Royal Society of Chemistry, www.rsc.org

systems. A second, interconnected initiative is the systematic attempt to define and quantify 'planetary boundaries', the state of earth system parameters that define a safe operating space for humanity.

Is there a community of research and practice that is better equipped to give leadership in connecting these two global interdisciplinary scientific initiatives to chemistry educational practice than the green chemistry community? Green chemistry philosophy and principles, formally articulated two decades ago, have been put forward out of concern that the everyday practice of chemistry be fundamentally transformed so as to start with sustainability and safety considerations. For green chemistry to take firmer hold, the next generation of educators, scientists, and citizens needs to own the philosophy and embed it into practice. To move ahead we need to understand where we are, and this volume presents an important snapshot of trends in world-wide green chemistry education.

Contributions to this title cover a wide range of green chemistry education initiatives on different continents, and include descriptions of formal and informal learning environments at secondary, post-secondary, and tertiary levels. Green chemistry education is appropriately situated relative to global sustainability education initiatives such as the Decade of Education for Sustainable Development, which ends the year this title is published. Connections are made to disciplines such as toxicology, and the crucial and often neglected area of assessment receives attention, with presentation of metrics for the 'greenness' of chemistry teaching.

The contributions in this book provide an important global snapshot of the progress being made in greening chemistry education practice, and point the way toward the important steps that are still needed to make mainstream chemistry education more relevant to the future of our planet.

Peter Mahaffy
The King's University, Alberta, Canada

Preface

Green chemistry education can be considered one of the hottest themes in our time. As is well known, green chemistry aims at the design, production and use of substances that are non-hazardous and at the design and use of environmentally benign production processes, in the perspective of sustainable development. This constitutes one of the most innovative and challenging tasks worldwide. Green chemistry education aims at incorporating information about green chemistry into chemical education, thus being called to design suitable options for all the broad educational areas—curriculum development, teaching, learning and outreach—and their specific components, from in-class activities to laboratory experiments to the dissemination of information to the public. A major objective of green chemistry education is to foster sustainable scientific literacy and to develop the corresponding skills among the present and future generations.

With this book, we aim at considering key issues of green chemistry education through the presentation of research, practices and theoretical reflections in different contexts, by educators from different countries and continents, *i.e.*, Austria, Brazil, Canada, England, Germany, Israel, Malaysia, Portugal, Russia, South Africa, Spain, Thailand and the USA. Our intention is that of offering a panorama of approaches and highlighting the connections between the general objectives of green chemistry education and the design of pedagogical options at different academic and school levels, apt for the characteristics of each individual experience and simultaneously interesting for other contexts. Presenting concrete didactic activities from different realities gives the opportunity to consider a variety of diverse possibilities for the incorporation of green chemistry education into chemical education. The book includes analyses of concrete experiences from the educational point of view, as well as general theoretical reflections on the approaches and on their suitability to promote the desired types of awareness in the young

Worldwide Trends in Green Chemistry Education
Edited by Vânia Gomes Zuin and Liliana Mammino
© The Royal Society of Chemistry 2015
Published by the Royal Society of Chemistry, www.rsc.org

generations, keeping in mind the importance of social and environmental sustainability (nowadays and in the future) and the role that Chemistry can play to promote sustainable development.

The first part of the book considers the significance of green chemistry, green chemistry education, sustainable development, education for sustainable development, and other crucial issues, and a variety of corresponding approaches. This is followed by the presentation of a number of current initiatives in, or designed for, secondary school level. The attention given to the teaching of the green chemistry and sustainability concepts at basic education level is presently inadequate, and this needs to change. Teacher training courses and other training initiatives constitute an excellent opportunity to raise the profile of secondary school green chemistry education, and can conveniently incorporate experiences from the undergraduate and postgraduate university levels, with suitable adaptations. We believe that, by presenting a panoramic of challenges and possible responses and offering an updated insight into the most recent trends in green chemistry education worldwide, this book may constitute a valuable resource not only for chemical educators specifically interested in green chemistry education, but also for scientists, students, professionals, industrialists and policy-makers. We really hope that the readers will enjoy the direct contact with the experiences presented.

We wish to express our sincere gratitude to Merlin Fox, Alice Toby-Brant, Rowan Frame and Marisa Sartori for their fruitful cooperation and dedicated efforts in supporting the preparation of this book.

Vânia Zuin and Liliana Mammino

Contents

Worldwide Trends in Green Chemistry Education
Edited by Vânia Gomes Zuin and Liliana Mammino
© The Royal Society of Chemistry 2015
Published by the Royal Society of Chemistry, www.rsc.org

A Great Challenge of Green Chemistry Education: The Interface between Provision of Information and Behaviour Patterns

LILIANA MAMMINO*[a]

[a]Department of Chemistry, University of Venda, Thohoyandou, South Africa
*E-mail: sasdestria@yahoo.com

1.1 Introduction

Green chemistry[1-3] aims at promoting environmentally benign patterns, a change that is essential for development to be sustainable. In line with the nature of chemistry as the science of substances, green chemistry is concerned with all the stages of the 'life' of a substance or a material: production, utilization and final disposal. For the production stage, green chemistry aims at designing inherently safer substances and less-polluting manufacturing processes. Pursuing these objectives falls within the technical domain of the design of substances and processes and, therefore, it concerns chemistry research and the chemical industry. After the production stages, the rest

Worldwide Trends in Green Chemistry Education
Edited by Vânia Gomes Zuin and Liliana Mammino
© The Royal Society of Chemistry 2015
Published by the Royal Society of Chemistry, www.rsc.org

of the *life* of substances and materials is in the hands of those who use them. Fostering informed and sustainable ways of handling them relies solely on education. Thus, green chemistry education needs to provide chemistry information in such a way that it may influence people's behaviour.

The importance of green chemistry education has been recognized since the *birth* of green chemistry.[4,5] Early recommendations already stressed the need for it to be 'both inside and outside academia'.[4] The last decade has witnessed enormous growth in approaches, projects and resource materials aimed at familiarizing pupils and students with the principles of green chemistry and with a variety of new, green industrial approaches. Their number is too high for a meaningful review within the space of a chapter. Several initiatives have also had an impact on behaviour patterns within specific communities (for instance, progressive *greening* of university campuses in some contexts). However, the extent to which the new messages have reached the general public, or have impacted on large-scale behaviour patterns, is still inadequate. This makes the interface between the provision of information and the actual promotion of sustainable behaviour patterns one of the major challenges currently facing green chemistry education. Meeting this challenge requires novelties in the educational approaches, with the objective of integrating the provision of information with a stimulation of awareness capable of influencing attitudes and behaviour patterns. The fundamental role of the provision of information goes hand in hand with the importance of stressing the meaning and role of chemistry. The main criteria in the design of educational (or dissemination-of-information) approaches may imply diverse aspects such as:

- Stressing the fundamental message that the handling of substances in everyday life is part of the broad domain of chemistry and, therefore, chemistry information is essential for proper handling, and green chemistry criteria apply to it. Recommendations concerning substances and materials (such as those written on their containers) are chemistry-based and, because of this, they need to be taken into account carefully, to ensure appropriate usage and appropriate disposal once they have finished their useful period.
- Efficiently highlighting the interplay between the two conceptual categories of 'general' and 'particular': the *general* (global) perspective of the environmental impact of certain actions and the *particular* perspective of the choices by individual persons or individual communities.
- Enabling sufficiently ample interfaces with ethics education, so as to provide motivations for sustainable behaviour patterns. This is an important pathway for trying to answer the often unspoken question of why an individual should care about what happens globally or what will happen in the future.
- Devoting attention to observable behaviour patterns. This implies observation of what occurs in one's surroundings, reflections on what is observed, and the design of approaches to foster the replacement of observed non-sustainable aspects with more sustainable ones.

The chapter considers concrete examples within both formal and informal education. The examples for formal education refer to efforts to integrate both industrial and everyday life green chemistry perspectives within chemical education, and are analysed in some detail. They comprise the integration of green chemistry perspectives into a process technology course at the University of Venda (South Africa), and the presentation of the interface between chemistry and ethics to secondary school pupils in Italy. The examples for which informal education needs to play extensive roles focus mostly on aspects for which the *outreach* to the public appears so far inadequate, and make extensive references to observations that can be made in one's surroundings. These examples suggest the importance of fostering chemical literacy and integrating green chemistry perspectives into information to the public. Some possible chemistry-oriented *outreach* options are outlined.

1.2 Green Chemistry Perspectives in a Process Technology Course

Green chemistry information and perspectives have been introduced for several years into the process technology course taught by the author at the University of Venda (UNIVEN). The context is an underprivileged one (what in South Africa is called a *Historically Black University*, HBU). Despite recent improvements in several respects, there are still difficulties related to the past (apartheid period) *lower* status of the university. Furthermore, the university mostly serves a poor rural community, which implies many of the disadvantages common to underprivileged communities. Students experience a variety of difficulties: general underpreparedness, difficulties related to poor language mastery and to the communication challenges typical of second language instruction; and the overall scarcity of learning skills and acquired mastery of essential learning tools, which goes under the comprehensive concept of inadequate *epistemological access*.[6,7] This ensemble of problems cumulatively results in generalized passive attitudes and a strong tendency to equate learning to passive memorization, both of which are also deeply rooted in the approaches of pre-university instruction.

The process technology course is a third year course providing the bases of chemical engineering. It has been considered the most apt course for the incorporation of green chemistry perspectives both for its content (directly related to the chemical industry) and because it is apt for explorative or pilot interventions, as it is not a large-enrolment course. The incorporation is realized in such a way as to engage students actively, which is considered essential for the acquired information to have an impact beyond the preparation of tests and exams. The practical approach is conceptually simple. It focuses on the twelve principles of green chemistry.[2] Students are invited to choose a principle (a different principle for each student) and to prepare a poster or a Power-Point presentation considering both industrial and everyday life implications of that principle. There are sessions during the semester, in which students can discuss the progress in the preparation of their presentation and ask for suggestions, so that guidance is provided for all the steps preceding the presentation. The posters are prepared individually, but the discussion sessions are common, to favour interchanges not only about practical challenges, but also about the content on which each students is working. The posters are presented at the end of the semester and are objects of assessment.

The overall approach has several advantages. It engages students actively, as they need to search for information, to design how to organize it and how to present it, and to be able to answer questions on it, after presenting. The request that they consider both industry and everyday life broadens the overall perspective and facilitates the recognition of parallelisms between the significance of the green chemistry principles for the industry and for everyday life.

The initiative has been implemented through the last ten years. In the UNIVEN context, it is so far the first occasion in which chemistry students encounter green chemistry. The impact has been different in different years (with different groups of students). In general, it has stimulated reflections on the relationships between chemistry knowledge and everyday handling of substances and materials and on the importance of considering the impacts of our actions on the environment. In some cases, the impact on students' perceptions and attitudes has gone beyond the recognition of the importance of these aspects, motivating students to search for ways to disseminate information beyond the campus, to the community and to younger (pre-university) pupils. It is interesting to note that this type of interest and commitment beyond the requirements of the course (*i.e.*, beyond doing a certain activity in order to pass the course) is perceived as something pertaining to the fact that they are (or are in the process of becoming) chemists. This is an important and desirable effect, as it links chemistry knowledge to sustainable behaviour and to a perception of a chemist's individual responsibility not only to comply with sustainable-behaviour criteria, but also to promote this attitude in their community.

1.3 Relating Ethics and Chemistry with Secondary School Pupils

An experience at presenting green chemistry to young pupils in the framework of the relationships between chemistry and ethics proved particularly successful. The school concerned was a *Scientific Lyceum* in Treviso (Italy), and the initiative involved senior pupils (16–19 years age). The overall initiative was a one-day conference on chemistry and ethics, titled *Ethics and Chemistry: a Feasible Dialogue*. It was organized by the chemistry teacher (Prof. Michele Zanata, assisted by the students themselves), and involved the participation of speakers from different backgrounds and countries, including academics (both chemists and a philosopher), representatives of chemists' professional associations and representatives from the industry. This enabled the consideration of the relationships between ethics and chemistry from a variety of diverse perspectives.

The author of this chapter contributed with a presentation titled 'Ethics and chemistry: the choices of research and the choices of citizens'. The title aimed at immediately highlighting the importance of two major conditions to enhance sustainability: chemical research, which can provide better substances and better processes; and citizens' behaviour, which determines other relevant aspects. The presentation itself aimed at stimulating awareness of the two essential aspects of ethical behaviour—wanting to do what is good and knowing how to do it[8]—and of the implications with regard to chemistry and to the production and use of substances and materials. These included the importance of chemical (and science) literacy to be able to make informed choices (knowing how *to do good*), and the importance of individual behaviours for global effects (a reason for *wanting to do good*). After an extensive introduction on the nature and purposes of green chemistry (including the presentation of its ten principles), the presentation focused on the sources of pollution (something in which pupils were specifically interested) and on the importance of choosing sustainable behaviour patterns. A number of images of environmental pollution were selected and combined with captions aimed at stimulating reflection, by conveying the main message in an expectedly impressive way. The major message was that pollution is not generated only by the industry, but also by the overall effect of the behaviour of a high number of individual persons. The selected images had the following subjects:

- A river polluted by industrial wastes
- A factory emitting huge clouds of black smokes from its chimneys
- An oil spill from an oil tanker
- A traffic jam, with a panoramic of a huge number of cars queuing from different directions at a cross-roads
- A river polluted by detergents
- An 'island' of plastic bags in the middle of the Pacific Ocean.

The captions for the first three images were 'This is due to industry'; the captions for the last three images were 'This is due to the choices of many normal citizens' (fourth image), 'This is due to the activities of many normal citizens' (fifth image) and 'This is due to the carelessness of many normal citizens' (sixth image). The aim was that of conveying the message that chemistry research can do something (hopefully a lot) to make industry more sustainable; but citizens also need to take responsibility for the ways in which they handle substances and materials.

The pupils' response was very positive. They showed active interest and asked many questions both in the question time after the presentation and informally later on. Several questions focused on chemical aspects ('What happens if...?'), showing that the main messages had gone through. Questions asked informally, after the sessions, showed pupils' remarkable prior exposure to the issue of chemistry and the environment: their chemistry teacher had put considerable efforts in this direction, stimulating curiosity and reflections as part of their overall attitude. The information on green chemistry added the information that it is possible to use chemistry to protect the environment, and also conveyed the message that a lot of research is still needed, and that sustainable behaviour is an ethical issue requiring adequate chemistry literacy to be pursued effectively. The general theme of the conference stressed the importance of cross-discipline and holistic thinking, a key contribution to the pupils' overall formation. A presentation of green chemistry within such a perspective is particularly suitable because it highlights a variety of cross-discipline aspects and their significance for sustainable behaviour patterns.

1.4 From Observations to Design: The Route to Effective Educational Approaches

1.4.1 Observation, Reflection and Design

Educational approaches need to be designed on the basis of observations and diagnoses, to respond more effectively to the characteristics of the target groups. This is true both for formal instruction (where the target groups are pupils or students) and for informal education (where the target groups may be specific groups of persons, or entire communities). When educational approaches, or approaches aimed at disseminating information, are meant for the general public, it is important to take into account existing attitudes and behaviours as the starting point.

Informal education is not delegated only to persons who are 'officially' in charge of it. Each person can make a number of observations/diagnoses by devoting careful attention to the surroundings, considering one's own choices and the choices of the persons around. Observations lead to reflections. Reflections provide the basis to design approaches, which can be implemented through direct communication (*e.g.*, talking between individual persons), or through inclusion into educational approaches and material development, if one is engaged in education. Two components are particularly important in such processes:

- The consideration that, in most cases, environmentally unfriendly choices are based on inadequate information, or inadequate awareness of the importance of the choices of each person
- The importance of underpinning any recommendation or suggestion on sound scientific information.

Many people still tend to consider that those who talk about the environment have mostly aesthetic and emotional motivations (liking nature as it is, loving trees, forests, and animals, or other similar reasons). These motivations, although important for those who perceive them, do not have a sufficiently significant impact on others, when communicated as such. Only scientific information can stimulate the awareness that environmental issues are important for our health, for the general economy in our society and for the wellbeing of the future generations. The relevant scientific information has mostly a chemical core, although significant interfacing contributions may come from mathematics, biology, medicine, economics, and other disciplines. A number of basic examples important for everyday life will be briefly considered in the next subsections, to highlight how chemistry information can be incorporated as the scientific basis to stimulate changes in behaviour patterns. The selection of the examples, and of the corresponding suggestions, is based on direct experience in different contexts. Therefore, the themes of the examples are not treated in an exhaustive way (what would require much more space than that of a chapter), but as a rather fast overview of possibilities.

1.4.2 The Selection of Transport

The selection of transport is a crucial issue because cars are currently the major source of air pollution. A huge number of cars on the roads implies the generation of huge amounts of pollutants and greenhouse gases. Using alternative transport responds to the criteria of reducing the generation of pollutants and utilising energy efficiently (the efficient use of energy is one of the principles of green chemistry). Public transport (buses, trains) constitutes the optimal choice for long distances. Bicycles are the best choice for sufficiently short distances, on non-rainy days.

In some countries, the use of bicycles is common and/or increasing. In other countries, it is viewed as a symbol of poverty and avoided altogether. Paradoxically, it may happen that a person actively involved in green chemistry research (in terms of green processes and syntheses) shows a total lack of understanding of why one would or should choose to move by bicycle, even for short distances, if one owns a car. The status-symbol perceptions in relation to transport overshadow considerations in terms of energy consumption, pollution or even simply personal health (riding a bicycle is surely healthier than driving a car). These perceptions are widely diffuse in countries where emerging economies are currently enabling people to emerge 'out of poverty'. 'Poverty' and what it implies remains the subconscious reference, and the wish to 'separate oneself' from the features typical of poverty becomes the dominant subconscious feature motivating choices.

It becomes important to disseminate information about the advantages of bicycles both to reduce the generation of air pollution and for our own health in general. The information needs to be based on scientific data, and to report and explain them in a way accessible to the audience. It may also be important to consider the subconscious motivations, an aspect that could envisage interesting collaborations between chemists, chemical educators and psychologists.

1.4.3   The Use of Air Conditioning

Air conditioning has high energy demands and is not friendly to our health. In some contexts/countries, it is utilized only when temperatures are extreme, and this can be considered reasonable usage. In other contexts, it is utilized throughout the year, including the long periods in which the outdoor temperature is comfortable and opening windows would be the ideal choice. The reasons behind this may be various, from a passive (unquestioned) acceptance of habits acquired since childhood (more frequent in some developed contexts) to the perception that air conditioning is one of the markers of getting 'out of poverty' (more frequent in developing contexts). There may be paradoxical situations, like periods in which, in some countries (*e.g.*, in southern Africa, including South Africa) the electricity supply is not adequate for the needs of the community, and yet people prefer to have periods without electricity (the so-called 'shadings'), with all the inconveniences that they involve, rather than opting for switching off an appliance (air conditioning) that consumes most of the power in normal (non-industrial) buildings.

Education requires dissemination of information, which would ideally target the following issues:

- The high energy demands of air conditioning and the impacts they have on the environment. This requires the explanation of the environmental impacts of energy production.
- The effects of air conditioning on our health. Many persons complain about the negative effects, realize that they are caused by air conditioning, but have not yet reached the stage for which they may decide to switch it off (they would be too different from the other persons in their surroundings), or to request the right to have natural ventilation (there have been cases in which some employees have opted in this way, but they are still rare).[9]
- Promoting a rational utilization, limited to when the outdoors temperature is extreme.
- Promoting the awareness of the importance of natural ventilation for the indoor air quality. It is the only way of getting rid of indoor pollutants and replacing the oxygen that is consumed by respiration. It is significant that, for instance, in a big city in Australia, a company is experimenting with an innovative double system: using air-conditioning during the day, because employees have a psychological need for it, and opening the windows during the night, to improve the indoor air quality and, thus, take better care of the employees' health.

The awareness of both the energy implications (huge consumption) and the health implications (negative impacts) should be the key for which scientific information may gradually change a highly non-sustainable behaviour pattern.

1.4.4 The Attitude Towards Trees

The attitude towards trees varies largely in different contexts and communities. Planting trees is recognized as one of the few effective and realistic ways currently available to fight climate change. However, many persons do not want trees in their neighbourhood, or chop down the existing ones, because of reasons as diverse as considering their leaves as something disorderly or untidy, or fearing that spirits might choose big trees as their residence (Figures 1.1 and 1.2).

The main information to be disseminated concerns the chemical nature of photosynthesis; it produces oxygen, something that many people know. It simultaneously traps carbon dioxide, removing it from the environment. This is something that is not always part of common awareness, and public

Figure 1.1   An attitude towards trees. In Italy somebody is insisting that this 52-year-old pine tree should be chopped down, because she considers trees to be untidy in an urban context.

Figure 1.2   A similar attitude towards trees. In South Africa some persons considered that the high number of weavers' nests on top of a tall tree made
the tree look untidy. Those persons managed to get the 50-year-old
large tree (more than 1 metre diameter) chopped down, although the
location was within a biodiversity preserve.

perception does not always associate it with the prevention of climate change.
The chemical equation of photosynthesis, $6CO_2 + 6H_2O \rightarrow C_6H_{12}O_6 + O_2$,
should become the main tool (even with the role of slogan) to stimulate the
awareness of the importance of planting new trees and protecting the existing ones to try and slow down climate change. It has the potential to become
a high-impact tool because of being chemistry-based and, simultaneously,
referring to nature. The overall objective is a huge increase in the number of
trees in all locations, including urban trees and woods.

1.4.5   The Attitude Towards Saving

The attitude towards saving varies widely in different contexts. Saving
resources is a sound economic principle (besides being a fundamental green
chemistry criterion). However, it is not always adhered to. Attitudes are more
negative in contexts currently coming out of poverty, because, then, saving is
perceived as a necessity associated with poverty, of which to get rid as poverty
decreases. For instance, the lack of attempts to save energy, even in contexts
where power failures are frequent because of inadequate electricity supply,
can be related to the general attitude towards saving. The dissemination of
information needs to focus on the economic soundness of saving resources,
on the limitedness of global resources, and on the overall impact of the sum
of many individual actions, although each of them may have an individually
tiny impact.

   Besides this attitude problem, there may also be a lack of awareness for specific cases. Recalling the importance of saving specific resources on specific

occasions may then have considerable impacts. For instance, from direct experience, the notice that 'One piece of tissue should be enough to dry your hands', in an airport in Brazil stimulated the wish to check whether that was true. It proved true (the size of the pieces was adequate) and it prompted the habit not to use more tissue than needed.

1.4.6   The Attitude Towards Garbage Disposal

The attitude towards garbage disposal also varies widely in different contexts. Separate collection and recycling is practised only in some contexts, mostly highly 'developed' ones. Sometimes, at the moment when it is initiated, it may be viewed as an unnecessary imposition or nuisance; soon after, the awareness grows and nobody would turn back to different, less rational ways of disposal. It is important to promote separate collection and recycling everywhere; to convince citizens of the importance of this practice; to find economic incentives (for their high convincing role); and to relate it to science-based motivations.

Recycling is one of the green chemistry principles. Thus, chemistry and everyday life may use the same perspective to work for sustainability.

1.4.7   The Handling of Substances and Materials

The handling of substances and materials has enormous impacts on the environment. We handle a lot of substances and materials: household products; products for the garden; pesticides, fertilizers and weed-killers in agriculture; medicines; and many others. Too often, substances are not handled or disposed of according to recommendations, or the main utilisation criterion seems to be 'the more, the better'. Some chemistry literacy is necessary to understand:

- The reasons why 'the more, the better' is not true, and how appropriate usage combines saving with best results, decreases environmental impacts and simultaneously lowers costs
- The reasons and importance of the usage recommendations usually given on the containers.

Safety is a key issue in the utilization of substances and materials. Using more than needed may decrease safety, *e.g.*, by increasing the risks of side effects from handling. But also utilizations contradicting recommendations may become a safety issue. For instance, it is not safe to mix even common household materials, if we are not in a position to predict the outcomes. Some kids enjoy making small explosions by mixing common household products; this should by itself be a warning symptom about the importance of being cautious when mixing products. However, some persons mix household materials in the hope of improving the overall action (*e.g.*, the disinfection action), without being in a position to predict which substances may form from the mixing, and openly contradicting the recommendations written on the containers. Being cautious is an outcome of basic chemical literacy, for which people become aware that adequate chemical knowledge would be needed to perform operations like mixing different products safely.

1.4.8   Relating Individual/Local and Global Perspectives

Promoting the awareness of the overall outcomes resulting from the sum of a high numbers of individual behaviours is a crucial issue. It is not easy to foster this awareness, because of the immediately perceivable disproportion between the massiveness of the global outcome and the tiny-ness of the impact of the individual action. Some persons say it straightforwardly: 'If I waste one A4 page, it is only few grams, it is negligible', or 'If I do not save one litre of water now, it is negligible'. It is true. The problem is that the sum of a very high number of negligible amounts results in a huge overall amount.

How to educate in this regard? Mathematics may help, for instance by calculating the sum of the effects of individual actions by multiplying the effect of one action by the number of citizens in a country. One A4 page is probably 5 g of paper. If, in a country of 50 million citizens, each person saves one A4 page in a day, it amounts to $(5 \text{ g}) \times (50\,000\,000) = 250\,000\,000 \text{ g} = 250\,000 \text{ kg}$ of paper. This is a huge amount to save. If, on the other hand, all those 50 million citizens decide that their contribution is not relevant, that amount cannot be saved.

Visualization may also contribute to stimulate awareness. For instance, the previously mentioned image of the 'island' of plastic bags in the middle of the ocean may have a strong visual impact, as it shows what can be the final outcome when familiar objects like plastic bags are not disposed of correctly by millions of persons.

1.4.9   Considering 'Protocols' Critically

Many practices end up acquiring the role of protocols that are passively followed as routines, without evaluating when they are reasonable and when they become unreasonable. The previously mentioned use of air conditioning whatever the outdoor temperature can be viewed as a telling example. Many others can be identified simply by 'looking around us'.

For instance, cutting the grass with weekly or two-week frequency during the rainy season in a garden in a tropical area has a meaning. Doing the same thing during the dry season becomes unfriendly both to the environment and to people's health. Figure 1.3 shows a case of this type: the gardener, instructed to follow the same protocol all year round, is performing the grass-cutting operation in the dry season, when the soil is mostly barren and exposed; there is no grass worth considering, and the grass-cutting tool lifts huge amounts of soil-dust for several metres into the air, often higher than

Figure 1.3 Non-sustainable behaviour related to uncritical implementation of protocols: a worker performs the operation of cutting grass in the dry season, when the soil is nearly bare and the machine lifts enormous clouds of dust.

the nearby houses. The unsustainability of the option is self-evident: unnecessary consumption of fuel and increase in the particulate level in the air.

1.5   Some Key Educational Features

The design of interventions aimed at stimulating environmentally sustainable behaviours in the young generation and in the general public needs to emphasize the interface between scientific information and behaviour patterns, so that the former motivates the latter. Issues like the appropriate handling of substances and materials, or the importance of saving materials and energy, are viewed as the most urgent focuses of such initiatives. The next paragraphs consider the design of possible approaches through the analysis of concrete examples.

Indirect invitations to environmentally friendly behaviours may take place in a variety of situations. The responses may be largely different in different contexts. For instance, at a meeting, referring to some material sent electronically, a person may say: 'Let us not print all this, so we save some trees'. If the other persons have already been sensitized to the need to save resources, they will agree immediately, because they are aware of the motivations. If the other persons have not yet been sensitized to the need to save resources, they will perceive the invitation as unnecessary and awkward. On the other hand, making environmentally friendly invitations and statements is a way to slowly convey important messages. The efficacy increases if the number of persons making such invitations and statements in a certain context increases with time.

1.6   The Issue of Ethics

One of the first analyses of the ethical connotations of the objectives of green chemistry was carried out by a professor of philosophy,[10] who showed how those objectives have an intrinsic ethical value. Very recently, the entire issue of climate change is viewed as an urgent ethical question of our times.

Within educational perspectives (whether formal or informal), ethics becomes the reference for questions relating to our behaviour. An answer to questions such as why an individual should be concerned about global effects, or about the wellbeing of future generations, can be provided only by ethics. Integrating the perspectives of chemical literacy and ethics may play key roles in prompting shifts to more sustainable behaviours. Chemistry and chemical literacy can indicate practical ways for pursuing the objectives related to sustainability, and motivate these objectives in terms of scientific information; in other words, they can provide the knowledge for a person to be in a position to pursue something that is good in an effective way. The choice of pursuing these objectives pertains to each individual, and can be motivated by ethical considerations (choosing to *do good*).

1.7   Discussion and Conclusions

The design of interventions aimed at stimulating environmentally sustainable behaviours in the young generation and in the general public needs:

- To be based on diagnoses about diffuse attitudes and about responses to already-attempted interventions
- To emphasize the interface between scientific information and behaviour patterns, so that the former motivates the latter.

To this purpose, green chemistry education links with other educational domains: chemical education in general, education aimed at fostering science literacy, and ethics education.

The most urgent focuses of interventions to foster sustainable behaviour patterns concern the appropriate handling of substances and materials, the importance of saving material resources and the importance of saving energy. All these components benefit the environment. The appropriate handling of substances and materials benefits also human health directly; so does pollution prevention. Saving resources and energy has fundamental economic value. Saving energy is one of the measures to try and decrease the rate at which climate change progresses.

Interfaces and integration with ethical considerations and ethics education are fundamental to link the dissemination/acquisition of information and knowledge with the adoption of environmentally friendly behaviour patterns.

Experience shows that changes are possible. Like for a new commercial product, the concept of *pioneers* who start a new type of behaviour, and whose example will eventually modify the behaviour of the others, is a key concept in the stimulation of environmentally benign behaviour patterns. For this specific purpose, the *pioneers* need not only to provide examples through their behaviour, but also to be able to explain the motivations of their choices, thus becoming educators who communicate science information and its implications. Therefore, pioneers wishing to foster sustainable behaviour patterns need to have basic scientific literacy, including basic knowledge of green chemistry and its principles.

References

1. P. T. Anastas and T. Williamson, in *Green Chemistry*, ed. P. T. Anastas and T. Williamson, American Chemical Society, Washington, 1996.
2. P. T. Anastas and I. C. Warner, *Green Chemistry: Theory and Practice*, Oxford University Press, New York, 1998.
3. P. Tundo and P. T. Anastas, *Green Chemistry, Challenging Perspectives*, Oxford University Press, Oxford, 2000.
4. World Commission on Environment and Development (WCED), *Our Common Future*, Oxford University Press, Oxford, 1987.
5. Recommendation 7, *OECD Workshop on Sustainable Chemistry*, Venice 15–17 October 1998.
6. L. Mammino, in *Chemistry as a Second Language: Chemical Education in a Globalized Society*, ed. C. Flener and P. Kelter, American Chemical Society, Washington, 2010, pp. 7–42.
7. L. Mammino, *ISTE International Conference Proceedings*, ed. D. Mogari, A. Mji and U.I. Ogbonnaya, UNISA Press, 2012, pp. 278–290.
8. B. Russell, *Perche' Non Sono Cristiano*, Feltrinelli, Milan, 1959.
9. L. Mammino, *29th ICOH, International Congress on Occupational Health*, Cape Town, South Africa, 22–27 March 2009.
10. J. B. R. Gaie, in *Green Chemistry in Africa*, ed. P. Tundo and L. Mammino, IUPAC & INCA, Venice, 2002, pp. 16–30.

CHAPTER 2

Education for Sustainable Development and Chemistry Education

FRANZ RAUCH*[a]

[a]Institute for Instructional and School Development, Alpen-Adria University Klagenfurt, Sterneckstrasse 15, 9010 Klagenfurt, Austria
*E-mail: Franz.Rauch@aau.at

2.1 Sustainable Development

> Humanity stands at a defining moment in history. We are confronted with a perpetuation of disparities between and within nations, a worsening of poverty, hunger, ill health and illiteracy, and the continuing deterioration of the ecosystems on which we depend for our well-being[1]

states the preamble to Agenda 21, the programme of action for the 21st century, which was adopted by the World Summit for Environment and Development in Rio de Janeiro in 1992 by virtually all countries of the world.

Sustainable development is to solve the above problems:

> However, integration of environment and development concerns and greater attention to them will lead to the fulfilment of basic needs,

Worldwide Trends in Green Chemistry Education
Edited by Vânia Gomes Zuin and Liliana Mammino
© The Royal Society of Chemistry 2015
Published by the Royal Society of Chemistry, www.rsc.org

improved living standards for all, better protected and managed ecosystems and a safer, more prosperous future. No nation can achieve this
on its own; but together we can – in a global partnership for sustainable
development.[1]

As early as 1987, the World Commission for Environment and Development
(WCEF) defined the concept of 'sustainable development' (in the Brundtland Report *Our Common Future*) as a 'development that meets the needs
of the present without compromising the ability of future generations to
meet their own needs.'[2] This also implies that environmental conservation
is no longer seen as a preferred means of preserving resources for future
generations, a tenet held predominantly and unilaterally by the Western
world, since:

> Sustainable development requires us to acknowledge the interdepen-
> dent relations between people and the natural environment. This
> interdependence means that no single social, economic, political or
> environmental objective be pursued to the detriment of others. The
> environment cannot be protected in a way that leaves half of humanity
> in poverty. Likewise, there can be no long-term development on this
> depleted planet.[3]

A fair and equitable distribution of capital and natural resources, and of living and development opportunities, among all people in the world was the
ambitious objective of the world community.

Homann[4] called this function a 'regulative idea', a term he borrowed from
Kant.[5] According to Homan, regulative ideas serve as heuristics for reflection. They:

> steer the searching, research and learning processes in a given direc-
> tion and direct it to a given focus; in this manner they keep us from
> poking about in a fog, incoherently and haphazardly. One needs at least
> an intuitive idea of what one is looking for. Without such pre-concepts,
> one cannot even formulate a reasonable question or identify a problem
> (...). Heuristics may help determine the agenda, keep it under a com-
> mon focus, attract attention to interdependencies in this field, but they
> cannot determine specific recommendations and proposals.[4]

The non-descriptness of sustainable development as a guiding principle can
be perceived as a lack; sustainability can be discounted as an empty formula,
even a container term.[6] Conversely, it may also be seen as an opportunity,
even a precondition, to fulfil its function.[7] The different interpretations to
which this guiding principle lends itself give it a broad range of points to tie
in with. The term's lacking precision, its non-descriptness, can make for a
highly creative, diverse, yet dynamic field, which is oriented to a given direction. In open societies, open notions are likely to meet with an echo, and this

is precisely what we are seeing in the current debate on sustainable development. Sustainable development forms a favourable backdrop for reacting to the complex issues, which modern-day society is facing in an adequate, manageably complex and not over-simplifying manner.[8,9]

2.2   Education for Sustainable Development: A Socio-Political Balancing Act

Education is perceived as the master key to achieving a sustainable society: 'It is widely agreed that education is the most effective means that society possesses for confronting the challenges of the future...' states UNESCO's policy report *Education for a Sustainable Future*,[10] and in 2000, the World Education Forum in Dakar noted that education constitutes the true basis for sustainable development.[11] In late 2002, the United National Plenary Assembly thus proclaimed the Decade of Education for Sustainable Development (2005–2014). Education must be geared to social visions and cannot be detached from society. One must be aware of the social dilemma in which education for sustainable development (ESD) operates. Yet, as educationalists, it is their very task to encourage and empower the next generation to partake in shaping society. Whenever facts are complex and controversial, whenever social and economic interests conflict, it is inadequate to 'settle the facts without strengthening the persons', Nagel and Affolter[12] stated, borrowing a quote from Hartmut von Hentig. It is only individuals with a sufficiently developed self-strength who can act self-confidently on the basis of their own reflection, especially when issues are contradictory and complex.[13]

ESD gives social concerns the appearance of social policy visions, an idea of a better world to which it can be directed. 'Education is about hope and therefore about strong and existential feelings of future':[14] In this context, Künzli-David[15] mentions three requirements, which pedagogical visions that ESD must meet:

- While recognizing social problems, a pedagogical vision must inspire optimism. By orientation to the notion of sustainable development, it is possible to convey complex facts while giving the feeling that the problems at hand can be tackled. Sustainable development does not deny problems, but presents them as fundamentally manageable. It can therefore generate and strengthen optimism about the future.
- Reality is complex and pluralistic; a pedagogical vision must not propose a one-sided view. Here, the regulative idea concept is an appropriate reference frame for sustainable development. What is sustainable depends on the conditions imposed by where and when stakeholders find themselves and requires a process of negotiation.
- Shifting social visions to the pedagogical level must not be the only measure by which to implement them. Education is only one measure that must go hand in hand with political and social transformations.

Education for sustainable development does not aim at changing people's lifestyles, but at 'empowering and encouraging them to participate in designing sustainable development and to critically reflect on their own action in this area.'[15]

2.2.1 The Role of Chemistry for Education on Sustainability

There is no doubt that the field of chemistry and the industries related to it are in the economic heart of every highly developed industrial society.[16] Industry provides the basic materials necessary for every other type of business. It also defines the basis of energy supply, modern agriculture, and innovative technologies. Unfortunately, many chemical industries around the world have not always been careful in the past. Quite often, they neither concerned themselves with the preservation of natural resources, nor did they give much thought to protecting the environment. Accidents both large and small have significantly contributed to the negative public image of industrial undertakings and chemistry as a science.[17] However, a change in attitude has slowly taken place, at least in Western societies.[18] This change goes hand in hand with a growing public awareness of both the finite nature of natural resources and the existence of limits, which regulate and determine feasible rates of growth.[19] Both being careful with our resources and avoiding damage to our ecology and health has aided in promoting a better image of chemistry in the past decades. Ever since the 1990s, sustainability and sustainable development have emerged as the core issues of today's chemical industry, its actions and its public image. The report by the European Communities Chemistry Council entitled *Chemistry for a Clean World* set the stage in 1993.[18] In the USA, the works of Anastas and Warner[20] presenting their concept of green chemistry began to expand and gain recognition in the mid-1990s. In the early years, a struggle to find a proper name for the concept occurred.[21] In Anastas and Warner,[20] the idea of a more environmentally friendly and resource-preserving synthesis in chemistry was expanded to include twelve fundamental principles. These principles became the generally accepted guidelines for the contemporary understanding of green, sustainable chemistry, which has been implemented by both research and industry worldwide. Limited resources and a constantly growing consciousness of the value of environmental protection were both among the driving forces of this movement. But, increasingly stringent legal restrictions for the handling of chemicals and the search for a better self-image for industrial chemistry in Western society also contributed to this development. Today, worldwide initiatives are focusing on a more environmentally responsible form of chemistry. Examples of this include the ACS Green Chemistry Institute (ACS-GCI) and the European Technology Platform for Sustainable Chemistry (SUSCHEM). The core role of chemistry and chemical industry for sustainable development in modern societies suggests a central role for chemistry education in ESD.[16]

2.2.2 Basic Models of Approaching Sustainability Issues in Chemistry Education

Adding sustainable development issues to the chemistry curriculum is not a new idea by any stretch of the imagination. For the last two to three decades, many pupils around the world have been faced in chemistry education with issues such as keeping water resources clean, dealing with the effects of acid rain, coping with the hole in the ozone layer, and searching for both renewable sources of energy and raw materials. These topics and others have been widely implemented as content in many chemistry curricula worldwide. Examples include *Chemistry in the Community* in the US,[22] the *Salters* chemistry curriculum from the UK,[23] chemical industry case studies in Israel,[24] or environmentally oriented chemistry education in Germany.[25,26] In any case, the question of how to deal with issues of sustainable development can take on different appearances and can follow different models. Although they partially overlap and can be integrated in different ways, four different basic models are presented by Burmeister and colleagues[27] when it comes to implementing issues of sustainable development into formal chemistry education.

2.2.2.1 Model 1: Adopting Green Chemistry Principles to the Practice of School Lab Work

The first model applies the philosophy of green chemistry[20] to the handling of chemicals and laboratory work procedures in chemistry classes. Student experiments can be shifted from the macro- to the micro-scale, dangerous substances can be replaced by less poisonous alternatives, and catalysts can be used to stimulate reactions.[28,29] The potential of ESD—at least when it deals with learning about chemistry's contribution to sustainable development—can be expanded, if students are able to recognize, compare and reflect upon the altered strategies. Students can learn how chemistry research and chemical industry attempt to minimize the use of resources, maximize the effects, and protect the environment. Karpudewan *et al.*[30] have already demonstrated that this strategy has the potential to change the attitudes and knowledge of student teachers. The strength of this approach is that chemistry education truly contributes to sustainability by reducing the amounts of chemicals used and by producing less waste. The weakness of the approach concerning ESD is that it is often less embedded into continuous self-reflection upon how society handles debates around changing technologies. In this case, students will not develop skills for contributing to society's decision-making on new or alternative technologies. Additionally, students will barely touch upon the controversial nature of developments in society and the real interplay between science, technology and society. In this case, the holistic approach of ESD will hardly be achieved in the manner outlined above.

2.2.2.2 Model 2: Adding Sustainability Strategies as Content in Chemistry Education

This model takes the strategies and efforts used to contribute to sustainability development into account when deciding which content to include in chemistry education. In this approach, the basic chemical principles behind sustainable and green chemistry and their industrial applications appear as topics within chemistry curricula.[31] Practical examples of this include the development of efficient processes in industrial chemistry in the fields of energy and raw materials conservation, research into the structure, properties and application of innovative catalysts, and the chemical considerations behind the production of fuels stemming from renewable materials.[26] Learning about green chemistry and chemical research's contributions to sustainable development can also offer a basis for a better understanding of various developments in wide-ranging fields. The strength of this approach is that it highlights the learning of the chemical principles disguised behind everyday processes and end products, thus making them more meaningful to students.[32] At any rate, a thorough understanding of the interplay between science, technology and society—ESD terms it the 'interplay of economic, ecological and societal impacts'—will never take place, if learners' concentration is primarily focused on (or even restricted to) the learning of chemical content behind its technological application. In such a scenario, the general skills necessary for participating in societal debates on socio-scientific issues will hardly have a chance to emerge. Making sustainability issues part of chemically based content in the proper context can provide the initial step, which offers learners access to sustainability issues as they exist in modern chemistry.

2.2.2.3 Model 3: Using Controversial Sustainability Issues for Socio-Scientific Issues Which Drive Chemistry Education

The third model integrates the chemistry learning using socio-scientific issues (SSI) having the tension of current societal debate behind them.[33] SSI teaching does not primarily focus the learning of chemistry as a subject or sustainability issues *per se*. Instead, lessons tend to mould sustainable development education by developing general educational skills in the area of an individual's actions as a responsible member of society. This model's approach varies from that of the second model in that it includes both the chemical basis of knowledge and reflecting society's debate about its practical application in technology as factors to be learned. Model 3 focuses primarily on learning exactly how developments in chemistry can be and actually are evaluated and discussed within society using all of the sustainability dimensions.[34] This approach not only constitutes the explicit learning *of* chemistry, but also includes the learning *about* chemistry as it is dealt with in society. Examples with respect to sustainable chemistry include the ongoing controversy about the use of biofuels,[35-37] the application of specific compounds and alternatives to them in everyday products, and the evaluation of innovative products from chemistry using a multidimensional approach. The aspects of understanding societal debates and developing appropriate skills to actively participate in them are systematically built into the lesson focus. Students learn how to take part in societal decision-making in order to contribute to shaping a sustainable future. The strength of this approach is that it is skill-oriented with a sharp focus on ESD. It closely mirrors the differentiation defined by Holbrook,[38] who has demanded more education through science instead of science through education. However, some socio-scientific issues of controversial nature have limited potential in the areas of individual and local action. Often, debate about new technologies is extremely complex and occurs primarily in expert committees at the political level. In such an arena, the influence of the individual is very limited. But in a truly democratic society, no individual is hindered from entering the political scene, if he or she wishes to. The type of teaching in this third model attempts to prepare students for this very eventuality.

2.2.2.4   Model 4: Chemistry Education as a Part of ESD-Driven School Development

The fourth model integrates chemistry education as part of ESD-driven school development.[39] Such an approach demands opening chemistry classrooms even further.[40,41] This model suggests that school life and teaching should become part of ESD. Educating children to become active citizens who have the ability to achieve sustainable lifestyles requires entire school process models. Such models include development, self-evaluation and reflection.[42] All shareholders in the school system are required to explore future challenges, to clarify values, and to reflect on both learning and actively taking part in society in the light of ESD. If we understand school development as changing schools to become learning organizations offering new experience, reflections, innovations, we necessarily need to change both the way people lead discussions and act.[40] Chemistry education should help contribute to such an altered teaching culture. Many opportunities exist for opening chemistry teaching to reflect how this domain influences us in the here and now, including our current lives inside and outside of school or other educational institutions. Chemistry teaching can actively contribute to saving resources (energy, clean water, *etc.*) in local environments, including school. It can also offer suggestions for treating waste in an efficient fashion suitable for later recycling. Chemistry education no longer needs to stop at the point where teaching is limited to describing the chemical theories and knowledge behind sustainability issues and potential avenues of action. Chemistry lessons and school life morph into an action-based pattern of living and learning. Students gain first-hand experience of how taking action can fundamentally change their lives. This experience includes pupils seeing how their personal contributions to in-school decision-making processes factor into both changed behaviour on their part and alterations in the learning process in which chemistry is an integral part. In teaching practice, all four of the above-mentioned models may overlap or even combine in order to place a stronger focus on sustainability issues connected to chemistry education.

2.3   Conclusion and Outlook

Based on the notion of sustainable development as a regulative idea outlined above, the link between sustainable development and education can be summarized as follows. Sustainable development is part and parcel of a general educational task, aimed at empowering the young generation to design their conditions of life on a more humane scale. It is based on an educational notion, which focuses on humans' self-development and self-determination as they interact with, and reflect on, the world, with others, and with themselves. Education relates to the ability to contribute to the design of society in a reflected and responsible manner in terms of sustainable future development.

In the context of sustainable development, learning is equivalent to addressing issues of how to sustainably shape the future in concrete fields of action, *i.e.*, chemistry and chemistry education. This includes observation, analysis, assessment and design of a given context in creative and cooperative processes. What is addressed and called for specifically are a critical assessment of knowledge in the light of the present-day information overload, the development of self-worth, self-determination and self-reliance, as well as social competencies such as the ability to participate.[8,43]

Chemistry dramatically influences everything from the life of the individual to society as a whole. Chemistry curricula and chemistry teacher education should more thoroughly reflect not only the importance of education and sustainable development, but support the development of human identity, which is interrelated with the environment, both individually and collectively,[44] the goal being to allow students to actively learn how to shape society in a positive, sustainable fashion.

References

1. UNCED, *Agenda 21*, UNCED, New York, 1992, retrieved from the World Wide Web, December 30, 2012 at http://www.un.org/esa/dsd/agenda21/.
2. *Unsere gemeinsame Zukunft. Der Brundtland-Bericht der Weltkommission für Umwelt und Entwicklung*, ed. V. Hauff, Eggenkamp-Verlag, Greven, 1987.
3. UNESCO, Education for Sustainable Development. From Rio to Johannesburg: Lessons learnt from a decade of commitment, *Report presented at the Johannesburg world summit for Sustainable Development*, UNESCO, Paris, 2002.

4. K. Homann, Sustainability: Politikvorgabe oder regulative Idee? in *Ordnungspolitische Grundfragen einer Politik der Nachhaltigkeit*, ed. L. Gerken, Nomos Verlagsgesellschaft, Baden-Baden, 1996, pp. 33–46.

5. I. Kant, *Kritik der reinenVernunft*, Felix Meiner, Hamburg, 1956.

6. H. Eblinghaus and A. Stickler, *Nachhaltigkeit und Macht. Zur Kritik von Sustainable Development*, Verlag für Interkulturelle Kommunikation, Frankfurt, 1996.

7. *Nachhaltige Entwicklung. Eine Herausforderung für die Soziologie*, ed. K. W. Brand, Leske+Budrich, Opladen, 1997.

8. F. Rauch, Education for sustainability: A regulative idea and trigger for innovation, in *Key Issues in Sustainable Development and Learning: A Critical Review*, ed. W. Scott and S. Gough, RoudlegeFalmer, London, 2004, pp. 149–151.

9. F. Rauch, What do regulative ideas in education for sustainable development and scientific literacy as myth have in common? in *Contemporary Science Education – Implications from Science Education Research about Orientations, Strategies and Assessment*, ed. I. Eilks and B. Ralle, Shaker, Aachen, 2010, pp. 35–46.

10. UNESCO, *Educating for a Sustainable Future. A transdisciplinary vision for concerted action. EPD-97/CONF.401/CLD.1*, 1997, Retrieved from the World Wide Web, December 30, 2012 at http://www.unesco.org/education/tlsf/mods/theme_a/popups/mod01t05s01.html#pre.

11. UNESCO, The Dakar framework for action. Education for All. Meeting our collective commitments, *Adopted by the World Education FORUM in Dakar, Senegal, 26 – 28 April 2000*, Retrieved from the World Wide Web, December 30, 2012 at http://unesdoc.unesco.org/images/0012/001211/121147e.pdf.

12. U. Nagel and C. Affolter, Umweltbildung und Bildung für eine Nachhaltige Entwicklung – Von der Wissensvermittlung zur Kompetenzförderung. *Beiträge zur Lehrerbildung*, 2004, **22**, 95–105.

13. M. Heinrich, J. Minsch, F. Rauch, E. Schmidt and C. Vielhaber, Bildung und Nachhaltige Entwicklung: eine lernende Strategie für Österreich. Monsenstein & Vannerdat, *Münster*, 2007.

14. J. Oelkers, Utopie und Wirklichkeit. Ein Essay über Pädagogik und Erziehungswissenschaft. *Zeitschrift für Pädagogik*, 1990, **1**, 1–13.

15. C. Künzli-David, *Zukunft mitgestalten. Bildung für eine Nachhaltige Entwicklung – Didaktisches Konzept und Umsetzung in der Grundschule*, Haupt, Bern, 2007.

16. J. D. Bradley, Chemistry Education for Development. *Chem. Educ. Int.*, 2005, **6**, Retrieved from the World Wide Web, December 30, 2012, at http://old.iupac.org/publications/cei/vol6/index.html.

17. M. R. Hartings and D. Fahy, Communicating chemistry for public engagement, *Nat. Chem.*, 2011, **3**, 674–677.

18. ECCC, *Chemistry for a Clean World*, European Communities Chemistry Council, The Hague, 1993.

19. D. Meadows, D. H. Meadows and J. Randers, *Limits of Growth. The 30 Year Update*, Earthscan, Milton Park, 2005.
20. P. T. Anastas and J. C. Warner, *Green Chemistry Theory and Practice*, Oxford University, New York, 1998.
21. O. Hutzinger, The greening of chemistry – Is it sustainable? *Environ. Sci. Pollut. Res.*, 1999, **6**, 123.
22. ACS, *Chemistry in the Community*, Kendall/Hunta, Dubuque, 5th edn, 2006.
23. J. Benett and F. Lubben, Context-based chemistry: The Salters-approach, *Int. J. Sci. Educ.*, 2006, **28**, 999–1015.
24. A. Hofstein and M. Kesner, Industrial chemistry and school chemistry: making chemistry studies more relevant, *Int. J. Sci. Educ.*, 2006, **28**, 1017–1039.
25. H. J. Bader, Less polluting technologies, regrowing resources and recycling: new topics in the teaching of chemistry, *Int. Newsl. Chem. Educ.*, 1992, **38**, 12.
26. H. J. Bader and R. Blume, *Environmental chemistry in classroom experiments*, IUPAC, Delhi, 1997.
27. M. Burmeister, F. Rauch and I. Eilks, Education for Sustainable Development (ESD) and chemistry education, *Chem. Educ. Res. Pract.*, 2012, **13**, 59–68.
28. P. Schwarz, M. Hugerat and M. Livneh, *Microscale chemistry experimentation for all ages*, Arab Academic College for Education in Israel, Haifa, 2006.
29. M. M. Singh, Z. Szafran and R. M. Pike, Microscale chemistry and green chemistry: Complementary pedagogies, *J. Chem. Educ.*, 1999, **76**, 1684–1687.
30. M. Karpudewan, Z. H. Ismail and M. Norita, The integration of Green Chemistry experiments with sustainable development concepts in pre-service teachers' curriculum: experiences from Malaysia, *Int. J. Sustainable High. Educ.*, 2009, **10**, 118–135.
31. A. Lühken and H. J. Bader, Energy input from microwaves and ultra sound – examples of new approaches to Green Chemistry, In *Green Chemistry*, ed. Royal Society of Chemistry, Royal Chemical Society, Cambridge, 2003, http://www.rsc.org/Education/Teachers/Resources/green/docs/microwaves.pdf.
32. A. Pilot and A. M. W. Bulte, Special issue: Context based chemistry education, *Int. J. Sci. Educ.*, 2006, **28**, 953–1112.
33. R. Marks and I. Eilks, Promoting scientific literacy using a socio-critical and problem-oriented approach to chemistry teaching: concept, examples, experiences, *Int. J. Sci. Environ. Educ.*, 2009, **4**, 231–245.
34. M. Burmeister and I. Eilks, An example of learning about plastics and their evaluation as a contribution to Education for Sustainable Development in secondary school chemistry teaching, *Chem. Educ. Res. Pract.*, 2012, **13**, 93–102.

35. I. Eilks, Teaching 'Biodiesel': A sociocritical and problem-oriented approach to chemistry teaching, and students' first views on it, *Chem. Educ. Res. Pract.*, 2002, **3**, 67–75.
36. T. Feierabend and I. Eilks, Teaching the societal dimension of chemistry using a socio-critical and problem-oriented lesson plan on bioethanol usage, *J. Chem. Educ.*, published online first July 10, 2011.
37. A. Hiramatsu, K. Takabay, R. Utsumi, H. Fujii and H. Ogawa, A lesson model fostering fine ideas in chemistry concerning biodiesel on the basis of "Education for Sustainable Development": potentialities for collaboration with social studies, *Chem. Educ. J.*, 2009, 13, retrieved from the World Wide Web, December 30, 2012, at http://chem.sci.utsunomiya-u.ac.jp/v13n1/indexE.html.
38. J. Holbrook, Making chemistry teaching relevant. *Chem. Educ. Int.*, 2005, **6**, retrieved from the World Wide Web, December 30, 2012, at http://old.iupac.org/publications/cei/vol6/index.html.
39. F. Rauch, The potential of Education for Sustainable Development for reform in schools, *Environ. Educ. Res.*, 2002, **8**, 43–52.
40. S. Breiting, M. Mayer and F. Mogensen, *Quality criteria for ESD-schools*, ENSI, Vienna, 2005.
41. G. De Haan, The BLK '21' programme in Germany: a 'Gestaltungskompetenz'-based model for education for sustainable development, *Environ. Educ. Res.*, 2006, **12**, 19–32.
42. *Creating Sustainable Environments in Our Schools*, ed. T. Shallcross, J. Robinson and P. Pace, Trendham Books, Staffordshire, 2006.
43. F. Rauch, Bildung für Nachhaltige Entwicklung als eine lernende gesellschaftspolitische Strategie, in *Demokratie Lernen Heute*, ed. G. Gruber and K. Stainer-Hämmerle, Böhlau, Wien, 2008, pp. 173–188.
44. V. Zuin, *Environmental Dimension in Chemistry Teacher Education*, Editora Átomo, Guanabara, 2012.

CHAPTER 3

Green Chemistry Education in Brazil: Contemporary Tendencies and Reflections at Secondary School Level

VÂNIA GOMES ZUIN*[a] AND CARLOS ALBERTO MARQUES[b]

[a]Department of Chemistry at the Federal University of São Carlos, Rodovia Washington Luiz, Km 235, Jardim Guanabara, 13565-905 São Carlos, SP, Brazil; [b]Department of Teaching Methodology at the Federal University of Santa Catarina, Campus Universitário Reitor João David Ferreira Lima – Trindade, 88040-900 Florianópolis – SC, Brazil
*E-mail: vaniaz@ufscar.br

3.1 Introduction

Why teach green chemistry (GC) to pupils at secondary school? For many of those working with GC, the answer is perhaps easier but certainly this is not the case for most chemists and chemistry teachers at schools. Therefore, answering this challenging question can be an important step towards defining pedagogic strategies for teaching it. This is one of the aims of this study which, based on a survey of academic research in GC in Brazil, discusses trends and possible impacts of this research, arguing in favour of it being included in chemistry teaching at the basic education level.

Worldwide Trends in Green Chemistry Education
Edited by Vânia Gomes Zuin and Liliana Mammino
© The Royal Society of Chemistry 2015
Published by the Royal Society of Chemistry, www.rsc.org

Before that, however, it is worth recognizing that at a certain level, quality, intensity and scope, Brazilian schools are already including subjects and content related to the environment, mainly by initiatives known as environmental education, whose academic outcome has theoretically guided new practices, attitudes and behaviours involving a whole range of factors included in environmental issues. The role of environmental education has been shown through research and scientific production, mainly by teacher training, which has grown over the years[1-4] and involves varied subjects, ranging from theoretical–methodological orientations to including different ones, such as the environmental crisis and degradation, recycling practices, issues involving socio-economic relations and climate change.

However, for some authors, environmental education initiatives developed at school take place in a very timely way, not very focused on the content, non-systematic, emphasizing naturalist conceptions of the environment and very much concerned with preserving and protecting fauna, flora and natural resources.[5,6] Nevertheless, if on the one hand these contributions are appropriate, mainly in that they help to raise sensitivity and awareness about everyday challenges concerned with the environment among young people; while, on the other hand, at schools, these activities are still focused mainly within the subject of environmental education. This implies recognizing that the approach of environmental issues within the subject of chemistry at secondary school level is still quite incipient, and is applied in few schools and only in few states of the country. From a different point of view, official curricular documents have already indicated, since the 1990s, that the teaching of chemistry should allow for:

the comprehension of chemical processes themselves, as well as construction of scientific knowledge in close relation with technological applications and their environmental, social, political and economic consequences.[7]

However, studies show that including environmental issues in chemistry teaching, when it occurs, is carried out from different theoretical perspectives and involves different contents.[2,4,8,9] On the other hand, there is little tradition of research about environmental education in chemistry teaching.[10] Reigota[1] carried out a study where it was observed that from 246 Master's theses related to environmental education, which were produced between 1984 and 2002, only one was about chemistry teaching. Records since then have shown that a large number of reports on work involving this issue in chemistry education are mainly concerned with the practice and training of chemistry teachers.[9] If, on the one hand, this shows the lack of studies and school experiences involving the issue, then, on the other hand, it shows a promising trend, as these professionals are being trained in universities concerned with the issue and have access to contents related to this subject.

Therefore, it is within this wider educational scope and linked to environmental education experiences that initiatives of approaches in GC can be present at the basic education level. Moreover, going back to the initial question as to 'Why teach green chemistry at high school level?' we are convinced that the main aspect to the answer is related to recognizing that young people need to be trained from an early age to think about social life and future professional work from an environmentally aware point of view, by developing not only values but also thinking about structures (scientific ones) related to the environment. Taking this into account, the environment should be seen and understood mainly from the perspective of the chemical sciences. To study the natural composition and the inevitable consequences of human actions *with* and *about* the environment might help our students to better understand it and, mainly, to protect it. Thus, chemistry will be understood as a science that is also concerned about nature and life. If this line of thought is valid, then *where should we start from?*, and, considering that chemistry is already being taught in schools, how and where from can we 'change' its teaching in order to incorporate the principles and contributions of GC, so that current chemistry teaching becomes greener chemistry education, aimed at socio-environmental sustainability?

To better answer these questions, we expect the following to support the thesis that it is necessary, and feasible, to teach (aspects of) GC at the basic education level. We argue the importance of chemistry education aimed at socio-environmental sustainability, one that involves, and surpasses proposals and experiences already developed with GC, in particular those aimed at schools, in such a way as to consolidate an environmental/green culture among Brazilian chemists. Concerning this, we briefly discuss the problem of risk involved in chemical activities, in light of the idea-concept of socio-environmental sustainability and sustainable development, since it is the permanence of this paradigm (of the risk) in the practical rationality of chemists—and of science in general—that seems to sustain the idea of being safe with chemical products when human life and environment are involved. Then, some considerations concerning GC in Brazil will be put forward, including this exposition within a larger movement concerning the answers that chemistry has given to contemporary challenges for the safeguard of the environment. Aspects can be seen from basic education to higher education, highlighting experiences gained on chemistry teacher training courses (Bachelor of Education degree, BEd), as these professionals are important disseminators of chemistry knowledge in Brazilian secondary schools. Finally, which is the most substantial part of this reflection, we discuss the current reality of the area in the country, based on a survey of scientific research (in theses, articles in scientific journals and conferences) of Brazilian authors that reflect, to some extent, the *status quo* of GC education in Brazil.

3.2 Sustainability and Development: The Risks in Chemical Activities and How the Area has Dealt with this Issue

Dealing with risks, as mentioned previously, aims to emphasize that the approach of hazards related to chemical activities for human health and the environment can be an important pedagogic strategy for the study of chemistry within the green/environmental perspective, in the widest socio-environmental context and concerned with issues of sustainability and development.

It is interesting to note that chemistry is seen as a science of matter, or better said, of transforming matter, and as such it has been developing enormously from centuries, contributing to nations' development, solving problems and providing wellbeing to people. Considering this, it has built up a rich body of postulates, theories, concepts and procedures that reaffirm it, not only as a field of nature's science, but rather as a basis for diverse areas of knowledge, such as chemical engineering, biochemistry, among others. Chemistry is present in different areas and processes of industrial and agricultural production, in goods and services. The efficiency in our processes of synthesis is expressed, for example, in the current catalogue of almost 74 million different chemical substances, some of them not readily available and others non-existent in nature. However, in most of these processes, the environmental cost involved is quite high, whether it is from the energetic point of view (generating entropy), or from the inevitable generation of material residues from processes and products (anthropogeny). We can state then that, through the entropic prism, making chemistry does pollute.

Without exhausting the subject here, it is worth recalling that, guided by the dilution and risk paradigms,[11] chemists have sought to find forms, techniques and ways to deal with the environment through safer alternatives for the chemical activities themselves, as well as for the disposal of residues from their production, in addition to information about care to be taken for all chemical products. Finally, these paradigms were (and still are) a means to approach and behave to face risk. Brunet *et al.* (citing Ewald)[12] describe three moments at which to approach them: 'risk' as a fatality, a destiny, something called *hazard*. The responsibility about it is individual and based on the moral virtue; it suffices for the individual to act with prudence and care in order to avoid it. Therefore, it is a local issue and its responsibility is exclusive to those directly involved in the activity (industry and workers). The second way of viewing it converts the hazard into foreseeable risks and it is from there that the idea of prevention arises. Guided by the positivist view of science and technology, methods are established in order to quantify the risk, assigning values and acceptability standards to it, since this would be inherent to the human activity. This culminates in a compensation system through indemnifications.[12] Thornton[11] calls these ideas *risk paradigm*. The third way of viewing the risk is based on searching for safety, not only social but also environmental, as well as on the need to create mechanisms about the scientific and technological activity, since the social and industrial society is based on goods and harms production.[13]

In the area of chemistry, the most traditional forms of minimizing these harms were based on reducing the 'limit of exposition, by controlling the so-called circumstantial factors' (p. 807),[14] such as the use, availability and treatment of chemical products. To counteract this normative standard still in force, Thornton[11,15] proposes adopting an *ecologic paradigm*, based on the

Caution Principle, which appears explicitly in Principle 15 of the final declaration of Rio-92 Conference.[16]

By highlighting briefly some conceptual and historical aspects concerning risk, we aim to establish an associative line between the Caution Principle with GC whose emphasis is shown by several authors who are precursors of it.[17] In spite of it not being claimed as a new paradigm[18] of chemistry, it is true that some of the principles of GC (that is, numbers 1, 3, 4, 5, 10 and 12)[19] indicate the search for a preventive attitude, based not on moral issues—although important as they are—but rather on essentially scientific basis concerned with the prevention of future damage to the environment. The incorporation of these and other principles of GC have made it possible to revise and develop new processes and techniques.[20-22]

Considering this, GC is a gaining place in the international societal effort to safeguard the environment, being related by many to the definition–concept of sustainable development and the idea–guide of seeking socio-environmental sustainability, found in the Brundtland report.[23] Studies indicate that different GC precursor researchers justify their work based on this socio-environmental sustainability purpose, even without problematizing it in light of the thermodynamic postulates, as observed by Marques and Machado.[17] In spite of this gap, GC can be considered one of the main answers that chemistry has given to safeguard the environment, through which Brazilian chemistry has, increasingly occupied a significant position in the world.

As pointed out at the beginning of this section, dealing with the risks associated to chemical activities—hazards to individuals and the environment in different everyday situations—from the preventive point-of-view, it can be seen as a promising pedagogic strategy for the 'green' approach in chemistry education as for example, in experimental activities throughout different teaching levels, in particular at secondary schools. Such an approach could work as a kind of 'open door' for teaching GC.

3.3 Considerations about Green Chemistry in Brazil: From Quick Receptiveness to Strategic Future

The aim here is to raise awareness about information concerning GC in Brazil, mainly that concerned with the education area in general, and the basic education level in particular, observing its occurrence, difficulties and possibilities of setting it up within these areas. Although it is not possible to pinpoint the exact date GC was founded, nor can the exact moment be defined when the dissemination of proposals and experiences related to GC teaching started at the basic level (secondary school) and higher education in Brazil.

However, the diffusion of GC in Brazil had the Brazilian Chemistry Association (ABQ) as one of its main sources and, more recently, the Brazilian Chemistry Society (SBQ),[24] through its scientific journals, networks and meetings. The pioneers of this diffusion were Lenardão and colleagues,[25] Prado[26] and

Sanseverino[27,28] who presented the 12 principles of GC,[19] although four years before there were comments on the existence of activities with that name.[29] Within the education sector, more formal proposals identified as GC activities have been registered since 2003, by a proposal to include the concept of atomic economy in the experimental organic chemistry subject in courses at a higher education level.[30] It was only in 2007 that proposals for the basic education level appeared in the journal *Química Nova na Escola*.[31] It was also in that year that, in its editorial, the SBQ stated the values and contributions of GC.[32] Even though there is not significant bibliographic research in books, as will be seen throughout the text, a more prominent and pioneer example is the book by Corrêa and Zuin,[22] *Green Chemistry: Fundamentals and Applications* (in Portuguese), which, in addition to describing the general principles, provides examples of GC applications in the country.

While commenting on the 20 years of its implementation, Faria and Fávaro[21] highlight the relevant contribution of GC for the progress of science; by describing an investigation they made on the CAPES website[33] on works published between 1991 and 2010, where more than 250 000 of them cited the expression 'green chemistry'. This is a real tool of the information available to Brazilian chemists, providing a positive influence for its adoption on work developed here. The study also revealed two relevant aspects. First, that the content of this work showed the catalysis area as heavily present, quite understandably due to its importance in the 'maximization of reactions and reduction of undesirable by-product formations during the process' (p. 1091). Second, nearly half of these papers are from North American institutions, but 'their fast worldwide dissemination is quite evident', and that:

> in Brazil there are already more than 20 research groups[34] in universities and research centres, working in different areas [...]: education, synthesis, catalysis and biocatalysis, alternative solvents, renewable materials, processes and life cycle analysis' (p. 1092).

Among the pioneering actions, a relevant one is from the Centre of Management and Strategic Studies (CGEE),[35] a social organization supervised by the Brazilian government, which plays an important role in defining strategies for national programmes concerned with sustainable development, whereby the contribution of GC for the green economy is highlighted. The study called 'Green Chemistry in Brazil' (CGEE) in a joint effort with universities, presented an analysis of the Brazilian potential in GC over the next 20 years (2010–2030). It assesses panoramas, presents results and suggests actions to be taken by governments and institutions in order to implement the practice of the GC concept in Brazil, emphasizing the privileged position the country has in sustainable development, as well as the potential to become a world leader in the integral usage of biomasses. These advantages result, among other factors, from its enormous biodiversity, plenty of water and climate diversity. Throughout its chapters, the book separates and presents each of the areas where GC is found: biorefineries (biochemical and thermochemical routes), alcohol chemistry, oil chemistry, sugar chemistry, phytochemistry, CO_2 conversion, renewable energies and bioproducts, biofuels and bioprocesses. As prospects, the study also pointed out the need for a national programme to promote the creation of partnerships with the industrial sector in order to boost economy and production. The work sets out the priorities and immediate needs in the country's industry, proposing the creation of a Brazilian Research, Development and Innovation programme in GC, which would include a Brazilian school to disseminate knowledge and activities of GC in companies, research centres and universities, in such a way as to provide support to promote a sustainable industrial model based on the new paradigms of the bioeconomy.

Regarding the dissemination of the main outcomes from activities related to GC in the country concerning all sectors—academy, government and industry—the 4th International IUPAC Conference on Green Chemistry, which took place in Foz do Iguaçu, Brazil, in August 2012 supported by the SBQ and IUPAC, should be highlighted. From the many activities in the programme, the sessions with the highest number of submissions, and whose presentations attracted more researchers, were those aimed at GC education (formal and informal).

In this area of chemistry related to the environment, the historical contributions of environmental chemistry should also be considered.[36] Officially created in 1994 as a division of SBQ, it is defined as the one that 'studies the chemical processes that take place in nature, whether they are natural or artificial, and which hinder not only human wellbeing, but also the planet's as a whole.' Therefore, within this definition, environmental chemistry is not the science for monitoring the environment, but rather one for elucidating the mechanisms that define and control the concentration of chemical species that are candidates for being monitored.[37] This definition is in consonance with the more classic ones, for example by Manahan[38] who defines it as the study of origins, transport, effects and processes of chemical species in water, on land and in air, as well as the influence of human activity on these processes.

Since the 1980s, environmental chemistry in Brazil has been expanding its scope and assessing the damage already caused to the environment (ecosystems) by chemical products and processes and thus trying to understand the complex relations among the environmental compartments, as well as the mechanisms of action, bioaccumulation and transformation, ways to remediate and control harmful substances to human beings and to the environment emitted by different sources of pollution. Silva and Andrade[29] observe that environmental chemistry has contributed to preventing and correcting determined situations of environmental degradation/pollution by accumulated knowledge and corrective technologies developed over the last years. Cortes Junior[39] understands that environmental chemistry implies the knowledge of chemical processes and the complex interactions among the different systems in the planet (physical–chemical, biological and social–human) in an interwoven web interlocked among the parts. Machado,[40] a researcher from Portugal who has published in Brazil, sustains that environmental chemistry is not characterized exactly by studying the processes to eliminate the production of pollutant and toxic residues, or to avoid the wide use of harmful substances. This kind of more proactive study belongs to chemistry *for the* environment and, according to the author is dealt with by GC.

3.4 Academic–Scientific Work on Green Chemistry Education in Brazil

Some researchers, mainly those working in Brazilian public universities, have investigated and defended the inclusion of GC principles into teacher training curricula (BEd) in different ways and intensity,[41,42] arguing in favour of its transversal to all chemistry subjects in those curricula. One of the first studies characterized as a case study about teacher training in chemistry in Brazil, and including the environmental dimension, in particular GC, was developed by Zuin and co-workers.[4,9,43] Based on data collected over four years from a BEd chemistry course in a higher education institution (IES) in the State of São Paulo, Brazil, showed that as well as the emergent GC, whose constructs can, erroneously, be transformed into slogans, there is a striking need to individually and collectively re-assess the literature and curricular documents by discussing current complex socio-environmental issues and their social and historical contexts of production in courses at a higher education level in the country, particularly in teacher training.

More recently, taking as a reference another experience from an undergraduate course in environmental chemistry, Goes[44] (based on Shulman)[45] suggests including GC from the concept of pedagogic knowledge of contents. Considering this, the author suggests three models: incorporating GC principles into experimental procedures; incorporating sustainable strategies as part of the contents in the curriculum of chemistry courses; and the use of sustainability issues related to socio-scientific aspects (p. 115). As pointed out by Zuin *et al.*,[9] environmentalizing the curriculum, or incorporating the environmental dimension in the curriculum of chemistry courses, means comprehending GC itself not only within the teaching activities, but also in the research, extension and management ones. Within this curriculum scope, the environmental chemistry undergraduate course, offered by the University of São Paulo, is an outstanding one, whose main characteristic is its structured form and, while it is focused on studying the environment, it has GC as an important axis.[46]

By highlighting the influence of research in GC in Brazil and Latin America in the last three years, mainly at the higher education level, Correa *et al.*[47] have shown increasing interest in GC in the area of organic chemistry. They defend the possibility of introducing it into the curricula of undergraduate and postgraduate courses in chemistry in subjects of a transversal manner so as to:

> increase the students' skills to think critically about the chemistry contents, as well as the related risks and as an engine of those specific areas concerning science, technology and the society (STS) (p. 9).

However, in the general scenario of proposals related to including GC in chemistry courses in the country, these proposals have actually been drawn up more on the idea of GC seen as a specific subject or as a re-organization of environmental chemistry, which is already present in most of the curriculum of chemistry BEd courses in the country. At the basic education level, studies have shown the possibility of adopting the principles of GC in experimental activities[42] and of these associated with environmental education actions.[48] Implementing these experiments has brought changes to the routine of the laboratory; for example, by reducing the use of reagents and chemical residues and implementing safety measures as indicated.

Regarding the role of manuals and didactic texts, these can play a relevant role in diffusing activities related to environmental issues in situations that involve chemistry in general and GC in particular, as well as emphasizing the pedagogic role for teachers and students. In some studies carried out by Drews[24,49] in manuscripts of the journals *Revista Química Nova na Escola* (QNEsc) and *GREEN*[50] which included environmental issues published between 1998 and 2009—directly or indirectly involving GC—and aimed at teaching chemistry and teachers of basic education level schools in Brazil and Italy, a certain level of predominance in the way of comprehending the relation between chemistry knowledge and environment, and limited these to study the physical–chemical and biological processes resulting from the interaction between the biotic, non-biotic and social–human systems (52% of the published texts), with the texts mentioning 'hydro-resources', 'terrestrial atmosphere' and 'soils'. Other ways of comprehending these relations, such as prevention, were identified in 19 texts (40%), where 12 of were published in the Italian journal *GREEN*.

Taking into account the context presented, we expect that this text concerning scientific research in GC related to education and published over the last decade in Brazil can contribute to a better understanding of the scenario described above, making clear the challenges and possibilities of GC rooted to chemical education and to chemistry teaching in schools.

3.5 Methodological Aspects of the Survey and Analysis of Scientific Research

The study consists of a survey and an analysis of scientific research from Brazilian authors published as Doctoral and Master's theses, articles in journals, books and conferences. Data was collected from 2002 to 2013, considering that 2002 was when academic publication on GC started in Brazil and 2013 was the year this chapter was written.

The publications were searched based on their titles, keywords, abstracts and, in some cases, within the text itself, expressing terms in Portuguese or English: GC, GC teaching and GC education. As the survey was concentrated on Brazilian authors, initially the research was carried out mainly on journals from SBQ:[5] *Química Nova* (QN); *Revista Química Nova na Escola* (QNEsc); *Journal of the Brazilian Chemical Society* (JBCS); and *Revista Virtual da Química*

(RVq). The survey also included work published in the annals of SBQ Annual Meetings (RASBQ). The search mechanisms varied for the various sources: in QN journal, SciELO platform (an electronic library covering a selected collection of Brazilian scientific journals) was used; in the case of JBCS and the RASBQ's annals, was the online SBQ platform. For the QNEsc journal, all volumes were read in order to identify the relevant articles. In addition to these data, publications from Brazilian authors in books and book chapters already known in academia were considered. Other sources were theses in the areas of chemistry, Teaching/education and scientific education, whose search was carried out through the CAPES[51] website; in this case, the analysed available material included that until 2012. Because some Brazilian researchers known to be involved in GC education research have had their work published in international journals, to eventually find their publications, a survey was carried out in their curriculum vitas in the Lattes platform (CNPq),[52] a Brazilian information system to manage information on science, technology, and innovation related to individual researchers and institutions working in Brazil. In these cases, the search was carried out directly in the journals. The final results pointed to 228 works, published in 13 different vehicles (Table 3.1).

Table 3.1 shows the data according to the publication vehicles—whole and proportional number of works in the same vehicles—expressing the main subject contents they are aimed at, as well as the work aimed at curricular aspects involving proposals for basic education teaching and training (aimed at BEd and Bachelor of Science degree, BSc). In the columns related to subject, 43 works (about 19% from the total) can be seen, of which 18 are aimed at teacher training, split into six for basic education and 12 for BEd training specifically, that is, at higher education level. The remaining ones make up 25 pieces of work aimed at chemist training, which in real terms includes the training for all courses indistinctively (BSc, BEd and, eventually, industrial chemist) (see this distinction in Figure 3.1).

On the other hand, the data showed a larger number of Master's than Doctoral theses; however, in both cases there is a clear increase in numbers from 2009. From the total number of this kind of work (65), eight were from postgraduate programmes from the teaching and education area whose subjects

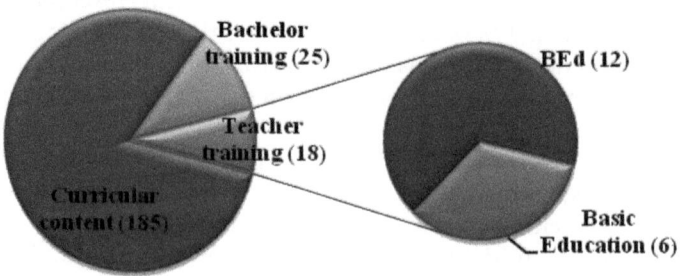

Figure 3.1   Distribution of production in terms of the focus.

Table 3.1 Summary of the distribution of publications in green chemistry, according to its vehicles and focus (curricular content/area and subject aim).[a]

Publication	No. of papers	Curricular content										Focus — Subject	
		ORG	INO	CAT	ANA	ENC	Integrative	Introductory	Practical	PHY	CMT	Teacher training	Bachelor training
RVq	19	13	—	2	—	—	—	—	—	1	—	1	2
JBCS	18	10	—	6	1	—	—	—	—	1	—	—	—
QN	29	9	1	2	4	1	—	—	—	—	—	2	10
QNEsc	4	—	—	—	—	—	—	—	—	—	—	3	1
PAC	1	—	—	—	—	—	—	—	—	—	—	—	1
Green Chemistry	1	—	—	—	1	—	—	—	—	—	—	—	—
Enseñanza	2	—	—	—	—	—	—	—	—	—	—	2	—
Educación	1	—	—	—	—	—	—	—	—	—	—	—	1
CERP	1	—	—	—	—	—	—	—	—	—	—	—	1
RASBQ	83	41	—	11	14	2	3	2	1	1	1	—	7
MSc theses	45	17	3	5	11	—	—	—	—	3	—	6	—
PhD theses	20	6	2	—	6	—	—	—	—	3	1	2	—
Books and chapters	4	—	—	—	—	—	—	—	—	—	—	2	2
Total	228	96	6	26	37	3	3	2	1	9	2	18	25

[a]RVq, *Revista Virtual da Química*; JBCS, *Journal of the Brazilian Chemical Society*; QN, *Química Nova*; QNEsc, *Química Nova na Escola*; RASBQ, Annual Meeting of the Brazilian Chemical Society; ORG, organic chemistry; INO, inorganic chemistry; CAT, catalysis; ANA, analytical chemistry; ENC, environmental chemistry; PHY, physical chemistry; CMT, chemistry of materials; PAC, pure and applied chemistry; CERP, *Chemical Education Research and Practice*; Educación, *Educación Química*; Enseñanza, *Enseñanza de las Ciencias*.

were concerned with those areas; the remaining ones (57) were from the chemistry area or another one related to it. Many of the papers in the RASBQ (59) were from scientific undergraduate projects and postgraduate programmes and, from a total of 21 aimed at education, only three were related to secondary school level. In the QN journal, there was a larger number of publications in 2012, 12 were in the education section; ten of them were divulgation ones and only two were aimed at basic education level teaching.

As can be seen in Figure 3.2, research aimed at curricular content, related to organic chemistry (96) was the most frequent, followed by that from the areas of analytical chemistry (37) and catalysis (26). Moreover, from the general survey of production, summarized in Table 3.1, an indication of the GC principles involved in the publications can be clearly seen (Figure 3.3). In some cases, many authors publish their work aimed at GC without making the involved principles explicit.

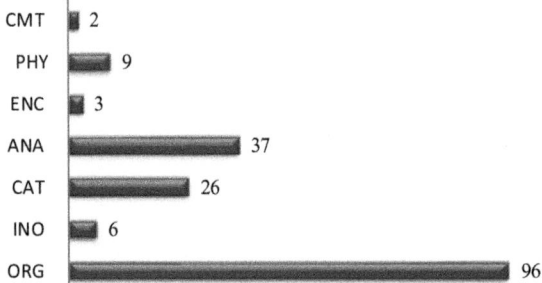

Figure 3.2   Publications according to curricular content.

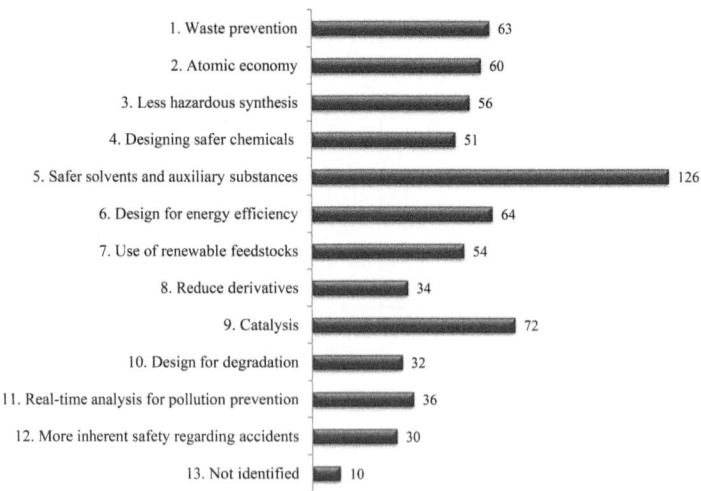

Figure 3.3   Principles of green chemistry identified in the publications analysed.

Therefore, the data presented reflects a scientific production that has a historical background. It basically started with the pioneering texts published on GC written in 2002 and 2003, including setting up the undergraduate course in environmental chemistry at the University of São Paulo, as well as the editorial of SBQ in 2007 which saw GC as a strategic measure for the development of chemistry in Brazil, and also the study developed by CGEE ('Green Chemistry in Brazil'). This last study establishes a wider scope about the state-of-the-art of GC in the country, pointing out strategic aspects for its advancement and consolidation over the next 20 years. Finally, there was choice of the IUPAC to hold its 4th International Conference on green chemistry in 2012 in Brazil, which was one of the main events currently dedicated to GC. Therefore, there are several aspects that can be analysed using the information obtained by the survey. However, among the most substantial ones, the first is that this volume of research involved a large number of people (about 600, among students and researchers) who published on GC in Brazil. A number that shows a significant interest for research, a relevant motivation from the community of chemists and encouraging intellectual vigour, shown mainly by the undergraduate students, especially in the RASBQ. Moreover, the Brazilian postgraduate programmes in the areas of chemistry and scientific education show a keen interest in including GC in teacher training and research concerning chemistry teaching, having publications, which in addition to diffusing them, have brought about the discussion of new subjects, contents and training processes. This signals a promising future for the training of new mentalities and professional abilities. Concerning teaching material, a relevant initiative was the journal QNEsc in 1994, which was a strategic action by SBQ to reach chemistry teachers at the basic education level. Bejarano and Carvalho[53] observed that:

> the journal [...] was not intended, since it started, to be a scientific journal of chemistry education; on the contrary, as shown by its language and sections structure, its aim was to reach chemistry teachers, especially at secondary schools, providing them with tools to improve their teaching (p. 164).

Still on books and didactic material, a relevant example is the book *Green Chemistry: Fundamentals and Methods*,[9] which is solely about GC and summarizes thematic indexes, academic initiatives and industrial applications of GC in Brazil.

Publications targeted at teaching and learning GC at the basic education level were also found that were more aimed at training teachers who will work at this level of education. Some milestones are worth mentioning, starting from the work that reflects the need and points to possibilities for including GC at the basic education level, considering that chemistry teachers working at that level already address environmental issues during their chemistry classes,[31,42] even considering the teaching challenges they have in their approach probably due to their university training.[54] Another study is about an experience in Italy with GC developed and aimed at the basic educa-tion level,[55,56] whose content and teaching focus of research were compared with Brazilian editorial productions. The major objective of the studies was to emphasize examples that show possibilities for GC to be adopted at this level of education and teacher training.

3.6   Final Considerations

To ask, *Why and for what do we need greener chemistry?*, is something quite often heard in the academic environment and whose meaning includes the assumption that chemistry being practised so far is (also) responsible for the current environmental problems. Therefore, changing it is an urgent chal-lenge. Here we are not analysing the degree of responsibility for the envi-ronmental crisis, or whether it in fact exists. The fact is that environmental issues are increasingly more present in the list of concerns of society and governments, as well as national and international bodies.

Therefore, chemistry should provide answers, and this will mean carrying out a thorough reassessment of the paradigms that have guided it through-out scientific history, even if such reassessment might not bring results that are ample, solid or even immediate to society concerning the solution of the environmental problems. It is not only a question of implementing technical changes, but rather of adopting another kind of scientific rationality, which consequently will require different professional training, not only for those starting out, but also from those already working. These changes will need to be supported by a new way of looking at the relationship between chemistry and the environment, expressed through chemistry education and aimed at socio-environmental sustainability.

Even though the first article related to GC in Brazil was published only in 2002, we should recognize that the success of GC depends on all the people who deal directly or indirectly with processes and chemical products care-fully adopting its principles and knowing the risks involved. It is precisely in this aspect that education plays a crucial role. Young people who are edu-cated with these principles will be adults tomorrow and will be more aware of their responsibilities about environmental issues; they will be much better informed and prepared to take decisions, being more proactive in exercising their citizenship.

This is something that can be obtained by strategic actions from govern-mental bodies of S&T (Science & Technology) and education, which could include, for example, supporting more circulation of scientific work of GC in academic and industrial environments; funding by national bodies of research in GC, adopting technological cleaning systems in industries, and curriculum changes in chemistry BEd and BSc courses.[4]

Finally, we reinforce the understanding that one of the biggest challenges in implementing GC at the basic education level in Brazil—in progress, but still having incipient results—involves teacher training. Which content, to what extent, and which approach (methodology) specific to GC all need to be dealt with and have to be better understood and improved. Moreover, this should not mean a passive attitude from chemistry educators, because adopting GC involves not only technical aspects, but also a change of mentality. The survey on GC in Brazil showed that for this teaching level (basic education), there are still few experiences or proposals of activities, teaching material and suitable practical work that could be included in the syllabus of a chemistry course or even that could be easily adapted for the classroom or in activities at the laboratory. From another angle, we reinforce – by reflecting on why and where to start implementing GC at the basic education level – where we show some examples, emphasizing the aspects about personal and environmental safety involving chemical products. In addition, we believe it is possible to use metrics in GC when the approach is concerned with environmental issues, and within this approach by using metrics, opening up the possibility of studying the concept of atom economy (expressed by the E factor),[57,58] an important parameter related to GC. These concepts can easily be included in laboratory activities involving some kind of synthesis, going beyond the classic determination of the reaction's outcome. This would allow students to analyse and determine 'how green' the chemical reaction is in comparison to these parameters. Analysing 'how green' chemistry is could also be part of the experimental activities proposed in teaching material, as already emphasized in the literature.[59] Then, these simple strategies could allow easy integration of relevant GC concepts with activities already existing in teaching programmes and materials. Along these lines, suggestions for experimental activities are already available in the literature,[60] which would strengthen the argument in favour of GC at the basic education level. Examples of experimental activities are found in the book *Introduction to Green Chemistry: Instructional Activities for Introductory Chemistry HS25*,[61] and, more recently, in an issue of the journal *Chemistry Education Research and Practice*[62] (published by the Royal Society of Chemistry) about environmental sustainability and GC in chemistry education. Finally, based on this survey, we attempted to point out the occurrence and recurrence of themes and proposals that Brazilian chemists and chemistry educators have offered to teacher training and schools in Brazil. These occurrences bring about, in general terms, trends of a positive synergy to adopt philosophical principles concerning GC to include it in all areas of chemistry, a contribution for the general reformulation of its classic postulates, such as risk and dilution, towards the safeguard of the environment in a critical historical moment expressed by the environmental crisis. Taking this into account, Brazilian chemistry joins international societary efforts with academic creativity and rigour.

References and Notes

1. M. Reigota, *Pesqui. Educ. Ambiental*, 2007, **2**(1), 33–65.
2. C. M. Kawazaki and L. M. Carvalho, *Educ. Rev.*, 2009, **25**(3), 143–157.
3. M. Tristão and R. P. Jacobi, *A educação Ambiental e os movimentos de um campo de pesquisa*, Annablume, São Paulo, 2010, pp. 13–29.

4. V. G. Zuin, *A inserção da dimensão ambiental na formação inicial de professores de Química*, Editora Átomo, Campinas, 2011, p. 179.
5. H. Fracalanza, I. A. Amaral, J. Megid Neto and T. S. Eberlin, *Ci. em Foco*, 2008, **1**(1), http://www.fe.unicamp.br/revistas/ged/cef/article/view/4458, accessed January 2014.
6. M. Sorrentino, *Debates Socioambientais*, 1997, **2**(7), 3–5.
7. *Parâmetros Curriculares Nacionais para o Ensino Médio*, Ministério da Educação do Brasil (MEC), 1999, http://portal.mec.gov.br/seb/arquivos/pdf/blegais.pdf, accessed January 2014.
8. E. F. Moradillo and M. C. M. Oki, *Quim. Nova*, 2004, **27**(2), 332–336.
9. V. G. Zuin, C. R. Farias and D. Freitas, *Rev. Eletrón. Enseñanza Ciênc.*, 2009, **8**(2), 552–570.
10. P. F. L. Machado, J. A. Baptista, J. A. Trindade and W. L. P. Santos, *Concepções de professores sobre Educação Ambiental no ensino de Química*, 2007, VI ENPEC, http://www.nutes.ufrj.br/abrapec/vienpec/CR2/p1122.pdf, accessed January 2014.
11. J. Thornton, *Int. J. Occup. Environ. Health*, 2000, **6**(4), 318–330.
12. S. Brunet, P. Delvenne and G. Joris, *Sociologias*, 2011, **13**(26), 176–200.
13. C. U. Luchesi, *Considerações sobre o Princípio da Precaução*, Editora SRS, São Paulo-SP, 2011, p. 142.
14. M. Poliakoff, J. M. Fitzpatrick, T. R. Farren and P. T. Anastas, *Science*, 2002, **297**(5582), 807–810.
15. J. Thornton, *Pure Appl. Chem.*, 2001, **73**(8), 1231–1236.
16. *Declaração do Rio sobre Meio Ambiente e Desenvolvimento*, Princípio 15, http://www.onu.org.br/rio20/img/2012/01/rio92.pdf, 1992, accessed January 2014.
17. C. A. Marques and A. A. S. C. Machado, *Found. Chem.*, 2014, **16**, 125–147.
18. T. S. Kuhn, *The Structure of Scientific Revolutions*, University of Chicago Press, Chicago- IL, 1970, p. 210.
19. The Twelve Principles of Green Chemistry: (1) Residues Prevention; (2) Atomic Economy; (3) Less Hazardous Synthesis; (4) Molecular Planning of Safer Products; (5) Safer Solvents and Auxiliary Substances; (6) Planning for the Energetic Efficiency; (7) Use of Renewable Raw-materials; (8) Reduction of Derivatisation; (9) Preference for Catalytic Reactions; (10) Planning for Degradation; (11) Real-time Analysis for Pollution Prevention; (12) More Inherent Safety Regarding Accidents.
20. *4th IUPAC Conference on Green Chemistry*, International Union Pure and Applied Chemistry, 2012, http://www.ufscar.br/icgc4/, accessed January 2014.
21. L. A. Faria and D. I. T. Fávaro, *Quim. Nova*, 2011, **34**(6), 1089–1093.
22. A. G. Correa and V. G. Zuin, *Química Verde: Fundamentos e Aplicações*, Ed. UFSCar, São Carlos-SP, 2009, p. 172.
23. WCED (World Commission on Environmental and Development), *Our Common Future*, Oxford University Press, Oxford, 1987, p. 383.
24. *Sociedade Brasileira de Química*, http://www.sbq.or.br, accessed January 2014.

25. E. J. Lenardão, R. A. Freitag, M. J. Dabdoub, A. C. F. Batista and C. Silveira, *Quim. Nova*, 2003, **26**(1), 123–129.
26. A. G. S. Prado, *Quim. Nova*, 2003, **26**(5), 738–744.
27. A. M. Sanseverino, *Ciência Hoje*, 2002, **31**(185), 20–27.
28. A. M. Sanseverino, *Quim. Nova*, 2000, **23**(1), 102–107.
29. L. A. Silva and J. B. Andrade, *Cad. Temáticos Quim. Nova Esc.*, 2003, **5**, 3–6.
30. L. M. O. C. Merat and R. A. S. San Gil, *Quim. Nova*, 2003, **26**(5), 779–781.
31. J. C. Coelho and C. A. Marques, *Quim. Nova Esc.*, 2007, **25**, 14–19.
32. Sociedade Brasileira de Química, *Quim. Nova*, 2007, **30**(2), 255.
33. Coordenação de Aperfeiçoamento de Pessoal de Nível Superior (CAPES), http://www.periodicos.capes.gov.br, accessed January 2014.
34. Diretório dos Grupos de Pesquisa no Brasil, http://dgp.cnpq.br/buscaoperacional/, accessed January 2014.
35. Centro de Gestão e Estudos Estratégicos, *Química Verde No Brasil: 2010–2030*, CGEE, Brasília, 2010, p. 438.
36. A. A. Mozeto and W. F. Jardim, *Quim. Nova*, 2002, **25**(1), 7–11.
37. Sociedade Brasileira de Química, *Divisão de Química Ambiental*, http://www.sbq.org.br/filiais/index.php, accessed January 2014.
38. S. E. Manahan, *Environmental Chemistry*, Lewis Publishers, Boca Ratón, 6th Edição, 1994, p. 811.
39. L. P. Cortes Jr, *As Representações Sociais de Química Ambiental: contribuições para a formação de bacharéis e professores de Química*. Dissertação (Mestrado em Ensino de Ciências) – Universidade de São Paulo, São Paulo, 2008.
40. A. Machado, *Bol. S. P. Q.*, 2004, **95**, 59–67.
41. *In Brazil, There Are Specific Undergraduate Courses to Teacher Training in All Scientific Areas Aiming the Basic Education.*
42. C. A. Marques, J. C. Coelho, F. P. Gonçalves, R. H. Lindemann, L. C. Mello, P. R. S. Oliveira and E. Zampiron, *Quim. Nova*, 2007, **30**(8), 2043–2052.
43. V. G. Zuin and J. L. A. Pacca, *Enseñanza Cienc.*, 2013, **31**(1), 79–94.
44. L. F. Goes, S. H. Leal, P. Corio and C. Fernandez, *Educ. Quím.*, 2013, **24**(1), 113–123.
45. L. S. Shulman, *Educ. Res.*, 1986, **15**(2), 4–14.
46. P. Porto, P. Corio, F. Maximiniano and C. Fernandez, *Enseñanza Cienc.*, 2009, V. Extra, 1527–1533.
47. A. G. Corrêa, V. G. Zuin, V. F. Ferreira and P. G. Vasquez, *Pure Appl. Chem.*, 2013, **85**(8), 1643–1653.
48. E. L. Silva, *Educação ambiental em aulas de química em uma escola pública: sugestões de atividades para o professor a partir da análise da experiência vivenciada durante um ano letivo*, dissertação de mestrado, Instituto de Ciências Biológicas, Física e Química da Universidade de Brasília, 2007.
49. F. Drews, *Abordagem de Temáticas Ambientais no Ensino de Química: um olhar sobre textos destinado sao professor da Educação Básica*, dissertação de mestrado, Programa de Pós-Graduação em Educação Científica e Tecnológica, UFSC, Florianópolis, 2009.

50. *Green - Revista Italiana Do Consórcio Interuniversitário Química Nacional Do Meio Ambiente (INCA), Em Colaboração Com a Sociedade Química Italiana (SCI)*, http://www.incaweb.org/green, accessed January 2014.
51. http://www.periodicos.capes.gov.br/ ou http://bancodeteses.capes.gov.br/
52. *Plataforma Lattes/CNPQ*, http://www.lattes.cnpq.br, accessed January 2014.
53. N. R. R. Bejarano and A. M. P. Carvalho, *Educ. Quím.*, 2000, **11**(1), 160–167.
54. A. L. Leal and C. A. Marques, *Quim. Nova Esc.*, 2008, **29**, 30–33.
55. F. Drews and C. A. Marques, in *Temas de Ensino e Formação e Professores de Ciências*, Editora da UFRN, Natal-RN, 2012, cap. 4, p. 73.
56. C. A. Marques, *Revista Electrón. Enseñanza Cienc.*, 2012, **11**(2), 316–340.
57. B. M. Trost, *Science*, 1991, **254**(5037), 1471–1477.
58. R. A. Sheldon, *Green Chem.*, 2007, **9**, 1273–1283.
59. M. G. T. C. Ribeiro and A. A. S. C. Machado, *J. Chem. Ed.*, 2013, **90**(4), 432–439.
60. http://jchemed.chem.wisc.edu.
61. American Chemical Society (ed.), *Introduction to Green Chemistry: Instructional Activities for Introductory Chemistry*, HS25, 2002.
62. *Chem. Educ. Res. Pract.*, 2012, **13**(2), DOI: 10.1039/C2RP90005J, 53–54.

CHAPTER 4

Learning about Sustainable Development in Socio-Scientific Issues-Based Chemistry Lessons on Fuels and Bioplastics

RACHEL MAMLOK-NAAMAN*[a], DVORA KATCHEVICH[a], MALKA YAYON[a], MAREIKE BURMEISTER[b], TIMO FEIERABEND[b], AND INGO EILKS[b]

[a]Weizmann Institute of Science, Rehovot, Israel; [b]University of Bremen, Bremen, Germany
*E-mail: Rachel.mamlok@weizmann.ac.il

4.1 Introduction

The central focus of education for sustainable development (ESD) is to prepare the younger generation to become responsible citizens in the future. Students should be able to participate in a democratic society and to help in shaping future society in a sustainable fashion. They should learn to take responsibility for both themselves and future generations, based on the concept of sustainable development.[1]

From the beginning the United Nations added a key role for sustainable development to education. The idea of education for sustainable development

Worldwide Trends in Green Chemistry Education
Edited by Vânia Gomes Zuin and Liliana Mammino
© The Royal Society of Chemistry 2015
Published by the Royal Society of Chemistry, www.rsc.org

(ESD) was suggested in the Agenda 21.[2] People at all levels of education were asked to contribute to ESD, among them students receiving secondary chemistry education. ESD is a skill-oriented educational paradigm that aims to prepare young people to become responsible future citizens.[3] ESD aims to develop skills in the learner in a way that they become able to make decisions and actively shape present and future society in a sustainable fashion.[4] For sustainable development of our present and future society decisions are to be made on, *e.g.*, consumption of resources, environmental protection, or new and alternative technologies. Many of these decisions are related to chemistry and chemical technology.[5,6] However, a thorough integration of ESD into secondary school chemistry education is still lacking.[3]

Investment in the development of a chemistry curriculum and teacher education is needed to develop general and domain-specific knowledge and skills in the learners that enables future students to assess and make decisions about chemistry-based processes, technologies and products.[3] All students need to develop corresponding skills irrespective of whether or not they will later embark on a career in science and technology, because all of them will be asked to act as responsible citizens in the future and to contribute to societal decision-making.[6,7]

Recently, Burmeister *et al.* suggested four basic modes of how to integrate ESD with chemistry education. The modes range from (1) an application of green chemistry principles in school science practical work, *via* (2) using sustainability issues to contextualize chemistry content learning, towards (3) addressing technological and environmental challenges in socio-scientific issues-based science education, and finally (4) innovating school life along sustainability principles.[3] Concerning the development of domain-specific educational skills for participation in societal debate and decision making about chemistry-related issues of sustainable development they suggested socio-scientific issues (SSI)-based science education as one of the most promising strategy.[3] Accordingly, this chapter discusses how secondary chemistry education can contribute to ESD based on the idea of SSI-based science education. Different examples will be discussed as they were developed in Israeli and German chemistry education. Alternative fuels and bioplastics will serve as examples.

4.2   Socio-Scientific Issues of Sustainable Development and Chemistry Teaching

Chemistry education can follow different curriculum orientations. These orientations range from rote following the structure of the discipline towards context-based and societal driven chemistry curricula.[8] Within this range, research suggests a stronger orientation of science learning along student-relevant contexts.[9] Many justifications are available to give societal issues among these contexts a more prominent place in science education in general[7,10–12] and in chemistry education in particular.[6,13]

Among the many underlying frameworks suggesting a stronger societal orientation of science education we can allocate for example Activity Theory[14,15] or the European tradition of Bildung.[7,16] The main reason for a more thorough societal orientation of chemistry education in all these theories is suggested in the over-arching goal behind any education of the young generation to helping them becoming self-standing and responsible citizens in the future.[7,17] All school education domains have to contribute to this aim, among them secondary chemistry education.[18] Accordingly, the societal dimension of science education should be an essential component of any relevant secondary chemistry curriculum.[12]

At this point curriculum theory meets the political suggested framework of ESD.[3] Also, ESD aims to develop skills in the learner for responsible citizenship and enable future generations to act in and shape society in a sustainable way.[4,19,20] Thus, societal driven science and chemistry teaching is essential if formal chemistry education at the primary and secondary school levels is asked to contribute to ESD.

For strengthening the societal dimension of science education different justifications and organizers have been suggested in recent years.[7,10,11,15,21] Models were developed that suggest appropriate curriculum structures starting from societal questions, leading over to science content learning, and coming back to understand the impacts of the societal issue.[15,22] Related innovative pedagogies were developed, *e.g.* mimicking societal practices of communication and decision-making in science education about controversial questions from science and technology, *e.g.* working like a journalist or role-playing of political debate.[23] Most of these works see themselves as being part of the SSI-based science curriculum movement.[11] SSI-based science teaching goes beyond just contextualizing science content learning by societal questions.[8,21,24] SSI-based science teaching makes the societal issue itself the content of science learning.[11] Science learning starts from a controversial societal issue that challenges the learner. Sadler[25] suggests that those issues have most potential to challenge the learners that 'encourage personal connections between students and the issues discussed, explicitly address the value of justifying claims and expose the importance of attending to contradictory opinions.'

Recently, Stolz *et al.*[26] suggested quality criteria for selecting and reflecting societal contexts on their potential to act as SSIs in SSI-based science teaching:

- *Authenticity*: The topic is authentic, when it is currently being discussed by society. For proof, common media can be checked for the presence of the topic, *e.g.*, in newspapers and magazines, on TV, or by advertising.
- *Relevance*: The topic is relevant if respective decisions will affect the current or future lives of the students. Scenarios are reflected to see which impact potential decisions will have on, *e.g.*, behavioural choices, consumer behaviour, or availability of products.

- *Evaluation is undetermined in a socio-scientific respect*: The socio-scientific issue allows for different points of view. Media can be analysed whether controversial viewpoints are represented, *e.g.*, by interest groups, the media, politicians, or scientists.
- *Allows for open discussions*: The topic must allow for discussion in an open forum. Thought experiments test arguments to make sure that no individuals, religious or ethnic groups would feel themselves to be insulted or pushed to the fringes of society by their use.
- *Deals with an issue based on science and technology*: This topic concerns itself with a techno-scientific query. Discourse in the media is analysed. The question is raised, whether scientific facts and concepts are addressed and either explicitly or implicitly used for argumentation.

If we now consider prototypical chemistry-related issues from the debate about sustainable development and green chemistry many of them meet these criteria. Chemistry-related issues of sustainable development, like a more intense production and use of alternative fuels or bioplastics, are discussed in mass media. Regulations on their use will potentially have an impact on the students' consumer choices. The use of alternative fuels or bioplastics is still controversial; pro and con arguments are given in societal debate and can be used in students' open debate. And, finally, these issues are definitely questions of chemistry and technology; corresponding arguments are used in public debate.

As a result, many chemistry-related issues of the sustainability debate well suit the criteria for feasible societal questions to be used in SSI-based chemistry teaching. Operated in a corresponding pedagogy they are suggested as having potential to motivate chemistry content learning and also to contribute to the development of general educational skills in the means of ESD.[8] The following cases will provide some ideas of how this can be done in class.

4.3   Issues of Sustainable Development in the SSI-Based Chemistry Classroom

Central issues of sustainable development concern, among others, a responsible use of natural resources. Challenges include raising living standards and economic prosperity while, at the same time, maintaining the environment, protecting fossil resources, or limiting climate change. Many of these challenges are connected to traditional and alternative practices in chemistry and chemical technology.[5] Chemistry's answer to these challenges is the search for a green or sustainable chemistry that has emerged in the last 20 years, starting from early ideas by Anastas and Warner.[27,28] One idea for more green and sustainable chemistry is suggested in the use of raw materials from renewable sources.[28] Fuels and materials from plants offer a chance to protect fossil resources and reduce the exposition of carbon dioxide into the atmosphere. Additional advantages can lie in better environmental compatibility, as well as investment in innovative technologies can provide positive economic and societal perspectives.

Developments in corresponding technologies have been intensified and become industrial practice in many countries in recent decades. On a broad base industry started to produce alternative fuels starting from vegetable oils or using ethanol. Also, the problem of growing amounts of plastic waste in the landscape and the oceans was taken into focus. The idea of developing bioplastics emerged. Bioplastics are made, at least partially, from renewable plant materials; some of them also have the advantage of being biodegradable and thus cause less waste to be deposited.

4.3.1 Teaching and Learning on Traditional and Alternative Fuels

In recent years, in Israel, a group of teachers developed a lesson plan which was called 'Can used oil be the next generation fuel?'.[29] This lesson plan focuses students learning about traditional and alternative sources of fuels. The students should learn about the advantages and disadvantages of each of different suggested technologies: fuels from crude oil, recycling of used oils, or producing biodiesel from vegetable oil.

The teaching–learning module uses a structure that comes from the socio-scientific issue, *via* learning about the content behind the issue, towards questions of evaluation and reflecting the issue from different perspectives and in the foreground of the societal discourse. This structure is parallel to those suggested in the curriculum models by Holbrook and Rannikmae[30] and Marks and Eilks[22] for societal-oriented or even SSI-driven science education, respectively.

The lesson starts with students being exposed to information about the world's energy crisis and its consequences. This information is discussed to activate prior knowledge and create questions. The idea that teachers should convey to their students is that the crisis is a worldwide problem and is not only another scenario for the science classroom. It is also made clear that there are several proposed solutions to this crisis, but often these solutions introduce new problems.

In order to make any decision regarding the various fuels students undertake different activities to investigate and compare the fuel types. One activity is that the students are requested to inquire into the chemistry of the use of different fuel types, one of which is biodiesel. Comparative activities require students to select criteria such as enthalpy of combustion values or the release of emissions. The teacher introduces the student to an experiment that compares the energy released by the combustion of different fuel types. By measuring the mass of the fuel needed to increase the temperature of a certain volume of water by 30 °C (Figure 4.1), students can compare the caloric value of different fuels. They also can investigate the level of pollutants emitted from the burning fuels with a special board called the Ringelmann Scale, which determines the concentration of soot particles accompanying the flame.

Figure 4.1   Which is the best fuel?

Following the comparison of chemical behaviour students are requested to decide: Which is the best fuel? Before making a final decision there is an attempt to involve students emotionally and involving an ethical perspective by creating a conflict regarding the use of biodiesel. This activity is based on viewing pictures that highlight the use of crops for fuels instead of using them as a food source in our world's growing population. Students' decisions should be based on arguments, but their decision first needs an agreement within the group: What do they suggest to be the meaning of 'the best' fuel? This discussion leads to understanding that a thorough comparison requires more criteria behind chemical behaviour. These criteria need to include price, environmental behaviour, production methods, or societal impacts. An open discussion about which technology has the most promising potential for sustainable development closes the lesson plan.

Within this lesson plan, the students learn about an authentic sustainability issue and the complexity of its solution. On the one hand, they learn that there is no 'best fuel', nor any 'best solution' to many sustainability problems. On the other hand, they learn that making use of used oil or biofuels is not 'the ideal solution', too. Nevertheless, both ways can offer a contribution to environmental protection because less waste is produced and fossil resources are also saved. However, the students also learn how complex are such evaluations and how many dimensions need to be taken into consideration if an overall decision is requested.

Also, in other countries, biofuels have become part of the secondary chemistry curriculum. In 2002, Eilks suggested a lesson plan on the use of biodiesel;[31] a couple of years later a quite similar lesson plan was published concerning the societal debate about the use of bio-ethanol by Feierabend and Eilks.[32] Both lesson plans follow the curriculum model suggested by Marks and Eilks[22] and include some innovative pedagogy. They start from authentic media from the public debate. In the case of biodiesel, advertising and brochures from pressure groups are used to open the context and to provoke questions. In the bio-ethanol example articles from news magazines are analysed on the concurrency of food and fuel production. In both cases, questions are derived from the material. Student questions regularly concern the science behind the issue. Starting from the public media consequently other questions are regularly also set up. The questions encompass aspects of consumer behaviour, but also on implications of the new technology. Implications are put into question stemming from the different fields of economy, ecology and society, the three most prominent dimensions in current models of sustainability.[3]

After a start from authentic media, the applied curriculum model suggests a phase of clarifying the basic chemistry behind the issues. This is the chemistry of fat and *trans*-esterification or of fermentation and alcohols, respectively. Learning the science background is done in a combination of theoretical learning and practical work. Similar experiments are done as suggested in the previous example from Israel. Additionally, biofuels are produced by *trans*-esterification of rape seed oil and distillation from fermented grapes, respectively. Practical work is embedded into cooperative learning techniques, especially the learning-at-stations method.[33]

Reflecting the chemistry content, learning makes clear that chemistry can only help understand the technological background behind the issue. However, balanced evaluations and decisions need to also include ecological, economic and societal aspects. For learning about how society is handling socio-scientific issues the curriculum model applied suggests the mimicking of an authentic societal practice of communication and decision-making.[22,23] Consequently, in both lesson plans role play activities were implemented. In the biodiesel example the students mimic a public panel discussion. Discussants are students representing stakeholders from crude oil industry research, a traditional gasoline company, a pressure group for biodiesel promotion, and an environmental protection agency. On bio-ethanol a role play is operated in the means of a business game. In this case the scenario is that a parliamentary committee has to suggest a decision for making 10% bio-ethanol in all the gasoline compulsory. This was an authentic scenario since Germany was discussing exactly such a law during the time the lesson plan was developed. In the business game a fictive parliamentary commission conducts a hearing with different student groups representing chemists, engineers, environmental protection activists, car manufacturers, *etc*.

In both cases of role playing, intense debate regularly emerges among the students. The students experience which perspectives contribute to respective decisions. They face the situation that it is not only chemistry, science and technology which provide relevant arguments in the chemistry classroom, but arguments stemming from economy, ethics, ecology and many other fields are also needed for a balanced view. The students also learn which arguments are selected by different societal groups and how they are put into the debate. The learners even see that science-related arguments are mainly introduced by non-scientists. Representatives of industry and pressure groups purposely select items from the available scientific information and use them in a way that supports their interests. The representatives take their choice from the available arguments and transform them as they like. Hofstein *et al.* called this learning 'filtered information'.[7] All these observations are reflected in the final part of the lesson plan to contribute to students' skills of understanding and providing critical reaction to public debate about sustainability-related socio-scientific questions.

4.3.2 Teaching and Learning on Traditional and Alternative Plastics

As is the case for fuels and biofuels, conventional and alternative plastics also provide an interesting field for SSI-based chemistry education. Similar to the example described above another teaching–learning module was developed in Israel. The module *Plastic: Reduce the use!*[34] is designed to expose the students to the convenient use of plastic products, on the one hand, and to the environmental impact caused by over-using those products, on the other. The module aims at developing, in the students, a sense of responsibility to the environment, by increasing their awareness of the environmental impact of non-biodegradable polymers, and by exposing them to ways of reducing damage, such as by recycling and the use of alternative biodegradable materials.

The lesson sequence follows a similar curriculum model to the examples described above. At the beginning of the lesson plan, the students are shown pictures of environmental pollution caused by plastic products and dead animals lying on a shore that is full of plastic bags. The goal of this activity is to affect students emotionally and get them ethically engaged. The students are requested to answer some questions about their feelings and thoughts regarding the pictures:

- What do you feel when you look at the picture?
- Does it bother you to see the picture? If so, why? If not, why?
- What would you like to do after seeing the picture? Write down at least two ideas.

After raising the issue, the students become acquainted with the basic and relevant scientific concepts related to polymers such as the nature of polymers, the repeating unit, recycling, and biodegradable polymers. The students need to know these concepts to make better decisions at the end of the module. This phase is inspired by another connection to the students' life and personal living environment. It starts by addressing the emotional aspects of the issues: a story about a sea turtle that ate a plastic bag and needed to undergo surgery to survive. At this point students have already acquired enough knowledge and understanding to reflect some questions that are asked in connection to the story focusing responsibility and potential action, and in which only part of them is to be answered by chemistry:

- Who is responsible for the damage to the sea turtle?
- Why is food not a problem when taken into the stomach, but plastic bags are?
- Why are plastics so resistant to decomposition?
- How can plastics bags be made harmless after use?

In order to understand why the turtle had to undergo surgery, students can refer to knowledge that they acquired during an investigation of the water solubility of two polymers, polyethylene and polyvinyl alcohol. The question of solubility is one of the important stages in the degradation process. Most of the students are not familiar with solubility of polymers, but this concept is needed in order to understand the chemical side of the problem.

The decision-making activity, which sums up the module, is a result of their emotional involvement and the cognitive change the students underwent. The students are asked to use a tool called 'Analysis of profit gain and loss'. The tool is a two-dimensional table. The first dimension shows different alternatives for reducing non-biodegradable plastic waste, such as consuming less, use of biodegradable bags, participating in campaigns to collect bottles for recycling, limiting the production of non-biodegradable plastics, taxing the production of polyethylene bags, and subsidizing biodegradable plastics. The second dimension shows criteria that may be affected by the first dimension for better or worse, such as space at waste sites, terrain conditions, employment of workers in factories for production of polymers, the environment, and the price of useful products. The students are asked to think about how each alternative can affect any given criterion. In addition, the students shall offer more alternatives and other criteria. At the end of this activity, each group is asked to decide which alternatives they suggest choosing in order to decrease the environmental damage caused by non-degradable plastic materials. At the end of the process each group presents to the class its decision and reflects why they chose it.

Also in this case, there are examples available in other countries. In Germany, Burmeister and Eilks recently suggested a lesson plan on evaluating conventional and alternative plastics in comparison.[35] As in the example from Israel, it also picks up the problem of the plastic waste in the environment. This problem of plastic waste in the oceans is contrasted with screening industry brochures on innovative bioplastics. The students are given single paragraphs from the industry brochures and a news magazine article on the plastic waste in the oceans. Their task is to develop headlines as if these paragraphs are to become short newspaper articles. Clustering the headlines opens the full scene of the issue from the chemistry, *via* alternative products, towards ecological and economic impacts. In addition, the socio-scientific context is used to provoke learning about polymers, their structures and synthesis. Experiments are made to investigate properties, make comparisons, and also to synthesize polymer materials. However, as described above, learning about the scientific content only helps to understand the properties, *e.g.*, questions of degradability. The question of whether alternative plastics are better than conventional ones, needs, first, an agreement about criteria and, second, a broad approach involving different dimensions.

To thoroughly focus on the three dimensions of most current sustainability models (ecological, economic and societal sustainability) a special pedagogy is applied: the consumer test method.[36] In this method the students mimic the authentic societal practice of a consumer test agency. Such agencies (like the Stiftung Warentest in Germany) provide tests of consumer products. However, to conduct a corresponding test it is the staff in the consumer test agency who decide first about the criteria to be tested and how they are weighted against each other. By applying sustainability-related criteria in the test, this method makes explicit the partially competitive nature of the different sustainability dimensions and reveals the need to balance them fairly in order to arrive at an equitable and holistic evaluation for a given product.

Within the product test method three plastics are to be evaluated: polyvinyl chloride (PVC), polyethylene terephthalate (PET), and thermoplastic starch (TPS). These particular plastics are chosen on purpose because of their significant (dis)advantages, which largely contradict one another when compared in a comprehensive analysis: PVC is a well-established, broadly applicable, and cheap material. However, PVC remains quite controversial. Even though modern use of PVC is hardly more problematic than other types of plastics, this fact is hardly ever heard in public discourse. The risks arising from unplanned combustion and those inherent in improper disposal, dominate widely the public perception. In contrast, the reputation of the 'bioplastic' TPS remains mainly untarnished. It is biologically degradable and is made from renewable sources. However, the uses of TPS are still few and its price is relatively high. TPS may help protect crude oil resources, but large-scale production in future might also increase the risk of overly intensive agricultural land use. PET is well-known for its use in drinking bottles. PET is largely ignored as a benign substance, even though recycling PET products successfully can be a highly volatile societal problem. Until recently, old bottles were seldom recycled in Western countries. Instead, they were shipped off to developing countries or to China, where they were disposed of or remanufactured into fleece pullovers and other clothing items under deplorable environmental and social conditions. Also the 'carbon footprint' of transporting the waste is an often neglected factor in evaluating PET.

The consumer testing phase begins by negotiating the evaluation dimensions. Technical properties and durability become logical factors in the evaluation. Availability and price are also easily derivable criteria. Environmental friendliness, societal compatibility, and recycling suitability become additional touchstones. Pragmatic criteria (properties and utility) are combined with economic and value-driven dimensions from the sustainability debate, namely economic (saleability, cost-price effectiveness, availability), ecological (environmental effects, health in production, disposal/recycling) and societal (production, recycling under appropriate social standards) impacts.

In the next step, the students first attribute weighting factors to the various dimensions. This step is guided by a respective worksheet (Figure 4.2). The students' suggestions regularly cover a wide spectrum. Some participants find environmental friendliness most important; others view technical properties as more crucial. Students need to search for compromises to come to joint weighting factors.

Experience shows that negotiations eventually lead to an overall rounding off of extreme values. A weighting factor of roughly 20–30% for each of the four categories is generally the end result, but distributions vary by 10–20% during each application of the method.

Each student is then assigned to individually rating one of the three plastics. Symbols ranging from ++ to – are used, which parallel the grades A, B, C, D and F in many English-speaking countries. Specially designed texts are provided, in which information about each of the plastics is tailored to the various criteria. Students carry out the evaluation individually. Then they negotiate the ratings for their particular sort of plastic in a group, and finally switch to new groups composed of at least one expert per plastic type. The learners discuss their various results and decisions, including their reasons for positively or negatively evaluating one of the plastics.

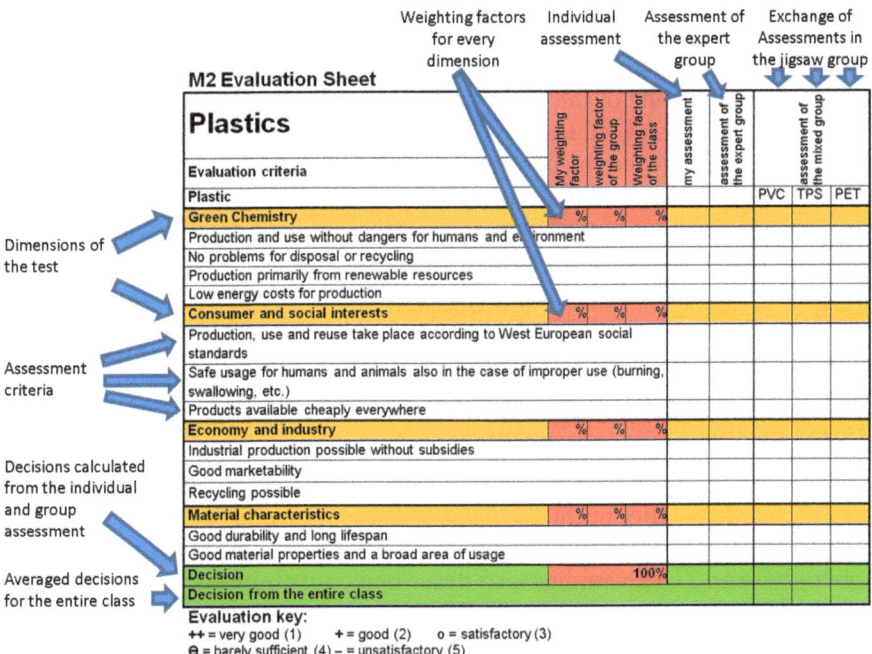

Figure 4.2 The consumer test activity.

In the reflection of the consumer test it quickly becomes clear that any advantages of one product are often negated by disadvantages in other areas. It also becomes clear that there is huge influence on the final result of both the weighting between the dimensions as well as of the individual testers' ratings. This exercise shows clearly that chemistry related products and processes need to be evaluated on a broader spectrum of criteria than those offered by chemistry alone. However, it explicates also that there needs to be balanced recognition between the different dimensions.

4.4   Effects on the Chemistry Classroom

Societal-oriented science education is suggested for many decades now and dates back at least to the 1980s.[7] Over the years, many cases and research works were published. The studies took different aspects into focus.[11,24] Positive results were found in questions of motivation and emotions,[37,38] perception of relevance of chemistry education,[39] the development of higher order cognitive skills,[31,40] or promoting environmental attitudes and activism.[19,41]

Also, the examples described in this paper were evaluated in different case studies.[29,31,32,34,35] All the cases agree the motivating potential of connecting environmental and sustainability issues with SSI-based science education and an ESD framework as also suggested by Swan and Spiro[42] or Robottom and Simonneaux.[24] A positive perception of the lesson cases among teachers and students were repeatedly described. This might be illustrated by quotes from two teachers who participated in developing the biofuels lesson plan from Israel discussed above:

'I felt that the chemistry became intriguing. Students dealt with social dilemmas. Indeed one of the main goals of Profiles [the project this module was developed in] is as 'education through science' and that was the feeling in the air. Pupils enjoyed the possibility of talking during class and even learn from it. I felt that this was a different experience for them. They can express their opinions and make decisions in a mature way and all the work is done by themselves.' (Teacher from Israel).

'I saw a lot of enthusiasm; students ran the debate and drew conclusions. Their seriousness was surprising. Each student had an opportunity to express himself clearly. Students divided their roles without any intervention on my part, and each group worked differently. A group of boys focused on the fuel type, and a group of girls focused more on the environmental consequences of burning fuel.' (Teacher from Israel).

All the cases described in this chapter also suggest a contribution to a perception of the relevance of chemistry education among the learners and a better awareness of environmental and sustainability issues, as also discussed by Mandler *et al.*[39] The studies also described contributions to general skill development, *e.g.*, how to react to politics, media or consumer choices. These claims will be illustrated by a quote from one student from Israel and three from Germany all after having learned about biodiesel in the first year of the senior high school cycle:

'I do not drive in the meantime but I already have the data on prices, the advantages and disadvantages of different kinds of fuel. Maybe I'll think about biodiesel as well. Regarding the dilemma: food or fuel, thinking and discussing the issue...they all contribute to an understanding of the complexity of the topic. It was interesting to understand how things can be easily understood.' (Student from Israel).

'We debated the use of biodiesel, so that now we can make up our own minds and argue for our own position using our knowledge (about production, properties,...). In addition, we learned something about self-organised learning, which I feel was extremely positive. Also, the group work encourages each member of the group to fulfil the requirements each time.' (Student from Germany).

'I learned a lot about the production, structure, use, advantages and disadvantages of biodiesel. Also, I consider it to be important that I learned about our environment and its protection. I especially learned about how companies sell environmentally friendly products and how naive we can be if there is the syllable 'bio' in it.' (Student from Germany).

'I have learned about the advantages and disadvantages of biodiesel, about interests of pressure groups and how to evaluate their opinions by considering their particular interests, and how to develop an opinion and make up my own mind.' (Student from Germany).

4.5 Conclusions

Sustainability issues are prevalent in the news these days, from global climate change to local concerns such as pollution in previously pure waterways.[39,43] A scientifically educated person is often appalled by the way scientific facts and theories are mis-stated (in the means of the idea of filtered information[23]) and how the roles of science and technology in both causing and curing these problems are misunderstood. Therefore chemistry education needs to prepare students to deal informed with these issues as chemistry-literate citizens in a world in which they are able to confront new problems intelligently.[6,23]

There are different effects of combining chemistry learning with sustainability issues.[3] Contextualizing chemistry by environmental issues suggests an effective method of motivating the students to learn chemistry and affecting their attitudes and emotions towards sustainability and environmental issues.[19,38,39] Adding sustainability and environmental contexts can increase students' perceptions of how relevant chemistry is to real-life problems and thus will raise motivation.[31,41,44] Motivating the students to study science, while providing a good understanding of chemical concepts, is still ranked among the most important concerns in chemistry education.[45]

The inclusion of sustainability and environmental contexts into chemistry education can have a significant impact on students' awareness of how chemistry is connected to the real world.[39,44] Students become more aware of the relationships between chemistry and the society in which they live.[22,32] Basic science knowledge is crucial for promoting sustainable development and improving the capacity of the people to address environmental and developmental issues, and thus the latter needs to be incorporated as an essential part of chemistry learning. Integrating chemistry learning with sustainability and environmental contexts is very much in line with the idea of ESD. This represents the context-based mode of integrating ESD with chemistry learning as suggested by Burmeister *et al.*[3]

However, learning chemistry-focused sustainability in a SSI-based mode should have a broader impact.[24] The NEETF environmental literacy report recommends that people should receive concrete suggestions on how to change their behaviour, with an emphasis on how they can join others who are doing the same thing.[46] The report notes a positive correlation between environmental knowledge and environmental behaviours. The more you know, the more likely you are to act to take positive steps to improve the environment.[46] Thus, the learning of chemistry should aim directly help citizens to better react to sustainability and environmental issues by providing data and theories that define the problem, skills to assess its seriousness and explain its causes, as well as to learn how to react to them.

It is important to stress that it is not only individual environmental behaviour that needs to be developed. Acting for a sustainable development also encompasses participation in societal discourse and decision-making.[4] The discussion in this chapter makes clear that raising learning success in pure chemistry knowledge is not sufficient to fully understand or to participate in societal debate on SSIs concerning questions of chemistry and chemical technology.[3,24] A broad skill-set is needed to become able to act responsibly in the societal context and respective pedagogies have to be applied.[4,35] Communication and decision-making skills need to be developed as well as skills in analysing how science and chemistry related arguments are used in society, and chemistry education should contribute to it.[23,47] Integrating chemistry learning with issues of sustainable development in an SSI approach offers great chances for creating learning environments that contribute general skill development beyond pure chemistry learning.[19,31,44] If operated in an SSI-based approach there is chance that chemistry learning thoroughly starts to contribute to the development of the skills that are important for shaping the society in a sustainable fashion.[3] The examples presented in this chapter show that corresponding lesson plans have great potential to both motivation and promoting chemistry learning on the one hand as well as general educational skill development on the other. This aim should be focused more thoroughly for chemistry teaching both at high school and undergraduate levels.[6]

Acknowledgement

Part of the work described here was funded by the European Community's Seventh Framework Program in the PROFILES Project under grant agreement no. 266589 and other parts were supported by two grants of the Deutsche Bundesstiftung Umwelt (German Federal Trust for the Environment).

References

1. G. De Haan, *Environ. Educ. Res.*, 2006, **12**, 19–32.
2. UNCED (United Nations Conference on Environment and Development), *Agenda*, vol. 21, 1992, http://sustainabledevelopment.un.org/content/documents/Agenda21.pdf (accessed February 2014).
3. M. Burmeister, F. Rauch and I. Eilks, *Chem. Educ. Res. Pract.*, 2012, **13**, 59.
4. G. de Haan, *Environ. Educ. Res.*, 2006, **12**, 19.
5. J. D. Bradley, *Chem. Educ. Int.*, 2005, 7, retrieved from the World Wide Web, July 01, 2011, at http://old.iupac.org/publications/cei/vol6/index.html.
6. S. A. Ware, *Pure Appl. Chem.*, 2001, **73**, 1209.
7. A. Hofstein, I. Eilks and R. Bybee, *Int. J. Sci. Math. Educ.*, 2011, **9**, 1459.
8. I. Eilks, B. Ralle, F. Rauch and A. Hofstein, in *Teaching chemistry – A studybook*, ed. I. Eilks and A. Hofstein, Sense, Rotterdam, 2013, pp. 1–36.
9. J. K. Gilbert, *Int. J. Sci. Educ.*, 2006, **28**, 957.
10. W. M. Roth and S. Lee, *Sci. Educ.*, 2004, **88**, 263–291.
11. T. D. Sadler, *Socio-scientific issues in the classroom*, Springer, Dordrecht, 2011.
12. M. Stuckey, R. Mamlok-Naaman, A. Hofstein and I. Eilks, *Stud. Sci. Educ.*, **49**, 1.
13. J. Holbrook, *Chem. Educ. Int.*, 2005, 7, retrieved from the World Wide Web, July 01, 2011, at http://old.iupac.org/publications/cei/vol6/index.html.
14. J. Van Aalsvoort, *Int. J. Sci. Educ.*, 2004, **26**, 1635.
15. J. Holbrook and M. Rannikmäe, *Int. J. Sci. Educ.*, 2007, **29**, 1347.
16. S. Elmose and W.-M. Roth, *J. Curr. Stud.*, 2005, **37**, 11.
17. A. M. Bodzin and R. Mamlok, *Sci. Teach.*, 2000, **67**(9), 36.
18. J. Sjöström, *Sci. Educ.*, 2013, **22**, 1873.
19. L. Bencze, E. Sperling and L. Carter, *Res. Sci. Educ.*, 2012, **42**, 129.
20. F. Mogensen and K. Schnack, *Environ. Educ. Res.*, 2010, **16**, 59.
21. J. Simonneaux and L. Simonneaux, *Res. Sci. Educ.*, 2012, **42**, 75.
22. R. Marks and I. Eilks, *Int. J. Environ. Sci. Educ.*, 2009, **4**, 131.
23. I. Eilks, J. Nielsen and A. Hofstein, in *Topics and trends in current science education*, ed. C. Bruguière, A. Tiberghien and P. Clément, Springer, Dordrecht, 2014, pp. 85–100.
24. I. Robbottom and L. Simonneaux, *Res. Sci. Educ.*, 2012, **42**, 1.
25. T. D. Sadler, *J. Res. Sci. Teach.*, **41**, 513.
26. M. Stolz, T. Witteck, R. Marks and I. Eilks, *Eurasia J. Math. Sci. Techn. Educ.*, **9**, 273.
27. G. Centi and S. Perathoner, in *Sustainable industrial processes*, ed. F. Cavani, G. Centi, S. Perathoner and F. Trifiro, Wiley-VCH, Weinheim, 2009, pp. 1–72.
28. P. T. Anastas and J. C. Warner, *Green chemistry: theory and practice*, Oxford University Press, Oxford, 1998.
29. L. Ezra, B. Skolnick, and G. Aghbariya, Can used oil be the next generation fuel? *Unpublished module developed in the framework of the PROFILES Project funded by the European Community´s 7th Framework Program*, 2012.

30. J. Holbrook and M. Rannikmae, in *Contemporary science education*, ed. I. Eilks and B. Ralle, Shaker, Aachen, 2010, pp. 69–82.
31. I. Eilks, *Chem. Educ. Res. Pract.*, 2002, **3**, 67.
32. T. Feierabend and I. Eilks, *J. Chem. Educ.*, 2011, **88**, 1250.
33. I. Eilks, *Sci. Educ. Int.*, 2002, **13**(1), 11.
34. S. Azulaos-Katz, A. Assi, and P. Kuzmin, Plastic: Reduce the use, *Unpublished module developed in the framework of the PROFILES Project Funded by the European Community´s 7th Framework Program*, 2012.
35. M. Burmeister and I. Eilks, *Chem. Educ. Res. Pract.*, 2012, **13**, 93.
36. M. Burmeister, J. von Döhlen and I. Eilks, in *Cases on pedagogical innovations for sustainable development*, ed. K. D. Thomas and H. E. Muga, IGI Global, Hershey, 2014, pp. 154–169.
37. M. Stuckey and I. Eilks, *Chem. Educ. Res. Pract.*, 2014, **15**, 156–167.
38. L. Tomas and S. M. Ritchie, *Res. Sci. Educ.*, 2012, **42**, 25.
39. D. Mandler, R. Mamlok-Naaman, R. Blonder, M. Yayon and A. Hofstein, *Chem. Educ. Res. Pract.*, 2012, **13**, 80.
40. T. D. Sadler and D. Zeidler, *J. Res. Sci. Teach.*, 2009, **46**, 909.
41. B. Robelia, K. McNeill, K. Wammer and F. Lawrenz, *J. Chem. Educ.*, 2010, **87**, 216.
42. J. A. Swan and T. G. Spiro, *J. Chem. Educ.*, 1995, **72**, 967.
43. M. L. Klostermann, T. D. Sadler and J. Brown, *Res. Sci. Educ.*, 2012, **42**, 51.
44. R. Marks and I. Eilks, *Chem. Educ. Res. Pract.*, 2010, **11**, 129.
45. U. Zoller and D. Pushkin, *Chem. Educ. Res. Pract.*, 2007, **8**, 153.
46. K. Coyle, *Environmental literacy in America: What ten years of NEETF/Roper research and related studies say about environmental literacy in the US*, NEETF, Washington, 2005, Retrieved from http://www.neefusa.org/pdf/ELR2005.pdf.
47. A. Hassan, H. Juahir and N. S. Jamaludin, *Am. J. Sci. Res.*, 2009, **5**, 50.

CHAPTER 5

Collaborative Development of a High School Green Chemistry Curriculum in Thailand

KENNETH M. DOXSEE*[a]

[a]Department of Chemistry, University of Oregon, USA
*E-mail: doxsee@uoregon.edu

5.1 Background

Faced with rising material acquisition, facility operation, and waste disposal costs associated with the undergraduate organic chemistry teaching laboratory, the University of Oregon joined numerous other institutions of higher education in moving from conventional scale to micro-scale experimentation in the early 1990s.[1] While this transition afforded us the opportunity to modernize both our curriculum and our laboratory facilities, it came with a significant unforeseen cost. With over 250 students enrolled each term, and with a laboratory that could safely accommodate no more than 17 students due to limited fume hood capacity, the laboratory instructor was faced with 17 scheduled laboratory sessions each week. At three hours per session, and with an additional one hour lecture per week, the resulting 52 student contact hours per week clearly represented an unsustainable situation.

A breakthrough in discussions of ways to ameliorate this problem came with the simple (in retrospect) realization that students working with

Worldwide Trends in Green Chemistry Education
Edited by Vânia Gomes Zuin and Liliana Mammino
© The Royal Society of Chemistry 2015
Published by the Royal Society of Chemistry, www.rsc.org

non-hazardous materials could perform experiments on open laboratory benches rather than in fume hoods (Figure 5.1).

By transitioning student work to open benches, amply available in our existing laboratory facility, each lab section could accommodate significantly more students, allowing us to schedule fewer sections and thereby reducing instructor contact hours to a more manageable level. The newly emerging field of green chemistry[2] appeared ideally suited to this notion. Finding few appropriate teaching experiments in existing laboratory texts, we developed and published[3] a set of experiments, designed to teach the essential techniques and concepts of the organic chemistry laboratory within the context of the principles of green chemistry.

While our green chemistry laboratory programme had its genesis in this very pragmatic need to address a critical scheduling problem, we quickly recognized that its value extended far beyond the solution of that local problem. Our focus turned naturally to the intrinsic importance of green chemistry in training the next generation of scientists, a generation that would be armed with the tools and information to help them avoid reproducing the periodic environmental and health catastrophes that are in many ways the hallmarks of 'traditional' chemistry. We have introduced the concepts and experimental materials developed at the University of Oregon to teachers from high schools, colleges, and universities across the United States and around the world. While the ambitions of our programme are lofty—educating a new generation of scientists that may help to 'save the world'—it is important not to lose sight of its grounding in addressing pragmatic realities. Beyond maintaining a reasonable instructor workload, the University is reaping hundreds of thousands of dollars in energy savings due to the decommissioning of numerous fume hoods. In addition, we are teaching record enrollments of students, and these students are moving on to successful employment and advanced study.

Figure 5.1   The green organic chemistry laboratory at the University of Oregon, with students working on open laboratory benches rather than in fume hoods.

In this chapter, I highlight the key role green chemistry can play in bringing a meaningful laboratory experience to students regardless of where they may be and regardless of what resources or facilities may (or may not) be available to them.

5.2   Introduction

During an official state visit in July 2006 to the University of Oregon by Her Royal Highness Princess Bajrakitiyabha Mahidol of Thailand, a new University of Oregon Distance Learning Program was announced, representing a significant expansion of a programme originated in 1996 for international teachers of English as a second language. A collaboration between the University of Oregon, the Thai Distance Learning Foundation,[4] and the U.S.–Thai Distance Learning Organization,[5] the expansion targeted the introduction of green chemistry as a tool for the teaching of chemistry. An introductory lesson for the Green Chemistry Program was presented at this time, carried *via* live satellite links to television channels throughout Thailand and neighbouring countries and reaching a potential audience of around 60 million. This lesson provided the conceptual underpinnings of the programme:

- Green chemistry serves as a platform for the introduction and discussion of the fundamentals of chemistry.
- The intrinsic and explicit connection of these fundamentals to real-life issues effectively engages students in new ways of thinking about chemistry.
- Modern pedagogical approaches, including small-group discussion, peer-led team learning[6] and guided inquiry,[7] are ideally suited to a green curriculum.
- Green chemistry furthers one of the central tenets of the Distance Learning Foundation, His Majesty King Bhumibol Adulyadej's educational policy of อบรมบ่มนิสัย (*ob rom bom nisai*): the teaching and training of a child to be good, honest, and economically sufficient.

While educators widely recognize the value of the laboratory in bring science 'to life',[8] the sad reality is that most schools do not have the facilities or the budget to allow safe and meaningful chemical experimentation. By designing intrinsically safe laboratory experiences, drawing from locally available resources, we can remove the safety and cost issues that prohibit traditional chemical experimentation. Thus, in the end, the most compelling justification for a green chemistry curriculum is that all these benefits are made available to students *everywhere*.

In February 2007, a second green chemistry lesson was presented, followed by a laboratory workshop conducted by Supawan Tantayanon (from Chulalongkorn University). The presentation focused on teaching methodologies for general chemistry and the 12 principles of green chemistry.[2] In the workshop, Tantayanon used the Small-Lab Kit[9] she designed to illustrate several principles of green chemistry, including the prevention of waste, energy efficiency, and minimization of the potential for accidents. During this lesson, teachers were surveyed regarding issues of particular interest, importance, and/or relevance to their students and teaching environments. Virtually all of the participants identified as key issues the connection to real-life issues and the ability to carry out experimentation safely, with available materials and at low cost.

A third green chemistry lesson was presented in December 2007, timed to coincide as closely as possible with the King's 80th birthday celebration in recognition of his strong commitment to education. In this lesson, an experiment from a typical laboratory manual was used as a platform for discussion of how one could determine if a chemical reaction was 'good' or 'bad', moving from simple concepts of theoretical yield through increasingly sophisticated analyses of reaction efficiency. The key message of this lesson was that, by systematically working through various ways of thinking about this issue, students may be led to *ask* to be taught the sometimes abstruse concepts that form a foundation of chemistry, such as balancing chemical equations, stoichiometry and the mole.[10]

5.3   Distance Learning in Green Chemistry

With this background, my attention was turned to creating new age-appropriate laboratory materials. Meetings with Thai teachers and exploration of the Thai national educational curriculum confirmed that Thai and U.S. expectations for high school chemistry course content were similar. I chose to focus initially on experiments with direct relevance to environmental issues, noting that while this does not guarantee that an experiment will be illustrative of green chemistry, it does maintain a concrete connection to intuitively relevant material. My collaborators, Dr Jorge Ibañez and his colleagues in the Mexican Institute for Microscale and Green Chemistry at the Universidad Iberoamericana in Mexico City, had recently authored a definitive textbook and laboratory guide for environmental chemistry.[11] Together, we evaluated selected experiments from this text for their relevance to teaching desired chemical topics, their illustration of multiple green chemical principles, and their reliability and reproducibility in the hands of inexperienced students. Other experiments were adapted from the literature and/or developed collaboratively, drawing especially from the creative developments of a uniquely international group of chemical educators informally known as the 'microscale family' (Figure 5.2).[12]

Final experiments (Table 5.1) were selected with consideration of the local availability of required materials. These experiments were successfully 'field tested' by 60 Kuwaiti high school teachers as part of a three-day Kuwait City workshop, presented in March 2009 under the coordination of the President of the Kuwait Chemical Society, whom I had serendipitously met the previous year at a conference (coincidentally in Bangkok). While in Kuwait, I had the opportunity to discuss green experimentation on the

Figure 5.2   Members of the micro-scale 'family' with Her Royal Highness Princess Chulabhorn Walailak, following a gathering in Ayutthaya, ancient capital of Thailand.

Table 5.1   Experiments for the distance learning programme in green chemistry.

Experiment	Relevant green principles[a]	Development notes[b]	Ref.
11 Yen battery (copper and aluminium)	1, 4, 6, and 12	2, 3	14
Aluminium–air battery	1, 4, 6, and 12	2, 3	13
The per cent of oxygen in air	1, 2, 3, 8, 10, 11, 12	1	—
Weight per cent of $CaCO_3$ in egg shells	1, 3, 7, 11, and 12	1	16
One-cent solar cell (Cu_2O on Cu)	3, 4, 5, 6, and 9	1	15
Metal ion recovery by cementation	1, 3, 4, 5, 10, 11, and 12	2, 3	17
Acetylsalicylic acid in an aspirin tablet	1, 3, 4, 5, 7, and 11	2, 3	18
Connecting solubility, equilibrium, and periodicity	1, 3, 11, and 12	2, 3	19
Greening the blue bottle reaction	1, 4, and 11	2, 3	20
Reaction rate (peroxide decomposition; combustion)	1, 4, 5, 7, and 9	2, 3	21
Reaction rate (clock reaction)	1, 4, 5, 7, and 11	2, 3	22
Polymer packing 'peanuts'	1, 3, 4, 7, and 10	1	—
Carbohydrates (biopolymers)	3, 4, 5, 7, and 10	2, 3	23
Analysis of charge with polymer gels	1, 4, 5, 10, 11, and 12	1	—
Reaction efficiency	1, 2, 3, 4, 8, 12	1	—

[a]The relevant green principles are taken from Anastas and Warner.[2]
[b]Experiment development notes: 1, new experiment; 2, modification to better illustrate green principles; 3, addition of green context.

live television programme *Good Morning Kuwait*. After making the point that green experimentation was so safe that we could perform it there in the television studio, the host enthusiastically asked, 'Oh, would you?' Two days later, Ibañez had the honour of demonstrating several experiments on live television (Figure 5.3).

Figure 5.3   Jorge Ibañez performing green experiments on the set of the *Good Morning Kuwait* television programme, joined by Abdulaziz Alnajjar, President of the Kuwait Chemical Society.

With experimental materials in hand, my undergraduate research assistant, John C. 'Jack' Niedbala and I presented a series of six 2.5-hour workshops from January through March 2010, with two-way audio and video *via* IP video-conferencing to two primary sites in Thailand—Bangkok and Hua Hin—and one-way audio and video connections to regional sites serving upwards of 30 000 Thai schools. Leading these workshops from the confines of a small, unventilated broadcast studio dramatically illustrated the intrinsic safety of the experiments, further highlighting their potential for bringing a safe and meaningful laboratory experience to students regardless of local resources or facilities.

Live translation was complemented by extended explanations in Thai, helping to bridge cultural differences between Thailand and the United States and ensuring real understanding of the curricular materials and of green chemistry in general. While the experimental procedures were for the most part risk free, each session began with a discussion of chemical safety. Rather than dogmatically insisting on protective gear even when it was not necessary (*e.g.*, when constructing a battery from aluminium foil, salt water, paper towelling, and charcoal)[13] we honestly appraised each situation and recommended *appropriate* safety precautions. Through this thoughtfulness, we were able to engage participants in the proper use of protective gear in those few cases where it was indeed called for.

The effective use of mobile cameras at the Thailand sites provided an experience that was remarkably similar to onsite supervision, allowing for real-time corrections and suggestions as readily as if all were present in the same laboratory. When Thai participants experienced difficulties in preparing a battery from copper and aluminium coins,[14] for example, we noted that an inadvertent short circuit had been formed, a problem that was quickly

Figure 5.4   Screen captures from a green chemistry distance learning programme workshop, illustrating a temporary loss of lights from Hua Hin followed by successful demonstration of a photocurrent (see text).

remedied. Not surprisingly, things did not always proceed as planned. These occasions served as additional opportunities for discussion and engagement with the participants. In the midst of one discussion about the apparent failure of a photocell preparation,[15] the lights suddenly went out at the Hua Hin site. After some frantic manoeuvring by the technical staff, the Hua Hin site reappeared, with smiling participants rejoicing about the success of their photocell. They had realized that the photocurrent created when they illuminated their photocell with a light bulb had been masked by the bright lights from the television cameras, so they simply turned off those lights, dropping from online visibility but allowing the successful measurement of a very respectable photocurrent[15] (Figure 5.4).

Following the conclusion of each experiment, we discussed interesting observations and carried out relevant data analyses, often using participant-generated data, helping to ensure that the participants were armed with the necessary information and background to transition the experiment to their own students. Finally, we concluded each experiment with a discussion of the green principles relevant to the experiment. This discussion proved particularly valuable to the participants, who were able to contribute intellectually to the workshops, and additionally allowed us to assess the extent of participant engagement.

Written materials complementing the workshops additionally included suggestions regarding related experimentation, encouraging participants to think beyond the materials presented and strive to offer their students the opportunity for open-ended investigation. As the photocell anecdote above highlights, when the materials are intrinsically safe, curiosity-driven independent investigation is not only tolerable but desirable.

Prior to the final workshop, one of the participants, Surapong Namnai, an instructor at the Wang Klaikangwon School in Hua Hin, taught a special class session focused on one of the experiments, carried live to Ratchaprachanukrao 21 School in Mae Hong Son Province on the border with Myanmar in northwest Thailand. This represented the first step in transmitting information from the workshops to the next generation of Thai science students (Figure 5.5).

Figure 5.5 Thai high school teacher Surapong Namnai teaching students at Wang Klaikangwon School in Hua Hin and, *via* video-conferencing, Ratchaprachanukrao 21 School in Mae Hong Son Province.

5.4 Assumption College, Thonburi

While initial momentum had developed, it was critical for there to be a 'next step'. This step was catalysed by Veerakarn Suebsang (co-Director, U.S.–Thai Distance Learning Organization), who introduced me to Wetchaiyan Jaturas, Director of the English Language Program at Assumption College, Thonburi (ACT). ACT, one of sixteen K-12 schools in Thailand's Assumption College system, enrolls roughly 5000 students, including nearly 1000 in the English Language Program, in which students receive bilingual instruction from teams of Thai and international teachers. Initial discussions revealed a theme common to science instruction worldwide; ACT students received only minimal laboratory instruction in chemistry. Following in-depth discussions of green chemistry and its implications, both practical and pedagogical, Jaturas enthusiastically proposed a partnership aimed at pioneering the introduction in Thailand of a green chemistry curriculum at the high school level.

During a subsequent visit to further the green project, ACT and the University of Oregon signed a Memorandum of Understanding addressing the creation of opportunities for student and teacher exchanges. Sensing some concern among ACT's teachers about our growing relationship and its implications for them, we convened an open discussion of the goals of the green chemistry project and the anticipated collaborative development of the programme, culminating in a clear sense of enthusiasm and the beginning of individual expressions of interest by teachers, both Thai and international. This interest has continued to grow, providing what is unarguably a critical element for the success of the project—local *intellectual* engagement—reaffirming our discovery of the great value of meaningful engagement by key stakeholders when developing the green chemistry programme at the University of Oregon.

Discussions and groundwork continued for the next several months, and in December 2012, I travelled again to ACT, carrying a bundle of reference books and a laptop computer and planning to sequester myself in my room to develop, draft and test a full set of experiments. Had I done so, I would have been ignoring an essential additional component of stakeholder engagement. Simply creating a curriculum without ensuring its acceptance *beyond ACT* would have led to the development of materials destined for a dust-covered shelf rather than for implementation. Instead, in a series of meetings throughout Bangkok and south central Thailand, Jaturas, Suebsang and I met with numerous influential scholars and administrators. While the concept received initial responses ranging from intrigued to sceptical, each meeting ended with enthusiastic endorsement of the effort. In the course of these meetings, an important additional consideration arose. While the Thai national educational curriculum could be relatively liberally interpreted in the context of a private institution like ACT, transition to a broader range of schools would require careful connection of the green materials to the expectations of the national curriculum.

During a subsequent meeting with the Director of ACT, we were delighted to hear him propose the construction of a green chemistry lab on the ACT campus. A promising site was rapidly identified, in the heart of campus and several feet above ground level (an important consideration given the devastating floods that had inundated much of Thailand, including the ACT campus, only a year earlier). Armed with a tape measure, note pad, and pencil, we quickly sketched several lab designs, and a few short months later, construction began. An interim meeting at the University of Oregon that included representatives from several other Assumption College campuses furthered the lab construction project while also engaging the St. Gabriel Foundation and Assumption College system more broadly.

As the lab took form, plans were laid for a grand opening ceremony in December 2013. While planning for the range of speeches customary for such an event, the hallmark was to be the laboratory demonstration of several experiments illustrative of green chemistry. Careful design of simple but engaging experiments was crucial, since many of the 60 to 70 dignitaries who would travel from throughout Thailand for the grand opening were not chemists. In the end, the operative principles paralleled those governing the creation of laboratory content for the green curriculum itself: teach important chemistry, illustrate the principles of green chemistry, and utilize inexpensive materials that are locally available. Two experiments were presented— one exploring super-absorbent polymers and their use in the separation of coloured materials,[24] the other examining halogenation and saponification of plant-derived esters accompanying the electrolysis of brine[25]—and both stimulated lively discussions (Figure 5.6 and Figure 5.7).

While we were celebrating the completion of a new, modern laboratory, my comments focused on the accessibility of experimental chemistry to students

Figure 5.6   The author performing green chemistry demonstrations at the grand opening of the new green laboratory at Assumption College, Thonburi.

Figure 5.7   Key partners gathered at Assumption College, Thonburi, in December 2013, for the grand opening of the new green laboratory.

not just at ACT, but throughout Thailand and beyond. Further, I stressed that while the demonstrations were seemingly simple, they in fact embrace a surprising range and depth of science, from food science to advanced materials chemistry. We spent some time discussing the notion of 'drillable' experiments, which can be introduced to students at an early age, then periodically revisited and explored in greater detail as students progress through their educational careers. In the electrolysis of brine, for example, young children can simply learn about the use of electrical energy to split water into its constituent parts, while advanced students can explore issues as complex as electrode over-potentials and coupled chemical reactions. Each message was clearly well-received, and a university professor in attendance advocated for continued 'drilling' into the experiments beyond the K-12 years and into university-level studies as well.

5.5　Next Steps

It is our intention that the ACT green laboratory will serve not only as a place of learning for ACT's students, but also as an international resource for the dissemination of green chemistry to teachers throughout the Federation of Asian Chemical Societies. We will continue to work collaboratively, designing and implementing new experiments that further complement the high school curriculum, tailoring them to the national curriculum while also setting the stage for possible modernization of that curriculum. We will also continue to develop our connections with the Assumption College campuses in Sriracha and Ubon Ratchathani, where considerable interest in replicating the green curriculum is apparent, and beyond as interest inevitably grows.

Teachers have been enthusiastic and vocal in their desire to play developmental roles as we move forward, clearly having left any initial trepidation behind. The transition from scepticism to advocacy among the teachers was dramatically demonstrated during a single one-hour demonstration in Ubon Ratchathani. Initially, one international teacher sat in the back of the laboratory, with a stony expression and arms crossed. As the demonstration progressed, his faced relaxed, his arms uncrossed … then he stood, and moved toward the front of the laboratory … and by the end of the lesson, he was animatedly translating and explaining details to the students.

5.6　Lessons Learned

In the hope that my experiences through the many years of development of this international green chemistry programme will prove valuable to others as they embark on similar adventures. I conclude with a summary of some of the key lessons learned.

5.6.1　Loss of Meaning during Translation

Be aware of the loss of 'nuance in translation.' Initial translation of 'green chemistry' into Thai proved entirely inappropriate, representing at best a baffling juxtaposition of the science of chemistry and the colour green, which does not, in the Thai language, convey any sense of environmentalism or safety. After numerous discussions about Thai terms that could convey the desired nuance, in the end, the solution was simple. When translating from English to Thai, we simply left the words 'green chemistry' in English, providing an instant 'call-out' and eliciting in the listeners the response, 'Oh, there are those English words that mean we are talking about chemistry that is safe for health and the environment.' Perhaps the common use in the United States of the Russian term 'glasnost' rather than the comparatively nuance-free translation 'openness' during the Gorbachev era is a similar example of an instance where 'non-translation' represents the best way of effecting a translation.

5.6.2   Differences in Teaching Methods

Find ways to address cultural differences in teaching and learning. The desire to engage students in the learning process through active discussion and questioning represented a second cultural barrier because Thai students, in general, believe it is disrespectful to ask questions of a teacher. Two approaches helped to overcome this obstacle. Simplest and perhaps most effective was the awarding of small prizes (*e.g.*, University of Oregon lapel pins) to participants who asked questions. Also effective was a brief but frank discussion of this cross-cultural issue, in which we noted that in the United States it can be considered disrespectful—a sign of lack of attention or engagement—*not* to ask questions of a speaker! In the end, the ability to formulate questions succinctly and clearly, and *simply to ask them*, may be one of the most valuable lessons the programme imparts. Perhaps these students—the ones who ask questions instead of simply writing down what they are told—will be the ones to solve the problems facing our world today.

By honestly appraising each situation and recommending *appropriate* safety precautions, we avoided contributing to the 'chemophobia' that is instilled by insisting that all chemical experimentation is dangerous and simultaneously highlighted a central lesson of green chemistry regarding the *assessment* rather than assumption of risk. Through this thoughtfulness, we were able to engage participants in the proper use of protective gear in those few cases where it was indeed called for.

5.6.3   Involve Stakeholders

Anticipate and engage all stakeholders. There is much at stake in major curricular reform, and many individuals in positions of responsibility, authority, and personal intellectual investment are available to help, or to hinder, if personal and professional relationships are not thoughtfully developed and nurtured. With all on board, difficulties become minor bumps in the road; with key players left wondering what is happening, the same difficulties can become insurmountable obstacles. Collaboratively developed materials are likely to transition successfully to the classroom; the same materials 'inflicted' on over-burdened teachers are likely to fail.

5.6.4   Be Realistic

Develop locally relevant and accessible experimentation. While this can represent a significant challenge, we can neglect it only at the risk of developing experiments that will never be used. All too often, I have observed workshop participants working in teams instead of independently so as to preserve a few milligrams of a reactant that they know they will not find when they return to their home institution. More than a challenge, this represents a key opportunity to engage teachers in the active development of new experiments. Begin with a 'brainstorming' session, exploring local and regional foods, dyes and pigments, traditional medicines, natural resources, and artistry. Each can suggest innovations. Red cabbage, popular in much of the world as a source of a pH-dependent extract, is difficult to find in Thailand. While peeling a wonderful local fruit, the mangosteen (mangkhud, มังคุด), I saw that my fingers were turned red from contact with the peel. We will soon be exploring the pH-dependent behaviour of extracts of mangosteen peel!

Acknowledgements

Each of the colleagues mentioned by name in this chapter played an integral role in the development of this programme. I express my most profound appreciation for each of them, including my international brothers—Muhamad Hugerat, Jorge Ibañez, and Wetchaiyan Jaturas—and the entire worldwide micro-scale family.

References and Notes

1. See, *e.g.*, D. L. Pavia, G. S. Kriz, G. M. Lampman and R. G. Engel, *A Microscale Approach to Organic Chemistry Laboratory Techniques*, Cengage Learning, Stamford, CT, 2012.
2. P. T. Anastas and J. C. Warner, *Green Chemistry – Theory and Practice*, Oxford University Press, New York, NY, 2000.
3. K. M. Doxsee and J. E. Hutchison, *Green Organic Chemistry: Strategies, Tools and Laboratory Experiments*, Cengage Learning, Stamford, CT, 2004.
4. In 1996, in commemoration of the 50th anniversary of His Majesty King Bhumibol Adulyadej's accession to the throne, the Thai Distance Learning Foundation was established with the goal of providing education to every child in the country, including those in remote areas where teachers and resources are scarce. The Distance Learning Foundation brings live lessons from Wang Klaikangwon School in Hua Hin to schools throughout Thailand through distance learning equipment, installed without charge by the Royal Thai Army, and toll-free telephone lines, provided by the Telephone Organization of Thailand. Twelve channels carry live broadcasts directed to grades 1–12 at some 30,000 schools throughout Thailand. An additional three channels carry vocational, language training, international documentaries, and technical college educational content. Live broadcasts are also provided asynchronously via satellite, conventional, and internet broadcasts throughout Southeast Asia, reaching Myanmar (Burma), Laos, Cambodia, Vietnam, Malaysia, and southern China. See: K. Vajarodaya, '*Twelfth Anniversary of the Distance Learning Foundation: Free and Open Low-Cost Distance Education via Satellite and Internet, Klaikangwon Model*,' *Fourth International Conference on eLearning for Knowledge-Based Society*, 18–19 November 2007, Bangkok, Thailand. Other key partners for the DLF within Thailand have included the Telecommunications Association of Thailand, the Institute for the Promotion of Teaching Science and Technology, the Thailand Ministry of Education, Advance Vision Systems Co., Ltd., Telesat Corporation Co., Ltd., and Cisco Systems (Thailand), Inc.

5. The U.S.-Thai Distance Learning Organization (DLO) is a non-profit organization based in Eugene, Oregon, co-directed by Richard Lindholm, Valaya Lindholm, and Veerakarn Suebsang. The DLO serves as the official U.S. affiliate of the Distance Learning Foundation. With the mission of fostering cultural and economic cooperation and facilitating the exchange of information and learning between Thailand and U.S. educational organizations and leaders, the DLO acts as a liaison between the Distance Learning Foundation, the Thai Grand Chamberlain, the Thai Consulate (in Los Angeles, California), the Thai Embassy (in Washington, D.C.), and U.S. collaborators.

6. D. K. Gosser, M. S. Cracolice, J. A. Kampmeier, V. Roth, V. S. Strozak and P. Varma-Nelson, *Peer-led Team Learning: A Guidebook*, Prentice Hall, Upper Saddle River, NJ, 2001.

7. See, *e.g.* J. J. Farrell, R. S. Moog and J. N. Spencer, *J. Chem. Educ.*, 1999, **76**, 570–574.

8. M. P. Clough, in *Using the Laboratory to Enhance Student Learning. Learning Science and the Science of Learning*, ed. R. W. Bybee, National Science Teachers Association Press, Arlington, VA, 2002, pp. 85–96.

9. S. Tantayanon, *Small-Scale Laboratory: Organic Chemistry at University Level*, http://www.unesco.org/science/doc/Organi_chem_220709_FINAL.pdf (accessed 27 February 2015).

10. S. Tantayanon, K. M. Doxsee, D. Nuntasri and J. C. Niedbala, *Chem. Int.*, 2011, **33**(4), 6–10. Portions of this article have been incorporated in this chapter.

11. J. G. Ibanez, M. Hernandez-Esparza, C. Doria-Serrano, A. Fregoso-Infante, and M. M. Singh, *Environmental Chemistry: Microscale Laboratory Experiments*, Springer, New York, NY, 2007.

12. The microscale family has evolved somewhat through the years. In its current configuration, with apologies to those who I have inadvertently omitted, it includes Abdulaziz Alnajjar (Kuwait), Ken Doxsee (USA), Marie DuToit (South Africa), Christer Gruvberg (Sweden), Muhamad Hugerat (Israel), Jorge Ibañez (Mexico), Angela Köhler-Krützfeldt (Germany), Mordechai Livneh (Israel), Metodija Najdoski (Macedonia), Kazuko Ogino (Japan), Peter Schwarz (Germany), Fortunato Sevilla III (Philippines), Supawan Tantayanon (Thailand), Robert Worley (UK), and N. H. Zhou (China), as well as our inspirational colleagues, the late Viktor Oberdrauf (Austria) and Mahmoud El Marsafy (Egypt).

13. M. Tamez and J. H. Yu, *J. Chem. Educ.*, 2007, **84**, 1936A–1936B, modified by J. Ibañez and coworkers. 'Local flavor' was added at a workshop in Mexico by replacing the activated carbon with carbon in the form of a tortilla that was intentionally over-cooked, burning it.

14. I. Otsuki, *Bussitsu no Henka (Matter and Change)*, Hyoronsha, Tokyo, 1973, (ISBN/ASIN:4566020045). This demonstration is particularly notable in that the functional battery produced from Japanese one yen (aluminium) and ten yen (copper) coins has the cathode and anode identified by the kanji characters for one (–) and ten (+).

15. J. G. Ibañez, A. Finck-Pastrana, A. Mugica-Barrera, P. Balderas-Hernandez, M. E. Ibarguengitia-Cervantes, E. Garcia-Pintor, J. M. Hartasanchez-Frenk, C. E. Bonilla-Jaurez, C. Maldonado-Cordero, A. Struck-Garcia and F. Suberbie-Rocha, *J. Chem. Educ.*, 2011, **88**(9), 1287–1289.
16. Adapted by J. G. Ibañez from Daniel Bartet & Anamaría Jadue, Departamento de Química, Universidad Metropolitana de Ciencias de la Educación, Santiago, Chile.
17. J. G. Ibañez, M. Hernandez-Esparza, C. Doria-Serrano, A. Fregoso-Infante, and M. M. Singh, *Environmental Chemistry: Microscale Laboratory Experiments*, Springer, New York, NY, 2007, pp. 199–203.
18. Adapted by J. Ibañez from K. B. Gusmão, E. M. A. Martini and S. T. Amaral, *Chem. Educ.*, 2005, **10**, 444–446.
19. K. L. Cacciatore, J. Amado and J. J. Evans, *J. Chem. Educ.*, 2008, **85**(2), 251–253.
20. W. E. Wellman and M. E. Noble, *J. Chem. Educ.*, 2003, **80**(5), 537–540.
21. Adapted from K. Schug, SMILE (Science and Mathematics Initiative for Learning Enhancement) (http://mypages.iit.edu/~smile/chbi2000.htm).
22. Adapted by J. Ibañez from M. E. Ibarguengoitia, *et al.*, *Microscale Chemistry (in Spanish)*, Universidad Iberoamericana, Mexico, 2005, For a related vitamin C clock procedure, see S. W. Wright, *J. Chem. Educ.* 2002, **79**(1), 40A–40B.
23. B. A. Watkins, *IFT Experiments in Food Science Series: Food Chemistry Experiments*, 7 pp., http://www.ift.org/~/media/Knowledge%20Center/Learn%20Food%20Science/Experiments/TeacherGuideCARBS.pdf (accessed March 12, 2014).
24. Adapted by J. G. Ibañez and R. M. Mainero-Mancera with invaluable input from, A. Köhler Romaine Rolland High School, Berlin, Germany, For related information, see: B. Criswell, *J. Chem. Educ.*, 2006, **83**(4), 576A–576B.
25. M. Hugerat, R. Abu-Much, A. Basheer and S. Basheer, *Chem. Educ. J.*, 2009, **13**(2) (http://chem.sci.utsunomiya-u.ac.jp/v13n2/13M_Hugerat/M_Hugerat.html, accessed 27 February 2015).

CHAPTER 6

On the Development of Non-formal Learning Environments for Secondary School Students Focusing on Sustainability and Green Chemistry

NICOLE GARNER[a], JOHANNES HUWER[b], ANTJE SIOL[c], ROLF HEMPELMANN[b], AND INGO EILKS*[a]

[a]Institute for Science Education, University of Bremen, Germany; [b]Institute for Physical Chemistry, Saarland University, Saarbrücken, Germany; [c]Institute for Environmental Research and Sustainable Technologies, University of Bremen, Germany
*E-mail: ingo.eilks@uni-bremen.de

6.1 Introduction

Green chemistry is one of the major current trends in chemistry research and industrial practice. However, green chemistry is not very old and only emerged in the last twenty years from the initial ideas of Anastas and Warner[1] towards a prominent field of scientific research and development.[2] Today, in many university chemistry departments worldwide ideas of sustainability and green chemistry have become part of the education of the new generation

Worldwide Trends in Green Chemistry Education
Edited by Vânia Gomes Zuin and Liliana Mammino
© The Royal Society of Chemistry 2015
Published by the Royal Society of Chemistry, www.rsc.org

of chemists. However, the situation in secondary school chemistry seems to be different.[3] Recent studies on chemistry student teachers, teacher trainees and experienced teachers in Germany showed that the teachers' knowledge about sustainability and green chemistry is not sufficiently developed in the context of secondary chemistry education. The application of respective curricula and pedagogies is still rare.[4,5] Related issues and content are also under-represented in many current chemistry school textbooks in Germany.[3] It can be suggested that the same is the case in many other countries.

From an analysis of the available literature, Burmeister and colleagues suggested that investment in chemistry curriculum development and teacher education is needed to allow secondary school teachers to consider sustainability issues and the ideas of green chemistry more thoroughly in their classrooms.[3] Ideas for corresponding pre-service chemistry teacher education modules are already available.[6] However, investment in pre-service teacher education will have impact only on future chemistry teachers. Help is also needed for those teachers that have already finished their teacher education; some of them completed this long before the idea of a green chemistry emerged.

This chapter presents a project in the intersection of formal and non-formal chemistry learning, curriculum development, and teachers' continuing professional development. The context is German secondary chemistry education. Teaching and learning modules are developed that combine learning in school with laboratory visits of whole classes to a university.[7] The development of the teaching and learning modules is used to also design new school-type chemistry experiments, as well as creating teaching and learning materials about sustainability issues and green chemistry principles for both the formal and non-formal educational sectors.[8] In inviting students and their teachers to implement respective modules and to conduct site visits to the university for learning in contact with authentic green science, the project implicitly also seeks to increase the motivation of the students studying science as well as contribute to the continuing professional development of their teachers.

6.2 Education for Sustainable Development and Chemistry Education

In response to growing environmental and global challenges, in 1987 the Brundtland Commission suggested the idea of orienting life along the ideas of a sustainable development. The United Nations defined sustainable development as, 'development that meets the needs of the present without compromising the ability of future generations to meet their own needs.'[9] In 1992, the United Nations agreed on sustainable development as a normative guiding principle of the international community, encompassing the global civil society, world economy, and politics. Accordingly, sustainable development was enshrined as the fundamental principle of the Rio Declaration and Agenda 21.[10]

More than many other fields, chemistry is considered to have a special responsibility in addressing a sustainable development for our common future. Effectiveness of industrial processes is to be increased; emissions, raw-material consumption and energy use have to be reduced. One answer of chemistry towards these challenges was set up by the ideas of a *green* or *sustainable chemistry* as a guiding framework for many fields of contemporary chemistry research, development, and industrial production.[2]

However, to have the goal of sustainable development accepted on a broad societal base more is needed than innovations in science and technology only. Investment in a change in the mind of the society is unavoidable.[11] Thus, from the 1990s the United Nations attached a key role for sustainable development to education. In Agenda 21, the idea of Education for Sustainable Development (ESD) was presented as a key element.[10] UNESCO even proclaimed a whole decade on ESD for promoting ESD-oriented innovations in all areas of education. The decade ranged from 2005 to 2014. Within this framework UNESCO defined ESD as skill-oriented education to enable pupils to act responsibly today and to actively contribute to developing their future in a sustainable way.[12] Within this educational paradigm, knowledge and skills are to be developed in the young generation to allow them to shape their society in a fair-minded and sustainable fashion.[13] These skills will enable students to participate in today's society dealing with sustainable issues and challenges and in the future.[14,15] The decade has led to numerous projects to integrate ESD in schools. Many countries committed even on a political base to integrate ESD into formal education.[16] Although there is a general will to integrate ESD into education,[17] research has shown that learning about sustainability issues is hardly represented in many domains of secondary school learning, among them chemistry education.[3]

Since chemistry is central for modern life and society[18] chemistry education receives a special responsibility for contributing to ESD in formal and non-formal education.[3] General and domain-specific knowledge and skills are to be developed to enable the individual to assess new chemistry-based products and technologies in their own life and in the society in which they live and operate, as well as to react appropriately.[19] All students need to develop corresponding skills irrespective of whether or not they will later embark on a career in science and technology. All of them will become future citizens and need well-developed abilities for participation in societal debate and decision concerning issues of sustainable development.

For achieving the aim of ESD for all students in chemistry, different suggestions of integrating sustainability with secondary chemistry teaching were published. These suggestions range from a change in content[20] towards a whole change in the curriculum emphasis behind chemistry learning.[21] Based on the literature, Burmeister and colleagues suggested four basic modes of how ESD and chemistry education can be merged.[3] The modes range from (1) an application of green chemistry principles in the school science laboratory, *via* (2) using sustainability issues to contextualize chemistry content learning, towards (3) addressing technological and environmental challenges in a socio-scientific issues-based curriculum, and finally

(4) innovating school life along sustainability principles. They suggested the latter two modes to be of most potential for secondary school chemistry education to contribute to ESD. Both ask for opening the chemistry curriculum and the classroom towards society, industry or research centres and thus encourage an integration of formal school education with informal and non-formal educational activities.

6.3 Non-formal Learning Environments as Catalysts for Innovation

For many decades laboratory work has been given a central role in any student-centred chemistry education.[22,23] However, the positive effect of laboratory work on students' learning is not self-evident.[24] In secondary school settings, practical work often takes place under difficult conditions: school laboratories are poorly equipped or even unavailable, time is limited, and teacher–student ratios are unfavourable for open and inquiry-type experiments. As a result, experiments are often limited to demonstrations. Experiments are used mainly for illustrating content, but do not challenge students' thinking or contribute to a better understanding of the nature of science.[22,25]

Non-formal educational settings provide an alternative. In 2012, the OECD defined non-formal education as out-of-school education that follows a specific programme and may be connected to a certain curriculum. In a non-formal learning environment, time and equipment often are less restricted; the teacher–student ratio is more comfortable.[26,27] If the out-of-school activities become a mandatory part of school chemistry learning in a class, the learning environment can be considered in the intersection of formal and non-formal learning.[27] Garner *et al.*[8] recently suggested using the term 'partially non-formal learning environment' in this case.

Werquin highlighted the main benefits of non-formal learning environments.[28] The environment can contribute to increasing motivation and having a positive perception of learning. It can support career orientation by identifying interests, and contribute to continuing professional development of the accompanying teachers. Eshach suggested that non-formal learning may offer enjoyable learning and create scientifically literate school leavers.[29] However, like doing practical work in class, advantages and benefits from non-formal learning are also not self-evident as it needs a good connection between in- and out-of-school learning. Thus, linking inner- and outer-school learning is the central challenge to make non-formal learning a success.[30,31]

For structuring non-formal learning activities the literature suggests different points in order to make optimum use of the learning environments in science education:[26,29,31–35]

- Individually adaptable programmes ease the integration of non-formal learning activities with formal school curricula.
- Working materials for the non-formal learning activities need to be adjustable to the current student's performance and knowledge level.

- The learning environment should be student-centred, inquiry-based, interactive, and provoke cooperative learning.
- A preparatory learning phase in school is necessary to raise effective learning in the non-formal setting.
- After the non-formal learning experience the contents and topics should be picked up in school again.
- National standards or governmental syllabi shall be met to bridge formal and non-formal education.

In the course of this project, we focus on developing partially non-formal learning environments that are structured according to the principles suggested in the literature. The learning environments are organized for visits of regular secondary school science and chemistry classes with their teachers to the university laboratory. The chosen topics are connected to the official science and chemistry syllabi. For each learning group, a specific programme is adapted that is both prepared before and continued after the laboratory visit in the regular school science lessons.

Apart from the offer for a more motivating learning environment to school students, there are other values of non-formal learning environments. Non-formal learning environments provide a platform for innovations in the curriculum and its related pedagogy.[8] The greater freedom and the open atmosphere of a non-formal setting allow innovative pedagogies to be applied. The non-formal setting offers a platform for testing open and inquiry-type practical work before transferring it to school classroom conditions. The better student-to-teacher ratio allows for implementing more challenging experimental tasks and procedures. The better equipment and the supervision by experienced academic staff can provide flexibility in testing out innovative experimental activities. All this can be used to design and evaluate new ideas of teaching and learning chemistry that after successful development and documentation might be adapted for and implemented in formal school education. Sight visits in the non-formal laboratory also have potential to contribute to teachers' continuing professional development since the teachers come into contact with new content and innovative pedagogies. One field to which this could be applied are chemistry-related issues of sustainability or learning about green chemistry.

6.4 Non-formal Learning on Sustainability and Green Chemistry

6.4.1 The Framework

In recent years, non-formal outreach laboratory environments for primary and secondary school students (in German: *Schülerlabor*) became widespread in universities, research facilities and large companies all over Germany. A large number of these laboratories were founded in order to motivate young people for science studies and to support science teaching by offering experimental set-ups which are not possible to implement in the school context.[36,37]

In a joint initiative of two of these *Schülerlabors* an attempt is made to re-focus student chemistry learning towards issues of sustainability and greener chemistry. The laboratories belong to the Universities of Bremen in the north and of the Saarland in the southwest of Germany. Within the joint project *Sustainability and Chemistry in Non-formal Student Laboratories* a thorough linkage is made between formal learning in school and partially non-formal learning in the university. Researchers in chemistry and environmental science work closely together with domain-specific educational researchers and curriculum experts from the field of chemistry education. Their joint aim is to develop secondary chemistry learning in the field of sustainable chemistry. Half- and full-day laboratory-based non-formal learning environments are designed and implemented that deal with sustainability-related issues from daily life, technology, and the environment.

The non-formal laboratory activities are embedded in modules encompassing pre- and post-laboratory activities in the formal school environment with the laboratory visit in the university. The topics of the learning environments range from usage of renewable raw materials, *via* water treatment, chemistry of the atmosphere, biofuels, to modern technologies and synthesis strategies in the chemical industry. The target groups are secondary science and chemistry classes of grades 5–13 (age range 10–19 years). All of the modules are based on the philosophy of socio-scientific issues-based science education[38] as has been suggested by Burmeister *et al.* as being one of the most promising pedagogies for contributing to ESD in science education.[3] Thus all modules incorporate reflections of the effects of chemistry on the environment, economy and society with respect to sustainability.

Ten learning modules, each five per partner, were to be developed and implemented in the course of the project. All modules combine formal chemistry learning in school with student laboratory visits to the universities in Bremen and Saarbrücken. The topics and contents to be learned are closely related to the German Science Education Standards and their respective application in the regional science and chemistry syllabi. The linkage with the syllabi ensures that students both possess the necessary knowledge to solve the given tasks and that the teaching and learning environments are worth visiting by the classes.

The project is funded by the Federal German Environmental Trust (*Deutsche Bundesstiftung Umwelt*) and, in 2014, was awarded an official project of the United Nations Decade on Education for Sustainable Development.

6.4.2 Design of the Formal/Non-formal Learning Environments

All teaching and learning modules follow the same structure and are composed of four phases (see Figure 6.1). All modules start with pre-laboratory activities in school, based on materials provided by the university laboratory staff. The materials focus upon increasing the motivation for the sight visit, on opening the context for the practical activity in the university, and to secure sufficient prior knowledge. The site visit to the university laboratory is mainly used for practical work. In a third phase a voluntary field trip into research laboratories in the university or branches of industry are suggested which fit the thematic issue of the module. These excursions are designed to increase the authenticity of what is learned. Finally, suggestions and materials for post-laboratory activities and assessment are also provided to the teachers.

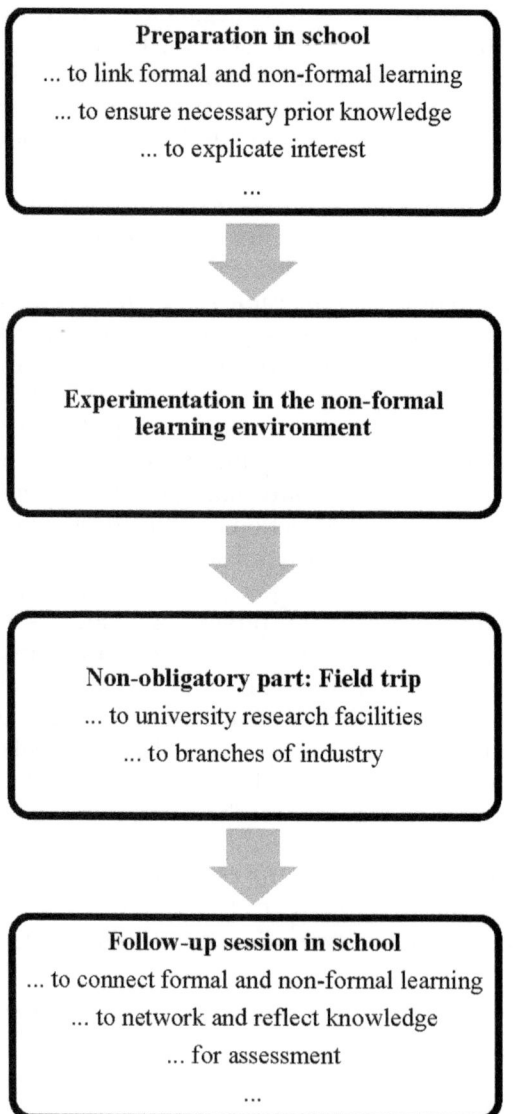

Figure 6.1   Design of the teaching–learning modules.

In the modules, a student-orientated[39] and inquiry-based[40] educational paradigm is applied. All teaching modules and tasks are embedded into meaningful contexts in order to allow for situated learning.[41] The inquiring nature of the activities aims at skill-oriented learning and contributing to an understanding of the nature of chemistry, as well as increasing motivation.[42] Students should be put in a situation where they can solve tasks autonomously. Therefore, experimental instructions are offered using varying levels of complexity and different degrees in openness of inquiry learning. Students are allowed the freedom to test out different methods and to make mistakes. Certain experiences of failure during inquiry-based experiments are considered essential for full learning.[43] Deliberately, the inquiries are split into small steps so that unsuccessful attempts will not become too time-consuming. By using graded learning aids students of different performance levels will be able to complete the tasks successfully on their own and avoid lengthy phases of unsuccessful working.

Students are asked to work in small groups of two or three to enable cooperative learning processes. Each student receives a 'researcher's booklet' that contains all working sheets and safety instructions. The students are guided through the whole teaching module by this booklet. All experiments start with a short delineation of a problem that needs to be solved experimentally. The booklets contain space to record hypotheses, ideas, sketches, observations, notes and experimental activities.

For every teaching–learning module a teaching handbook is available to the teachers. This handbook includes general information on the entire project and students' requirements to join the learning environment. Short descriptions are given about the scientific background, the connection between the topic and the science and chemistry syllabus, and how to embed the laboratory visit into the school curriculum; the teacher guide also provides master copies for all experiments and working sheets.

6.4.3 One Example in Practice: Natural Vanilla or Synthetic Vanillin?

Vanilla flavouring compounds belong to the most widely used additives in food, drinks and cosmetics. Natural vanilla extract is used in perfumes, soaps, medicines, soft drinks, and sweets. The basic flavouring compound in natural vanilla is vanillin (4-hydroxy-3-methoxybenzaldehyde). However, natural vanilla contains only 1–3% vanillin.[44] There are over 130 more compounds which contribute to the unique flavour and aroma of natural vanilla, among them are vanillic acid, 4-hydroxybenzaldehyde and 4-hydroxybenzoic acid.[45]

In cosmetics, food and drinks industry, the demand of additives that cause the smell and taste of vanilla is far larger than natural methods of vanilla production can accommodate. This is why synthetically manufactured vanillin is often used as a cheap and readily available alternative. The current worldwide production of synthetic vanillin is roughly 10000 tons per year, with most of the processes starting either from lignin, eugenol or guaiacol.[46–49]

Today many attempts are made to make the synthesis of vanillin as sustainable as possible and in line with green chemistry principles.[50-53]

Several suggestions have been made so far to satisfy the synthesis of vanillin in chemistry education.[44,54-56] In the framework of situated learning, contention with vanilla/vanillin might start from industrial processes since industrial chemistry can provide an authentic and motivating context for chemistry education.[57] Industrial chemistry is constantly seeking to make their practices more sustainable. This is done not only because of responsibility for the future, but also for simple economic reasons. Thus, learning about different routes for vanillin synthesis provides an authentic and relevant field for learning about sustainability and green chemistry.

Different routes to gain vanillin flavouring compounds can be analysed and compared in the foreground of sustainability and green chemistry principles. Vanilla flavour is produced both naturally and synthetically. Both processes have advantages and disadvantages. Amongst the advantages of natural vanilla the unique smell and taste are the most important, but these have to be balanced with the growth conditions, transport and availability. As an alternative, students may analyse ways of producing synthetic vanillin. Different procedures are available. Some consider aspects of green chemistry more thoroughly than others. Economic and societal aspects may be also considered.

As a result of the relevance and motivating character of substances like natural vanilla and artificial alternatives, a non-formal learning environment was developed. The learning environment focuses on comparing natural vanilla and artificial vanillin. It deals with analysing food and drinks to determine whether they contain natural vanilla or synthetic alternatives. It also provides the opportunity to learn how the synthesis of artificial vanillin can be improved taking the philosophy of green chemistry into consideration.

The student learning environment in the university laboratory comprises up to a total of six experiments (Table 6.1). These experiments are supplemented by several other activities that engage with more advanced chemical background knowledge, the societal discussions around the topic, and the principles of green chemistry applied. If it is not possible for a learning group to perform all activities, the teachers negotiate with the accompanying university staff the most useful experiments and materials that fit best the students' abilities and school curriculum. The teaching and learning module

Table 6.1   Experiments in the module on vanilla/vanillin.

Experiment no.	Title
1	Synthesis of vanillin from isoeugenol
2	Synthesis of vanillin from lignin
3	Extraction of vanillin from vanilla
4	Detection of vanilla and vanillin in food
5	Analysis of the molecular structure of vanillin
6	Structure–property relationships of aromatic fragrances

was developed for grades 11–13 (age range 17–19 years) upper secondary school students. The lesson content is related to organic chemistry topics in the governmental syllabus for this age range.

In one of the experiments, vanillin is produced from isoeugenol. Eugenol can be derived from clove oil, a quite cheap and widely available substance. Thus a renewable source of raw materials is applied. Since the isomerization of eugenol to isoeugenol takes a lot of time and students often become impatient with long waiting times, this step is not carried out. Nevertheless, the students are able to apply essential steps of the vanillin synthesis starting from isoeugenol.

The synthesis of vanillin was developed from experiments already existing in the literature.[58,59] Many suggestions for vanillin syntheses made for the undergraduate university level are no longer allowed in Germany when working with secondary school students. This is even the case when the procedures are characterized as 'green' like some of the suggested syntheses of vanillin by biotechnology.[49,60] Other procedures use additions of functional groups on 4-hydroxybenzaldehyde or guaiacol that may convert into substances which are not allowed in the hands of school students (such as brominated compounds). The synthesis route from the literature example was improved in accordance with some of the central ideas of green chemistry. It was only as a result of changes in the preparation route, that the synthesis became feasible under German safety policies for science experimentation with school students.

Changes that were made to given procedures and taking ideas of green chemistry into consideration are as follows. In order to generate vanillin the C–C double bond needs to be split. Oxidative cleavage is carried out in the presence of an oxidizing agent. Potassium permanganate is used herein is in contrast to other proposals from the literature.[51,57] A protecting group, which is added before the oxidative cleavage and subsequently cleaved off again, prevents the formation of waste by-products. The synthesis proceeds in part without the input of energy as well as with the use of non-hazardous solid-state acids as catalysts. Water and methyl *tert*-butyl ether are used as harmless and less toxic solvents, respectively. The synthesis of vanillin on the basis of isoeugenol thus corresponds to some basic aspects of green chemistry. The total time necessary is roughly 4 hours and the process employs only chemicals that are considered not to be harmful and which are allowed for use with secondary school students under German law.

Within this teaching learning module, students gain a holistic view to the issue of using vanilla or vanillin. This includes manufacturing processes, analytical chemistry tasks, as well as issues of green chemistry and sustainable development. The experiments motivate and contextualize students' learning in the field of organic synthesis. Reflection upon the synthesis of vanillin is possible in the foreground of sustainability issues. Beside the educational contents, students may obtain experience in preparing experiments as well as in documentation of chemical investigations. Problem-solving activities, cooperative working, and evaluation skills are also encouraged.

6.5   Findings

Since 2012 various modules have been developed in the course of the project. So far more than 30 teachers with a total of nearly 2000 students have visited the different modules in the laboratories in Bremen and Saarbrücken. Different cases were described about the teachers' and students' learning experience.[7,8]

The development of the teaching and learning modules orients itself on the principles of action research triggered innovation of classroom practice.[61,62] The development is cyclical. Data within the cyclical development are collected focusing on the points of view of the accompanying university staff, the teachers and the students. In all visiting groups, both teachers and students are invited to contribute to a survey prior to and after visiting the university laboratory. The pre-questionnaire for the teachers consists of 15 Likert-type items and five open questions, while the student's questionnaire consists of 16 Likert-type and two open items. The questions focus on the prior expectations of the participants concerning the visit in the non-formal university student laboratory. Parallel structured questionnaires inquire into the teachers' and students' personal reflections after the visit. On the basis of this assessment, the materials and experiments are optimized before they are finally implemented and fully evaluated.

So far, 70 upper secondary chemistry students from six different learning groups have visited, with their teachers, the non-formal laboratory environment on vanilla/vanillin in the university laboratory in Bremen. Prior to the visit, almost all students indicated that they look forward with great interest to the outreach laboratory experience. Almost 90% of students agree totally, mainly or at least partially with the statement that it would be important to do something different to the format given at school. This is a typical ratio like most other data on expectations fit those data gathered in the other modules.

The students connected their positive expectations mainly with their hope to do many experiments, especially those that cannot be done in schools. Also, in the open part of the survey, the students mentioned that doing interesting experiments during the experimental session in the university laboratory is the main reason for their positive anticipation. Accordingly, 75% of the students mentioned that it is important to do a lot of experiments during the laboratory session. Students seem to be reflectively aware that school conditions are far from being optimal for doing practical work. The missing availability of materials and chemicals were criticized by many students, also the 45 minute timing of the science lessons was believed to impede experimental work. It were exactly these aspects that the students expected from the outreach university laboratory. Hope was expressed for better conditions, less pressure to perform, and more time. The expectations to learn something new were enumerated by the students concerning the visit to the university laboratory.

The students also expressed their perception that there is a lack of open and problem-based experiments in school and their hope for a different experience in the university laboratory. The students also expected to gain a better understanding of chemical issues in school and later better grades by having visited the non-formal education environment. Sixty per cent of the students expected improvement in their grades by visiting the university laboratory. The majority of students did not want to see the non-formal chemistry laboratory separated from formal learning in school. They expected learning that will help them gain better marks in school.

After the visit, a large majority of the students were very satisfied and happy with their visit. Ninety per cent of the students gave thorough positive feedback. A quote like 'I'm happy that we were able to do an experiment which would have not been possible under school conditions' was typical for responses to the open questions. The students emphasized the importance of experiments for their learning. The students enjoyed the practical work and emphasized the intense and student-centred atmosphere of working in small groups of two or three. A respective statement was supported again by over 90% of the students. It was also the very promoting atmosphere of being supported by the university staff in a very good student-to-teacher ratio, which was positively emphasized.

The teachers' expectations and experiences were mostly in line with those of their students. The teachers hoped that their students will benefit from the visit of the university learning environment. As the students suggested, also the teachers felt a need for more intense experimentation in science classes. As their students experienced, the teachers indicated that for them it is difficult to conduct experiments in their school environment in the number and quality they want because of insufficient equipment, time, facilities, and increasing restrictions of handling hazardous materials in schools. Nearly all of the teachers suggested that it is important that students can do experiments in small groups, that chances in school are limited, and that the university laboratory can help to overcome the situation. The teachers also agreed that learning about the scientific way of thinking, experiencing scientific methods, and developing problem-solving thinking might be promoted by the visit to the university. The teachers expressed hope that the visit to the university laboratory would contribute to and enrich their practice of teaching in their classes.

The teachers also expected that the university laboratory visit would increase the motivation in science learning. In most modules, on average, three quarters of the teachers agreed with a respective statement completely. The teachers attributed motivational potential to the societally relevant aspects of the experience, such as providing students with insights into university education as well as into chemistry which is relevant to everyday life. Their point of view was that a visit to a university learning environment should have many more benefits beyond motivation and learning the course content. For the teachers it was more important to use the university laboratory context to make the students aware of the relevance of chemistry, than to fulfil part of their formal curriculum. Nearly 90% of the teachers visiting the different modules agreed at least partially with this statement.

Although the teachers mainly emphasized students' skills development beyond pure subject matter learning by the university visit, they nevertheless wanted the contents to be linked with those of the school curriculum and the official syllabus. In the questionnaire, this aspect was mentioned by almost all teachers. Also here there is overlap with the students' point of view, in that they expected better marks after visiting the university laboratory. The teachers clearly recognized that the topics offered in this project go beyond the traditional elements of the school curriculum. The inclusion of sustainability perspectives, learning about green chemistry, and encompassing a socio-scientific point of view was highly appreciated because it was on the one hand connected to but on the other hand also went beyond the pure content of the syllabus.

After the visit, the teachers gave similar feedback as their students did. The teachers placed great emphasis on the quality of care: 'The students worked well and with great interest. Very good, friendly and professional mentoring.' The teachers also followed their students' behaviour with great interest. Several teachers mentioned that they saw their students from a completely different angle. The lower-achieving students, in particular, surprised the teachers with their working behaviour. The teachers also gave positive feedback on the organization and on the materials. All teachers rated the experiments positive.

However, teachers described benefits not only for their students, but also for themselves. Almost 80% of the teachers visiting the different modules agreed at least partly with the statement 'It is important for me to be able to find ideas and inspiration from this excursion for my own teaching'. Teachers stated that they had become aware of new topics, experiments and teaching–learning materials. Most of the teachers stated that they have received inspiration for their own teaching. Most of the teachers indicated that they will implement working sheets, experiments and other ideas from sustainability issues and green chemistry into their classrooms. Burmeister *et al.* stated that a lack of materials and availability of feasible experiments for secondary chemistry classroom might be one of the major reasons why ESD is rarely implemented in school chemistry teaching.[3] The project provides the opportunity to present new and innovative experiments and materials to teachers and to bring them into schools. The willingness of the teachers is given, as long as appropriate materials are provided.[5] Thus, the project seems to offer direct potential to overcome this shortcoming on a regional level and, *via* publication of the materials and in-service teachers' continuing professional development, also on a larger scale.

The feedback from both the students and the teachers indicates that partially non-formal learning environments, as described here, have the potential to increase students' attitudes and motivation towards chemistry and science learning. Although there were many high expectations, the expectations of the teachers and students seemed to have been fulfilled. Most teachers expressed themselves positively after the laboratory session. The results of the other units that were developed in the project were reflected similarly by the students and their teachers.

6.6 Conclusions

Orion and Hofstein[63] suggested that the development of a more positive student attitude towards learning science could be fostered by visiting informal and non-formal learning environments. We found that the students enjoyed the unfamiliar, non-formal atmosphere of visiting the university chemistry laboratory. Hardly any student was not looking forward to the visit or was disappointed after it. The positive feedback is a promising sign that school students will develop more positive attitudes towards chemistry, science and technology. Our approach of connecting science learning to authentic and innovative issues from the sustainability debate embedded into the partially non-formal learning experience seems to be motivating and meaningful to the learners.

Also, the teachers reacted very positively to the programme. The teachers followed their students' behaviour in the laboratory with great interest. They had both a focus on the students' behaviour as well as on the students' activities and tasks. Through these observations they gained new knowledge about sustainability and green chemistry as well as becoming familiar with new experiments, which, at least in part, can also be carried out in regular school science classrooms. They also saw how motivating the issues of sustainable development can be for their students if they are integrated with chemistry learning and practical work.

Limitations in the initiative lie in the limited regional range of the two university laboratories. It is also suggested in the literature that the effects of such laboratory visits are short-term if they occur only once.[64] So far; there is limited research about whether a repeated visit to a non-formal learning environment will have more durable effects. A recent study by Zehren and colleagues showed that gaining corresponding effects is possible.[65] While offering modules for all grade levels in the lower and upper secondary level with connection to chemistry-related issues of sustainability and sustainable development the described effects may be stronger if classes visit the university laboratory more than once. However, more research is needed to show how often the visits should occur in order to contribute to long-term effects on motivation, attitudes and cognitive gains by the non-formal chemistry learning environment, in this case about sustainability and green chemistry.

What definitely proved to be true was the intention to use the partially non-formal student laboratory as a catalyst for innovations in practical work in chemistry education. New experiments and teaching–learning scenarios were developed, successfully tested by the visiting students and finally published. With these materials a growing body of classroom materials on chemistry-related sustainability issues became available which now offer a chance to be implemented by teachers even beyond this project.

Acknowledgement

We gratefully acknowledge the funding and support of the German Environmental Trust (*Deutsche Bundesstiftung Umwelt*) and the contributions of the non-formal laboratory staff in Bremen and Saarbrücken, the students and their teachers towards this project.

References

1. P. T. Anastas and J. C. Warner, *Green Chemistry: Theory and practice*, Oxford University Press, Oxford, 1998, ch. 2, pp. 11–19.
2. G. Centi and S. Perathoner, in *Sustainable industrial processes*, ed. F. Cavani, G. Centi, S. Perathoner and F. Trifiro, Wiley-VCH, Weinheim, 2009, pp. 1–72.
3. M. Burmeister, F. Rauch and I. Eilks, *Chem. Educ. Res. Pract.*, 2012, **13**, 59.
4. M. Burmeister and I. Eilks, *Sci. Educ. Int.*, 2013, **24**, 167.
5. M. Burmeister, S. Schmidt-Jacob and I. Eilks, *Chem. Educ. Res. Pract.*, 2013, **14**, 169.
6. M. Burmeister and I. Eilks, *Centre Educ. Pol. Stud. J.*, 2013, **3**, 59.
7. N. Garner, M. de Lourdes Lischke, A. Siol and I. Eilks, in *Handbook of research on pedagogical innovations for sustainable development*, ed. K. Thomas and H. Muga, IGI Global, Hershey, 2014, pp. 229–244.
8. N. Garner, S. M. Hayes and I. Eilks, *Sisyphus J. Educ.*, 2014, 2(2), 10.
9. World Commission on Environment and Development, *Our common future*, University Press, New York, 1987.
10. United Nations Conference on Environment and Development (UNCED), http://sustainabledevelopment.un.org/content/documents/Agenda21.pdf (accessed February 2014).
11. M. van Eijck and W.-M. Roth, *PLoS Biol.*, 2007, DOI: 10.1371/journal.pbio.0050306.
12. United Nations Educational, Scientific and Cultural Organization (UNESCO), http://unesdoc.unesco.org/images/0014/001486/148654e.pdf (accessed February 2014).
13. G. de Haan, *Environ. Educ. Res.*, 2006, **12**, 19.
14. R. McKeown, *Education for Sustainable Development Toolkit*, http://www.esdtoolkit.org/about.htm (accessed February 2014).
15. H. Gresch, M. Hasselhorn and S. Bögeholz, *Int. J. Sci. Educ.*, 2013, **35**, 2587.
16. United Nations Educational, Scientific and Cultural Organization (UNESCO), http://www.unesco.org/education/justpublished_desd2009.pdf (accessed February 2014).
17. P. Jones, C. J. Trier and J. P. Richards, *Int. J. Educ. Res.*, 2008, **47**, 341.
18. J. D. Bradley, *Chem. Educ. Int.*, 2005, **6**, 11.
19. M. Karpudewan, Z. Ismail and W.-M. Roth, *Chem. Educ. Res. Pract.*, 2012, **13**, 120.
20. M. Eissen, *Chem. Educ. Res. Pract.*, 2012, **13**, 103.

21. M. Burmeister and I. Eilks, *Chem. Educ. Res. Pract.*, 2012, **13**, 93.
22. K. G. Tobin, *Sch. Sci. Math.*, 1990, **90**, 403.
23. I. Abrahams, *Practical work in secondary science*, Continuum, London, 2011.
24. A. Hofstein, M. Kipnis and I. Abrahams, in *Teaching chemistry – A studybook*, ed. I. Eilks and A. Hofstein, Sense, Rotterdam, 2012, pp. 153–282.
25. A. Hofstein and P. M. Kind, in *Second international handbook of science education*, ed. B. J. Fraser, K. G. Tobin and C. J. McRobbie, Springer, Dordrecht, 2012, pp. 189–207.
26. L. J. Rennie, in *Handbook of research and science education*, ed. S. K. Abell and N. G. Ledemran, Lawrence Earlbaum Associates, Mahwah, 2007, pp. 125–167.
27. R. K. Coll, J. K. Gilbert, A. Pilot and S. Streller, in *Teaching chemistry – A studybook*, ed. I. Eilks and A. Hofstein, Sense, Rotterdam, 2012, pp. 241–267.
28. P. Werquin, *Recognising non-formal and informal learning. Outcomes, policies and practices*, OECD, Paris, 2010.
29. H. Eshach, *J. Sci. Educ. Technol.*, 2007, **16**, 171.
30. R. W. Bybee, in *Free-choice science education: How we learn science outside of school*, ed. J. H. Falk, Teachers College Press, New York, 2001, pp. 44–63.
31. N. Orion and A. Hofstein, *J. Res. Sci. Teach.*, 1994, **31**, 1097.
32. J. Gallacher and M. Feutrie, *Eur. J. Educ.*, 2003, **38**, 71.
33. A. Hofstein and S. Rosenfeld, *Stud. Sci. Educ.*, 1996, **28**, 87.
34. L. J. Rennie and T. P. McClafferty, *Stud. Sci. Educ.*, 1996, **27**, 53.
35. M. Braund and M. Reiss, *Int. J. Sci. Educ.*, 2006, **28**, 1373.
36. F.-J. Scharfenberg and F. X. Bogner, *Eurasian J. Math. Sci. Technol. Educ.*, 2014, **10**, 329.
37. O. J. Haupt, J. Domjahn, U. Martin, P. Skiebe-Corette, S. Vorst, W. Zehren and R. Hempelmann, *Math. Naturwiss. Unterr.*, 2013, **66**, 324.
38. T. D. Sadler, *Socio-scientific issues in the classroom*, Springer, Dordrecht, 2011.
39. I. Eilks, G. T. Prins and R. Lazarowitz, in *Teaching chemistry – A studybook*, ed. I. Eilks and A. Hofstein, Sense, Rotterdam, 2012, pp. 183–212.
40. L. Luehmann, *Int. J. Sci. Educ.*, 2009, **31**, 1831.
41. J. G. Greeno, in *Complex information processing: The impact of H. A. Simon*, ed. D. Klahr and K. Kotovsky, Lawrence Erlbaum, Hillsdale, 1988, pp. 285–318.
42. R. Berg, V. C. B. Bergendahl, B. K. S. Lundberg and L. A. E. Tibell, *Int. J. Sci. Educ.*, 2003, **25**, 351.
43. V. N. Lunetta, A. Hofstein and M. P. Clough, in *Handbook of research on science education*, ed. S. K. Abell and N. G. Lederman, Lawrence Erlbaum, Mahwah, 2007, pp. 393–441.
44. M. B. J. Hocking, *J. Chem. Educ.*, 1997, **74**, 1055.
45. A. Pérez-Silva, E. Odoux, P. Brat, F. Ribeyre, G. Rodriguez-Jimenes, V. Robles-Olevera and M. A. García-Alvarado, *Food Chem.*, 2006, **99**, 728.

46. R. Plaggenborg, J. Overhage, A. Loos, J. A. C. Archer, P. Lessard, A. J. Sinskey, A. Steinbüchel and H. Priefert, *Appl. Microbiol. Biotechnol.*, 2006, **72**, 745.

47. C. Fargues, A. Mathias and A. E. Rodrigues, *Ind. Eng. Chem. Res.*, 1996, **35**, 28.

48. T. Unno, S.-J. Kim, R. A. Kanaly, J.-H. Ahn, S.-I. Kang and H.-G. Hur, *J. Agric. Food Chem.*, 2007, **55**, 8556.

49. C. Brazinha, D. S. Barbosa and J. G. Crespo, *Green Chem.*, 2011, **13**, 2197.

50. J. Schrader, M. M. W. Etschmann, D. Sell, J. Hilmer and J. Rabenhorst, *Biotechnol. Lett.*, 2004, **26**, 463.

51. W. A. Herrmann, T. Weskamp, J. P. Zoller and R. W. Fischer, *J. Mol. Catal. A: Chem.*, 2000, **153**, 49.

52. T. X. T. Luu, T. T. Lam, T. N. Le and F. Duus, *Molecules*, 2009, **14**, 3411.

53. J. Hu, Y. Hu, J. Mao, J. Yao, Z. Chen and H. Li, *Green Chem.*, 2012, **14**, 2894.

54. R. T. Winter, H. L. van Beek and M. W. Fraaije, *J. Chem. Educ.*, 2011, **89**, 258.

55. F. Taber, S. Patel, T. M. Hambleton and E. E. Winkel, *J. Chem. Educ.*, 2007, **84**, 1158.

56. A. Hofstein and M. Kesner, *Int. J. Sci. Educ.*, 2006, **28**, 1017.

57. M. Branan, J. T. Butcher and L. R. Olsen, *J. Chem. Educ.*, 2007, **84**, 1979.

58. M. Lampman and S. D. Sharpe, *J. Chem. Educ.*, 1983, **60**, 503.

59. M. Lampman, J. Andrews, W. Bratz, O. Hanssen, K. Kelley, D. Perry and A. Ridgeway, *J. Chem. Educ.*, 1977, **54**, 776.

60. S. Wenda, S. Illner, A. Mell and U. Kragl, *Green Chem.*, 2011, **13**, 3007.

61. I. Eilks, in *Action Research, innovation and change*, ed. T. Stern, F. Rauch, A. Schuster and A. Townsend, Routledge, London, 2014, pp. 156–176.

62. I. Eilks and T. Feierabend, in *Educational design research: Introduction and illustrative cases*, ed. T. Plomp and N. Nieveen, SLO Netherlands Institute for Curriculum Development, Enschede, 2013, pp. 321–338.

63. N. Orion and A. Hofstein, *Sci. Educ.*, 1991, **75**, 513.

64. J. DeWitt and M. Storksdieck, *Visitor Stud.*, 2008, **11**, 181.

65. W. Zehren, H. Neber and R. Hempelmann, *Math. Naturwiss. Unterr.*, 2013, **66**, 416.

CHAPTER 7

Green Catalysts for Producing Liquid Fuels from Lignocellulosic Biomass

DEQUAN XIAO*[a] AND EVAN S. BEACH[b]

[a]Department of Chemistry and Chemical Engineering, University of New Haven, West Haven, CT 06517, USA; [b]Center for Green Chemistry and Green Engineering, Yale University, New Haven, CT 06511, USA
*E-mail: dxiao@newhaven.edu

7.1 Introduction

Green chemistry is a philosophy of chemical research and engineering that encourages the design of products and processes that reduce or eliminate the use and generation of hazardous substances.[1] In accordance with the goals of green chemistry, lignocellulose (or non-food) biomass is increasingly recognized as a valuable renewable resource that has been estimated to have the potential to produce 3 billion barrels of liquid transportation fuel annually in the US alone.[2] Cheap and efficient methodologies for converting biomass solids to liquid fuels *on a large scale* will not only help to reduce dependence on petroleum, but will also alleviate CO_2 emissions. However, such technologies are not yet fully developed. Designing effective, low-cost, robust, and sustainable biomass catalysts is an indispensable step

Worldwide Trends in Green Chemistry Education
Edited by Vânia Gomes Zuin and Liliana Mammino
© The Royal Society of Chemistry 2015
Published by the Royal Society of Chemistry, www.rsc.org

for developing the next generation technology for converting lignocellulosic biomass to liquid fuels.

Biomass is predominantly composed of biopolymers, for example cellulose, hemicellulose, and lignin. From a molecular viewpoint, converting non-food biomass to liquid fuels is a process of transforming solid-phase macromolecules into liquid-phase small molecules. Simultaneously, it is necessary to reduce the oxygen content of the small molecules to improve the energy density.

In this chapter, we focus on current efforts to apply various catalysts in the conversion of lignocellulosic biomass molecules into liquid fuels. From the viewpoint of the fundamental chemistry, we discuss the main chemical processes and reaction mechanisms for specific biomass conversion methods, as well as the typical catalysts that arc uscd. Since these biomass catalysts are used to perform green chemistry (*i.e.*, produce renewable energy) we can also call them green catalysts. We hope that this chapter will serve as a tutorial for green chemistry education, and inspire researchers to design more efficient, robust, low-cost and sustainable catalytic approaches. This chapter does not intend to serve as a comprehensive account for all the catalysts used for lignocellulosic biomass conversion.

The overall goal of converting lignocellulosic biomass to liquid fuels includes two tasks: (1) degrading biomass polymers by breaking C–C or C–O bonds; and (2) minimizing the oxygen content to improve the quality of the biomass liquid fuel. Conventional fuels derived from petroleum refinery typical contain less than 1 wt% of oxygen. However, biomass feedstocks usually contain 40–60 wt% oxygen. In the following sections, we will compare the chemical structures of three important components of lignocellulose biomass: cellulose, lignin, and hemicellulose, and discuss three main paths for converting the biomass polymers into liquid fuels: solid → gas → liquid (*i.e.*, S → G → L), solid → liquid (*i.e.*, S → L), and solid → gas/liquid → liquid (*i.e.*, S → GL → L). For each path, the chemical processes and catalysts used for specific approaches have been reviewed.

7.2   Biomass Polymers

Lignocellulosic biomass is composed of mainly three types of polymers: cellulose (40–50%), hemicellulose (25–35%), and lignin (15–20%).[3,4] Cellulose is a type of crystalline polymer composed of glucose units linked *via* β-glycosidic bonds (see Figure 7.1). Cellulose is typically isolated within the complex lignin/hemicellulose matrix. In addition, strong hydrogen bonds exist within the polymer chains and between polymer chains. Hence, cellulose is resistant to the solvation by water or direct attack by enzymes, making it difficult to hydrolyse untreated biomass.

The hemicellulose fraction of lignocellulosic biomass is an amorphous polymer that is generally comprised of five different sugar monomers: D-xylose, L-arabinose, D-galactose, D-glucose and D-mannose (see Figure 7.2).[5] The hemicellulose is usually branched polymers, and contains fewer hydrogen bonds between polymer chains, as compared to cellulose. Hence,

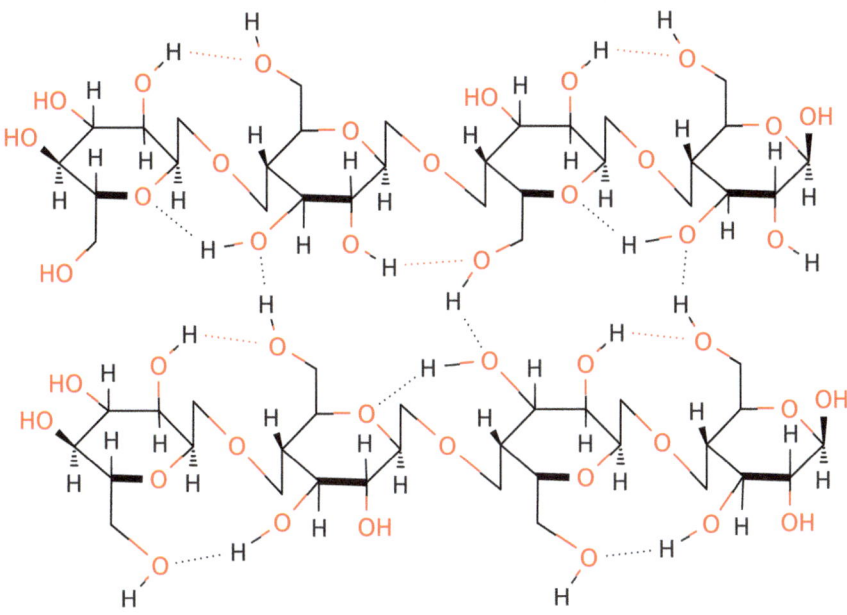

Figure 7.1   Typical chemical structure of celluloses. The structure of cellulose is composed of glucose monomers linked by β-glycosidic bonds. Strong hydrogen bonds link glucose units between different polymer chains, leading to crystalline structures that are resistant to water.

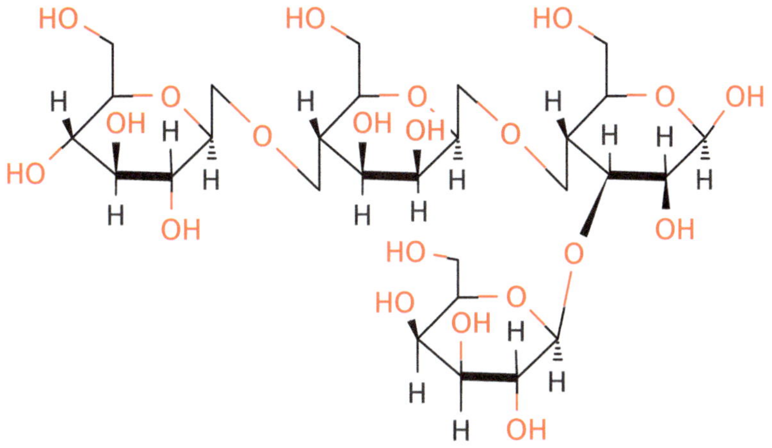

Figure 7.2   Typical chemical structure of hemicelluloses. The structure of hemicellulose is typically composed of five different sugar monomers: D-xylose, L-arabinose, D-galactose, D-glucose, and D-mannose.

hemicellulose is a non-crystalline polymer that is susceptible to the solvation of water and enzymatic attack. At temperatures above 180 °C, hemicellulose can dissolve in water fully.[6]

Lignin is an amorphous polymer based on coniferyl alcohol, sinapyl alcohol, and coumaryl alcohol,[2,7] linked by aromatic ether bonds (see Figure 7.3). Due to the presence of aromatic rings, lignin provides structural rigidity to

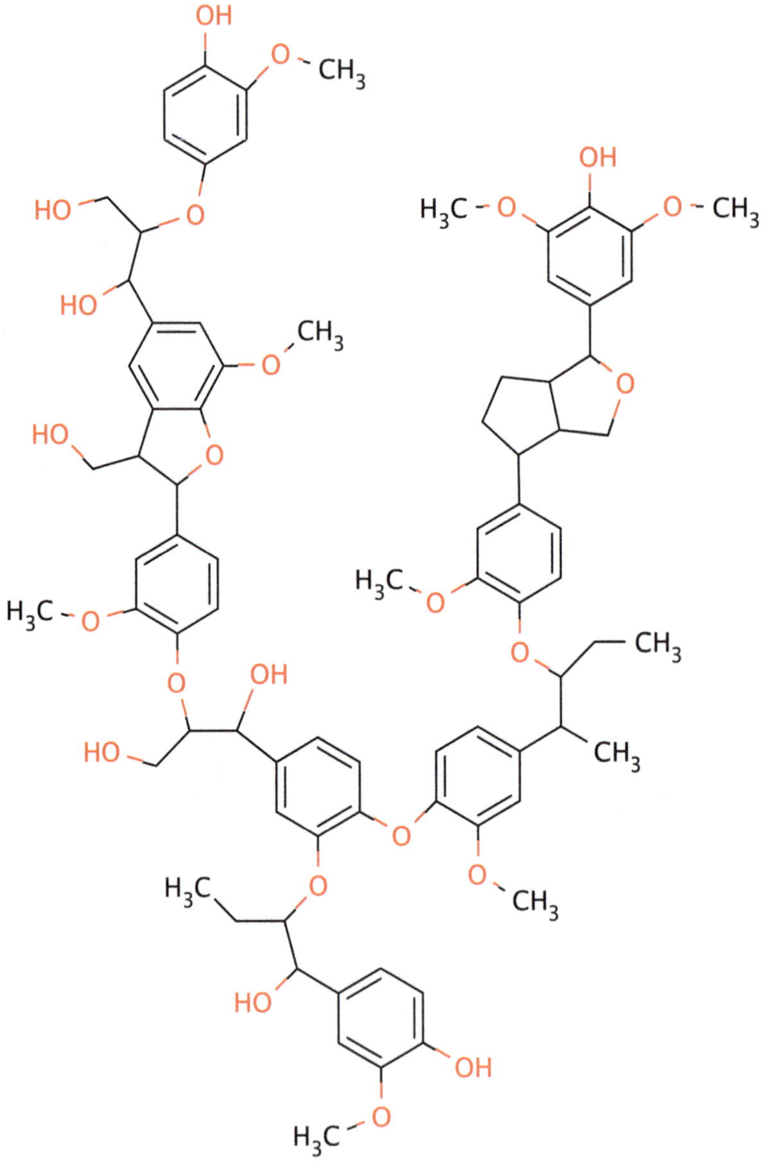

Figure 7.3 Typical chemical structure of lignin. The structure of lignin polymers is typically composed of coniferyl alcohol, sinapyl alcohol, and coumaryl alcohol linked by aromatic ether bonds.

plants, enabling the transport of water and solutes.[8] However, lignin is highly random in structure and resistant to hydrolysis.

7.3 Three Paths for Biomass Conversion

Convention liquid fuels derived from crude oils usually have 5–14 carbons in the molecules. In Table 7.1, we list common liquid fuels such as gasoline, kerosene, and fuel oils along with natural gas and petroleum ether for comparison.

As shown in Figure 7.4, there are three general paths for converting biomass polymers (solid phase) to small-molecule fuels (liquid phase): (1) solid → gas → liquid (S → G → L), (2) solid → liquid (S → L), and (3) solid → gas and liquid → liquid (S → GL → L). The reactions in path 3 are equivalent to the combined reactions from paths 1 and 2. In the following sections, we review specific biomass conversion methods that fall in the categories of path 1 and path 2.

7.3.1 Solid → Gas → Liquid

The solid → gas → liquid (*i.e.*, S → G → L) path denotes a general route of converting biomass solid into gas-phase molecules first, and then the gas molecules are combined to form liquid fuels. The approaches following the S → G → L path include gasification, aqueous phase reforming, and photocatalytic conversion.

Table 7.1 Common liquid fuels from petroleum fractions.

Fraction	Boiling range (°C)	Number of carbon atoms
Natural gas	<20	C_1 to C_4
Petroleum ether	20–60	C_5 to C_6
Gasoline	40–200	C_5 to C_{12}, mostly C_6 to C_8
Kerosene	150–260	Mostly C_{12} to C_{13}
Fuel oils	>260	C_{14} or higher

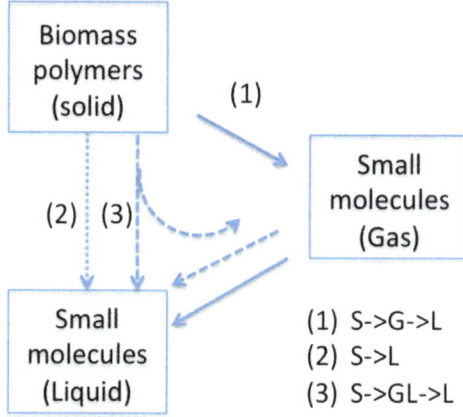

Figure 7.4 Three general paths for converting biomass solid to liquid fuels.

7.3.1.1 Gasification

Gasification is a thermal process to decompose the biomass polymers into small molecule gases, predominantly CO and H_2 (syngas).[2] Syngas is then used to produce liquid fuels *via* Fisher–Tropsch reactions.[9,10] This is the most developed technology in biomass renewable energy and has been commercialized by Choren, Coskata, South Africa (SASOL), and Range Fuels, among others.[11,12]

The gasification process is usually performed using four main units: (1) a biomass gasifier, (2) a gas clean-up unit, (3) a water-gas shift (WGS) reactor in certain cases, and finally (4) a syngas converter. The gasification unit converts the biomass at high temperature (600–900 °C) in the presence of oxygen or steam.

Depending on the catalysts or reaction conditions, syngas with different components can be generated. In the last three decades, the gasification process has been conducted in mainly two ways: (1) producing H_2-rich syngas under high temperature (>500 °C) without using any catalysts or using non-metal catalysts; and (2) producing methane-rich syngas at lower temperatures (from critical temperature to 500 °C or below) with catalysts.

Catalytic gasification of biomass was pioneered by Elliott *et al.* in the 1980s.[13-16] They found that Ru, Rh, and Ni are effective catalysts, using supports such as ZrO_2, Al_2O_3, TiO_2, and C. Sealock *et al.* used Harshaw Ni as the catalyst under the conditions of 450 °C and 34 MPa.[15,16] Bond breaking during catalytic gasification usually follows a radical mechanism.[17] The catalysts facilitate the breakage of C–C bonds (*e.g.*, opening phenol rings), and breaking the H–OH bond of water to produce O and OH radicals. The O or OH radicals attack biomass polymer fragments to generate CO and CO_2, and the H radical abstracts H from the fragments producing H_2 gas.

To control the H_2/CO ratio, several reactions are employed. Most important is the water gas shift reaction, which provides a source of hydrogen at the expense of carbon monoxide:[18]

$$H_2O + CO \rightarrow H_2 + CO_2 \tag{7.1}$$

For Fischer–Tropsch plants that use methane as the feedstock, another important reaction is steam reforming, which converts methane into CO and H_2:

$$H_2O + CH_4 \rightarrow CO + 3H_2 \tag{7.2}$$

Syngas can be converted to liquid fuels *via* the Fischer–Tropsch synthesis (FTS).[10] The FTS process involves a series of chemical reactions that produce a variety of alkanes, ideally having the formula (C_nH_{2n+2}):

$$(2n+1)H_2 + nCO \rightarrow C_nH_{2n+2} + nH_2O \tag{7.3}$$

where n is typically 10–20. For liquid fuels, the formation of methane ($n = 1$) is not desired. The synthesized alkanes are mainly straight-chain, and can be used as diesel fuel. As side products, small amounts of alkenes, alcohols and other oxygen-containing hydrocarbons can be generated as well. For the catalysts, Fischer and Tropsch originally used cobalt in an oil medium to perform FTS initially.[10] Later, iron catalysts were used for CO-rich syngas,[19] and Ru catalysts were developed for the syngas of CO and H_2.[20]

One of the challenges in gasification is to develop efficient technology to prevent or reduce the formation of tars, ammonia, hydrogen sulfide, and particulates. This problem is often resolved by designing effective catalysts. For example, using perovskite-structural La–Ni–Fe catalysts is reported to significantly reduce the formation of tars, and has been applied to biomass from almond shells.[21] Schmidt *et al.* have reported rhodium–cerium catalysts that can significantly reduce tar formation in outlet streams.[22,23] Further understanding of the mechanisms of tar formation will help aid rational catalyst design.[24]

7.3.1.2 Aqueous Phase Reforming

Aqueous phase (or liquid phase) reforming is a new approach for converting biomass feedstocks into oxygenated species in the presence of water, and then subsequently converting the oxygenates into H_2, syngas, or alkanes with the aid of catalysts in the aqueous phase (eqn (7.4)). This approach was pioneered by Dumesic and co-workers in 2002.[25] Since then, this approach has been used to convert biomass feed stocks into a variety of liquid fuels and chemicals.[26,27]

The advantages of using a liquid-phase process include low heating cost, the feasibility of using various catalysts, and easy separation of oil products from the aqueous phase. The drawback is that the biomass feedstocks need to be purified. A comprehensive description of reactions that occur in liquid phase processing can be found in others reviews.[12]

$$\tag{7.4}$$

The generated H_2 can be converted to liquid fuels through Fischer–Tropsch reactions, or used in the upgrading of bio-oils.

The first catalyst used was Pt (3 wt% on Al_2O_3), applied to renewable feedstocks such as sorbitol, glycerol, ethylene glycol, and methanol to produce hydrogen gas.[25] Pt black and Pt supported on TiO_2 or ZrO_2 have also been developed as active catalysts for generating hydrogen.[28] More recently, metal catalysts like Sn–Raney Ni have been shown to enhance the production of H_2 from sorbitol, glycerol, and ethylene glycol.[29] Conversion of glycerol has also been demonstrated with several first–row transition metals.[30] The aqueous phase reforming method was used to convert lignocellulosic biomass into liquid fuels using a Pt/Al_2O_3 catalyst.[31]

7.3.1.3   *Photocatalytic Conversion*

Photocatalytic conversion uses sunlight to convert biomass feedstocks into H_2 or liquid fuels, and is usually performed at room temperature. The cost could be potentially lower than other methods, since abundant sunlight is used as an energy source for the conversion. However, the photocatalytic conversion efficiency from biomass to H_2 is still low (1%). New photocatalysts need to be developed to improve the efficiency:

$$(C_6H_{12}O_6)_n + 6nH_2O \xrightarrow{\text{photocatalysts}} 6nCO_2 + 12nH_2 \qquad (7.5)$$

One of the first reports using TiO_2 photocatalysts to convert biomass into H_2 was reported by Kawai and Sakta in 1980.[32] They used $RuO_2/TiO_2/Pt$ photocatalysts to convert cellulose, protein, fat, and waste materials into H_2 under the irradiation by a Xe lamp. Recently, M/TiO_2 catalysts (M = Pt, Rh, Ru, Au or Ir) have been used to convert glucose to H_2, and Rh/TiO_2 was found to have the maximum catalytic efficiency.[33] Verykios found that adding alcohols as hole scavengers can improve the photocatalytic conversion efficiency.[34] Water-splitting chemistry has been coupled with photo-oxidation of biomass in a Pt/TiO_2 system operating under mild conditions.[35] As shown in Figure 7.5, the proposed mechanism of a typical metal-doped TiO_2 photocatalyst includes the migration of electron and hole to the catalyst surfaces. On the surface, biomass molecules can lose H^+ to produce a radical when interacting with the hole, and the H^+ can gain electron on the surface and combine with another H atom to generate H_2 gas.

7.3.2   Solid → Liquid

The solid → liquid (*i.e.*, S → L) path denotes a general route of converting biomass solids directly into liquid-state small molecules, usually called bio-oil. The bio-oil is then upgraded to generate high-quality liquid fuels. Approaches following the S → L path include pyrolysis, liquefaction, chemical hydrolysis, enzymatic hydrolysis, and methanol-phase reforming.

7.3.2.1   *Pyrolysis*

Pyrolysis entails thermal decomposition of biomass molecules in the absence of oxygen, usually at the temperatures up to 650–800 K. To produce liquid oils, the heating process should be short (*i.e.*, short residence time), thus this process is usually referred to as fast pyrolysis. At high temperature, the biomass is vaporized and then condensed upon cooling to produce a liquid oil mixture which may be comprised of more than 300 compounds such as alkanes, aromatic aliphatic, sugars, alcohols, ketones, aldehydes, acids and esters. If the residence time is longer (slow pyrolysis), the product mixture is likely to produce more solid coke than liquid fuels. An advantage of fast pyrolysis is that it is economical for use on a small scale (*i.e.*, 50–100 tons biomass per day). Yields of bio-oil production in excess of 70% have been

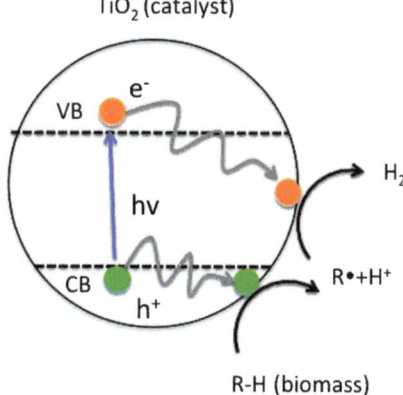

Figure 7.5 Proposed mechanism for photocatalytic conversion of biomass feedstocks to hydrogen on a Pt/TiO$_2$ photocatalyst.[36]

reported.[37] The most favourable conditions for maximizing the yield of bio-oils are rapid heating, high heat transfer rate, short residence time, moderate reaction temperature (*ca.* 500 °C), and rapid cooling of pyrolysis vapours.

The bio-oils have a dark-brown color and the composition includes organics (*ca.* 75–80 wt%) and water (*ca.* 20–25 wt%).[38–40] The crude bio-oil is rarely used as a liquid fuel directly, as it tends to have a low heating value that is less than half that of petroleum fuels (32–38 MJ kg^{-1}), strong corrosiveness (pH 2–3), high viscosity (30–1000 cp at 40 °C), and poor chemical stability.[38–40] Thus, bio-oil is commonly upgraded in order to be suitable for use as a liquid fuel. We will discuss the upgrading process in the later sections.

The following reactions (eqn (7.6)-(7.11)) are typical of the pyrolysis process:

$$\text{biomass} \xrightarrow{\text{heating}} \text{volatile vapour} + \text{char} \xrightarrow{\text{cooling}} \text{bio-oil} + \text{tar} \quad (7.6)$$

For the pyrolysis of cellulose, it first is degraded to methyl glyoxal:[41]

$$(C_6H_{10}O_5)_x \xleftrightarrow{535\,K} xC_6H_{10}O_5 \quad (7.7)$$

$$C_6H_{10}O_5 \leftrightarrow H_2O + 3CH_3-CO-CHO \quad (7.8)$$

This is followed by the hydrogenation of methyl glyoxal to produce isopropyl alcohol, propylene glycol or acetal:

$$2CH_3-CO-CHO + 2H_2 \leftrightarrow 2CH_3-CO-CH_2OH \quad (7.9)$$

$$2CH_3-CO-CH_2OH + 2H_2 \leftrightarrow 2CH_3-CHOH-CH_2OH \quad (7.10)$$

$$2CH_3-CO-CH_2OH + 2H_2 \leftrightarrow 2CH_3-CHOH-CH_3 + H_2O \quad (7.11)$$

For the pyrolysis of lignin, the aromatics and phenols and their alkyl substituted fractions may be formed by recombination and cyclization reactions, *via* Aldol condensation, and from C2, C3 and C4 fragments that occur as initial degradation products.[42] Further reaction may yield furans, aldehyde and ketones.

During pyrolysis, small molecules may be formed *via* radical mechanisms (eqn (7.11)–(7.14)). For example, a biomass substrate is decomposed into two radicals $2R^\bullet$[43] then the free radicals can attack the solvent molecules (DH_2), high molecular weight fractions (M), or recombine with other free radicals produced by these steps, leading to stable products:[44]

$$Biomass \rightarrow 2R^\bullet \qquad (7.12)$$

$$R^\bullet + DH \rightarrow RH + D^\bullet \qquad (7.13)$$

$$R^\bullet + MH \rightarrow RH + M^\bullet \qquad (7.14)$$

$$M^\bullet + M^\bullet \rightarrow M-M \qquad (7.15)$$

The remaining challenge for pyrolysis is to remove or reduce the production of cokes. Catalysts can play a critical role and catalyst development represents a significant new direction in the field. Recently, the Huber group used a zeolite-based catalyst (HZSM-5.57) and showed a high-yield of aromatics (20–30%) for the fast pyrolysis.[45] Follow-up studies have demonstrated the potential for producing commodity chemicals,[46] and the possibility of tuning the product stream by control of catalyst morphology.[47] These are promising steps toward producing high-value liquid fuels using catalysts.

7.3.2.2 *Liquefaction*

Liquefaction is the process of thermal decomposition of biomass by mixing the biomass with water and basic catalysts like sodium carbonate, usually carried out at a lower temperature than pyrolysis (300–400 K), at high pressure (120–200 atm) and longer residence time.[48,49] The liquefaction process is potentially more expensive than pyrolysis due to the high pressure requirements. However, the bio-oil produced from liquefaction has less oxygen content (12–14%) than that obtained from pyrolysis.[50] No pre-drying of biomass is required for the liquefaction process.

During the liquefaction process, the biomass is decomposed into small molecules, and then re-polymerized to form liquid fuels:

$$biomass + H_2O \xrightarrow{\text{catalysts}} small\,molecules \rightarrow bio\text{-}oil \qquad (7.16)$$

A Na_2CO_3-catalysed mechanism for biomass liquefaction in the presence of carbon monoxide was proposed by Appell *et al.* (eqn (7.16)–(7.20)).[51] First, the Na_2CO_3 reacts with water and CO to form sodium formate:

$$Na_2CO_3 + 2CO + H_2O \rightarrow 2HCOONa + CO_2 \qquad (7.17)$$

Dehydration of vicinal hydroxyl groups in carbohydrate structures produces an enol, and then the enol isomerizes to produce ketones:

$$-CH(OH)-CH(OH) \rightarrow -CH + C(OH) \rightarrow -CH_2-CO- \qquad (7.18)$$

The carbonyl group in the ketone is reduced to the corresponding alcohol in the presence of formate ions and water:

$$HCOO^- + CH_2-CO \rightarrow -CH_2-CH(O^-) + CO_2 \qquad (7.19)$$

$$-CH_2-CH(O^-)- + H_2O \rightarrow -CH_2-CH(OH)- + OH- \qquad (7.20)$$

Hydroxide reacts with another CO molecule to generate formate:

$$OH^- + CO \rightarrow HCOO^- \qquad (7.21)$$

According to this mechanism, deoxygenation occurs through decarboxylation of esters formed from the hydroxyl group (in eqn (7.20)) and formate ion [derived from the carbonate, in eqn (7.21)].

7.3.2.3   Chemical Hydrolysis

Chemical hydrolysis decomposes cellulose or hemicellulose into glucose and sugars, using the chemical methods (*e.g.*, mineral acids) or enzymes. Lignin is not decomposed under these conditions and typically needs to be separated from the decomposed products.

Concentrated sulfuric acid has been used to dissolve and hydrolyse native cellulose (see Figure 7.6).[52] The concentrated acid can disrupt hydrogen bonding between the cellulose chains and thus decrystallize the cellulose. Then, water is added to rapidly hydrolyse cellulose into glucose. The diluted sulfuric acid is re-concentrated for the next cycle of decrystallization and hydrolysis steps. The final products include a mixture of C5 and C6 sugars. The hydrolysis process is generally more complex than pyrolysis or liquefaction. However, hydrolysis enables selective decomposition of the biomass polymers and thus provides access to useful platform chemicals that are unavailable from pyrolysis or liquefaction techniques.

The disadvantages of using mineral acids such as concentrated HCl or H_2SO_4 to hydrolyse biomass is that they are toxic, corrosive, hazardous and difficult to recycle. The use of heterogeneous solid acids can ease product separation and provide better catalyst recyclability. For example, mesoporous transition metal oxides have been used in biomass transformations.[53,54] Polymer-based acids have been employed for the hydrolysis of various organic substrates.[55–57] In particular, carbon-based solid acids made by sulfonation of carbonized polymers, such as the solid acid shown in Figure 7.7, have shown promise.[55] Sulfonated bio-char has been similarly used.[58]

7.3.2.4   Enzymatic Hydrolysis

Enzymatic hydrolysis uses cellulose enzymes to perform hydrolysis of cellulose or hemicellulose under relative mild conditions (pH 4.8, 40–50 °C). The enzymatic methods avoid the use of corrosive acids. However, the hydrolysis reactions catalysed by enzymes are significantly slower than chemical hydrolysis, typically requiring days rather than minutes.

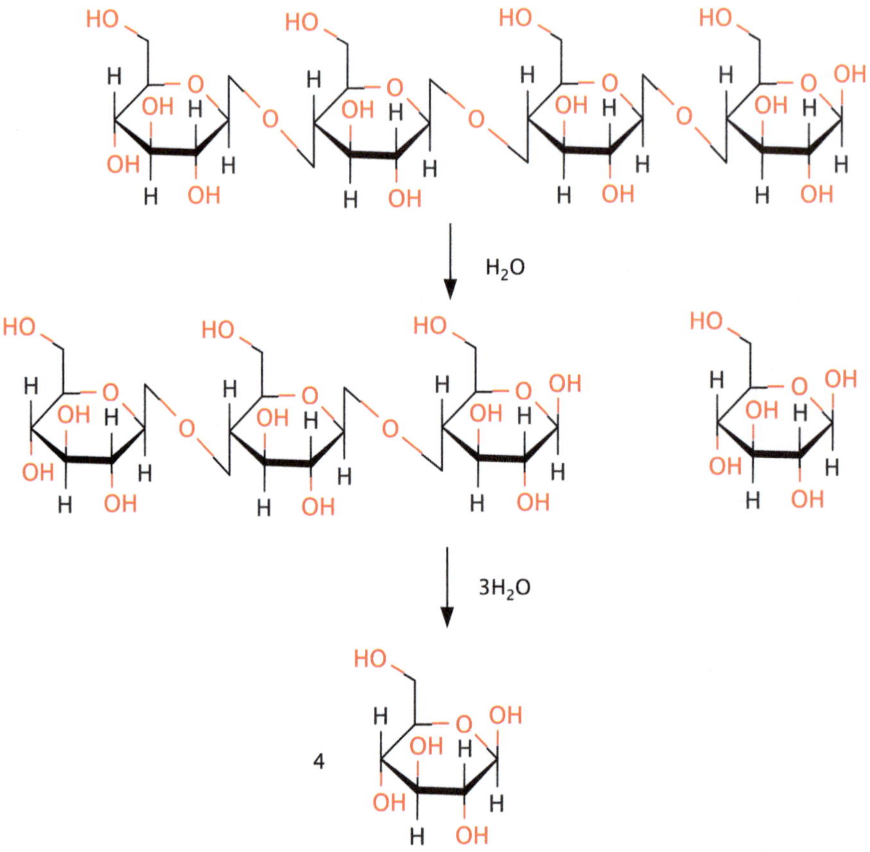

Figure 7.6   Hydrolysis of celluloses.

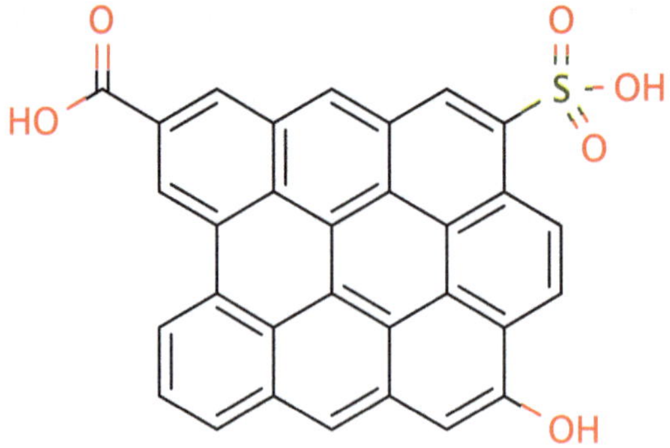

Figure 7.7   A form of carbon-based solid acid.

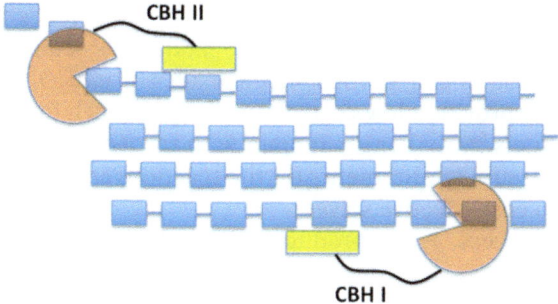

Figure 7.8 Illustration of the working mechanism of cellulase for enyzmatic hydrolysis of cellulose. CBH I and CBH II denote two primary cellobiohydrolases.

A variety of organisms are specialized for biomass degradation, including fungi, bacteria and protozoa. They are all potential sources of biomass-degrading enzymes. However, most current commercial cellulases are derived from fungi, because fungi can secrete the cellulases into the growth medium, providing a cost-effective means of separating the active cellulase enzymes for use in hydrolysis reactors.[59]

The fungus *Trichoderma reesei* is the source of one of the mostly widely used commercial cellulases.[60] *T. reesei* was first isolated from decaying cotton tents during World War II.[60] Since then, numerous mutants have been developed, increasing the productivity of the strain by over 20-fold.[61,62] The *T. reesei* cellulase includes three classes of enzymes: exoglucanases comprised of two primary cellobiohydrolases (CBH I and CBH II), a number of endoglucanases, and β-glucosidases. CBH I and CBH II account for roughly 60% and 20% of the secreted protein mix, and hydrolyse the cellulose chain progressively from the reducing and non-reducing ends, respectively, and release the glucose disaccharide cellobiose (see Figure 7.8).[63] Engdoglucanases account for 15% of the secreted protein mixture and hydrolyse β-1,4 linkages within the cellulose chains, providing new reducing and non-reducing ends for the attack by CBHs. β-Glucosidases account for roughly 0.5% of the secreted protein mixture and hydrolyse cellobiose and other short cellodextrins, producing glucose.

7.3.2.5   *Methanol Phase Reforming*

Methanol phase reforming is a process of using methanol as the solvent to degrade biomass polymers with the aid of catalysts, under conditions of high pressure and relatively mild temperature. Methanol can be reformed into CO and H_2, and the hydrogen is used to fragment biopolymers *via* hydrogenolysis and hydrogenation reactions. Such processes have been used to degrade biomass polymers into useful chemical monomers.[64,65] Methanol-phase reforming can be conducted as a simple process, in a one-pot fashion to produce liquid fuels. Depending on the catalyst employed, char formation can be minimized. Recent reports describe catalysts based on non-precious metals, such as Cu-doped porous metal oxides ($Cu/Al_2O_3/MgO$). Further development may provide a low-cost, simplified approach for converting biomass to liquid fuels.[66] Challenges remain with respect to reducing the temperature and pressure requirements for the reaction, and whether methanol can succeed as a cost-effective solvent.

In a typical process, H_2 is produced through the methanol reforming reaction:

$$CH_3OH \leftrightarrow 2H_2 + CO \tag{7.22}$$

Hydrogen is also generated by the water-gas shift reaction:

$$CO + H_2O \leftrightarrow H_2 + CO_2 \tag{7.23}$$

The hydrogen gas can then react with C–O or C=C bonds in the biomass substrate to form aliphatic alcohols. Depending on the catalyst, temperature, and time scale, the alcohols can be further deoxygenated:

$$\text{lignin or cellulose} + H_2 \xrightarrow{Cu/Al_2O_3/MgO} \text{alcohols} + CO_2 \tag{7.24}$$

7.4 Upgrading Bio-Oil

Upgrading bio-oil is an important step toward to the production of high-quality liquid fuels. Several routes are commonly used for bio-oil processing. Here we will focus on fluidized catalytic cracking, hydrotreating and decarboxylation.

Fluidized catalytic cracking is a process well known for breaking long-chain hydrocarbon molecules in petroleum feedstocks,[67,68] and is also applicable in biomass applications. The process applies a fluidized, powered catalyst after the molecules are vaporized:

$$C_6H_8O_4 \rightarrow C_{4.5}H_6 + H_2O + 1.5CO_2 \tag{7.25}$$

Hydrocracking uses hydrogen gas to remove S, N, O, and metallic contaminants from the bio-oil, through hydrodesulfurization, hydrodenitrogenation, hydrodeoxygenation and hydrodemetallization reactions, respectively.[69] Commercial processes for hydrocracking reactions are typically carried out at 300–600 °C, 30–170 atm using sulfided $Co/Mo/Al_2O_3$ and $Ni/Mo/Al_2O_3$ catalysts.[70] Other catalysts, such as $Pt/SiO_2/Al_2O_3$,[71] vanadium nitride,[72] and Ru,[73] have also been used for hydrodeoxygenation.

Decarboxylation removes carboxylic acid groups from bio-oil without the use of hydrogen gas, decreasing the oxygen content of the bio-oil substrate. Catalysts such as zeolites ZSM-5 and USY have been employed.[74,75] A significant challenge is that coke can be formed easily during the decarboxylation reaction. Therefore new catalysts are needed to avoid coke formation in this process.

7.5 Perspective

Designing effective, low-cost, robust, and sustainable catalysts for converting biomass polymers into liquid fuels remains a major challenge and opportunity area for applications of biomass in renewable energy.

Conventionally, the overall reactions for biomass conversion have been rationalized based on experimental observations, mainly through identification of the final products. However, the detailed mechanisms need further exploration. In particular, improvement of catalytic processes will depend on a better mechanistic understanding, and this represents a significant challenge due to the complexity of biomass systems. There is increasing recognition that sustainable catalytic processes must avoid the use of precious, scarce or supply-restricted metals.[76] In the near future, this will require extensive research on reactivity of first–row transition metals as well as biochemical approaches.

Currently, existing methods each present advantages and drawbacks in biomass conversion. In practice, a combination of different approaches can be applied to successfully convert specific biomass varieties to liquids with tailored fuel properties. However, application at large scale will require low-cost catalysts and effective performance under mild temperature and pressure, using environmental friendly process designs. In most cases, the cost of current biomass conversion technologies is still higher than producing liquid fuels from crude oils, and beyond economics the energy return on investment needs improvement. One area of catalyst research that is showing promise is the application of computational methods based on quantum chemistry methods. Especially as computational power continues to grow, and the methods continue to be developed as a valid complement to experimental research, the combined computational–experimental approach will play a critical role in identifying novel, high-quality catalysts. A recent example is the application of 'inverse molecular design',[77] developed by Xiao *et al.* for discovery of solar cell materials.[78] The inverse molecular design is a comparatively efficient and effective method to search for optimum materials and inform experimental projects. Such methods would be highly applicable to the biomass-to-fuel challenge.

References

1. P. T. Anastas and J. C. Warner. *Green Chemistry: Theory and Practice*, Oxford University Press, New York, 1998.
2. G. W. Huber, S. Iborra and A. Corma, *Chem. Rev.*, 2006, **106**, 4044.
3. C. E. Wyman, B. E. Dale, R. T. Elander, M. Holtzapple, M. R. Ladisch and Y. Y. Lee, *Bioresour. Technol.*, 2005, **96**, 1959.
4. P. Maki-Arvela, B. Holmbom, T. Salmi and D. Murzin, *Catal. Rev. Sci. Eng.*, 2007, **49**, 197.
5. L. R. Lynd, J. H. Cushman, R. J. Nichols and C. E. Wyman, *Science*, 1991, **251**, 1318.

6. O. Bobleter, *Prog. Polym. Sci.*, 1994, **19**, 797.

7. F. S. Chakar and A. J. Ragauskas, *Ind. Crops Prod.*, 2004, **20**, 131.

8. R. Vanholme, K. Morreel, J. Ralph and W. Boerjan, *Curr. Opin. Plant Biol.*, 2008, **11**, 278.

9. H. H. Storch, N. Golumbic and R. B. Anderson. *Fischer-tropsch and Related Syntheses*, Wiley, New York, 1951.

10. F. Fischer and H. Tropsch, *Brennst. Chem.*, 1932, **13**, 61.

11. Y.-C. Lin and G. W. Huber, *Energy Environ. Sci.*, 2009, **2**, 68.

12. D. M. Alonso, J. Q. Bond and J. A. Dumesic, *Green Chem.*, 2010, **12**, 1493.

13. R. Butner, L. Sealock Jr and D. Elliott, *Biotechnol. Bioeng. Symp.*, 1986, **15**, 3.

14. R. Butner, D. Elliott and L. Seaklock Jr, *Biotechnol. Bioeng. Symp.*, 1986, **15**, 169.

15. L. Sealock Jr and D. Elliott, In *US Pat.* 1991, Vol. 5019135.

16. D. Elliott, L. Sealsock Jr and E. Baker. In *US Pat.* 1997, Vol. 5616154.

17. A. A. Peterson, F. Vogel, R. P. Lachance, M. Fröling Jr, M. J. A. and J. W. Tester, *Energy Environ. Sci.*, 2008, **1**, 32.

18. T. Kaneko, F. Derbyshire, E. Makino, D. Gray and M. Tamura, in *Ullmann's Encyclopedia of Industrial Chemistry*, Wiley-VCH, Weinheim, 2001.

19. H. Kolbel and M. Balek, *Catal. Rev.*, 1980, **21**, 225.

20. E. Iglesia, S. C. Reyes and R. J. Madon, *J. Catal.*, 1991, **129**, 238.

21. S. Rapagná, H. Provendier, C. Petit, A. Kiennemann and P. U. Foscolo, *Biomass Bioenergy*, 2002, **22**, 377.

22. P. J. Dauenhauer, B. J. Dreyer, N. J. Degenstein and L. D. Schmidt, *Angew. Chem., Int. Ed.*, 2007, **46**, 5864.

23. J. R. Salge, B. J. Dreyer, P. J. Dauenhauer and L. D. Schmidt, *Science*, 2006, **314**, 801.

24. C. Font Palma, *Appl. Energy*, 2013, **111**, 129.

25. R. D. Cortright, R. R. Davda and J. A. Dumesic, *Nature*, 2002, **418**, 964.

26. G. W. Huber, R. D. Cortright and J. A. Dumesic, *Angew. Chem., Int. Ed.*, 2004, **43**, 1549.

27. E. L. Kunkes, D. A. Simonetti, R. M. West, J. C. Serrano-Ruiz, C. A. Gartner and J. A. Dumesic, *Science*, 2008, **322**, 417.

28. R. R. Davda, J. W. Shabaker, G. W. Huber, R. D. Cortright and J. A. Dumesic, *Appl. Catal., B*, 2005, **56**, 171.

29. G. W. Huber, J. W. Shabaker and J. A. Dumesic, *Science*, 2003, **300**, 2075.

30. G. Wen, Y. Xu, H. Ma, Z. Xu and Z. Tian, *Int. J. Hydrogen Energy*, 2008, **33**, 6657.

31. M. B. Valenzuela, C. W. Jones and P. K. Agrawal, *Energy Fuels*, 2006, **20**, 1744.

32. T. Kawai and T. Sakata, *Nature*, 1980, **286**, 474.

33. G. P. Wu, T. Chen, G. H. Zhou, X. Zong and C. Li, *Sci. China B*, 2008, **51**, 97.

34. A. Patsoura, D. I. Kondarides and X. E. Verykios, *Catal. Today*, 2007, **124**, 94.

35. D. I. Kondarides, V. M. Daskalaki, A. Patsoura and X. E. Verykios, *Catal. Lett.*, 2008, **122**, 26.

36. X. Fu, J. Long, X. Wang, D. Y. C. Leung, Z. Ding, L. Wu, Z. Zhang, Z. Li and X. Fu, *Int. J. Hydrogen Energy*, 2008, **33**, 6484.

37. D. J. Hayes, *Catal. Today*, 2009, **145**, 138.

38. A. V. Bridgewater, *Catal. Today*, 1996, **29**, 285.
39. A. V. Bridgewater, *J. Anal. Appl. Pyrolysis*, 1999, 3.
40. S. Czernik and A. V. Bridgewater, *Energy Fuels*, 2004, **18**, 590.
41. A. Demirbaº, *Energy Convers. Manage.*, 2000, **41**, 633.
42. A. Vuori and J. B-son Bredenberg, *Prepr. Pap. ACS Div. Fuel Chem.*, 1985, **30**, 366.
43. J. D. Adjaye, R. K. Sharma and N. N. Bakhshi, *Fuel Process. Technol.*, 1992, **31**, 241.
44. C. A. Koufopanos. In *Comm. Eur. Comm., Final Report of the Grant Period 1983–1986*, 1986.
45. T. R. Carlson, T. P. Vispute and G. W. Huber, *ChemSusChem*, 2008, **1**, 397.
46. T. P. Vispute, H. Zhang, A. Sanna, R. Xiao and G. W. Huber, *Science*, 2010, **330**, 1222.
47. J. Jae, G. A. Tompsett, A. J. Foster, K. D. Hammond, S. M. Auerbach, R. F. Lobo and G. W. Huber, *J. Catal.*, 2011, **279**, 257.
48. P. D. Patil and S. Deng, *Fuel*, 2009, **88**, 1302.
49. D. A. Simonetti, J. Rass-Hansen, E. L. Kunkes, R. R. Soares and J. A. Dumesic, *Green Chem.*, 2007, **9**, 1073.
50. D. C. Elliott, *Energy Fuels*, 2007, **21**, 1792.
51. H. R. Appell, in *Fuels from Waste*, ed. L. Anderson and D. A. Tilman, Academic Press, New York, 1967.
52. E. E. Harris, *Wood Saccharificaiton Advances in Carbohydrate Chemistry*, Academic Press, New York, 1949.
53. C. Tagusagawa, A. Takagaki, A. Iguchi, K. Takanabe, J. N. Kondo, K. Ebitani, S. Hayashi, T. Tatsumi and K. Domen, *Angew. Chem., Int. Ed.*, 2010, **49**, 1128.
54. C. Tagusagawa, A. Takagaki, A. Iguchi, K. Takanabe, J. N. Kondo, K. Ebitani, T. Tatsumi and K. Domen, *Chem. Mater.*, 2010, 2035.
55. X. Qi, M. Watanabe, T. M. Aida and R. Lee Smith Jr, *Green Chem.*, 2008, **10**, 799.
56. M. T. Sanz, R. Murga, S. Beltran and J. L. Cabezas, *Ind. Eng. Chem. Res.*, 2002, **41**, 512.
57. Y. Uozumi and K. Shibatiomi, *J. Am. Chem. Soc.*, 2001, **123**, 2919.
58. Y. Wu, Z. Fu, D. Yin, Q. Xu, F. Liu, C. Lu and L. Mao, *Green Chem.*, 2010, **12**, 696.
59. S. T. Merino and J. Cherry, *Adv. Biochem. Eng./Biotechnol.*, 2007, **108**, 95.
60. B. Nidetzky, W. Steiner, M. Hayn and M. Claeyssens, *Biochem. J.*, 1994, **298**, 705.
61. M. Mandels, J. Weber and R. Parizek, *Appl. Micobiol.*, 1971, **21**, 152.
62. S. Mishra and K. S. Gopaldrishnan, *J. Ferment. Technol.*, 1984, **62**, 495.
63. P. L. Suominen, A. L. Mantyla, T. Karhunen, S. Hakola and K. M. H. Nevalainen, *Mol. Gen. Genet.*, 1993, **241**, 523.
64. T. D. Matson, K. Barta, A. V. Iretskii and P. C. Ford, *J. Am. Chem. Soc.*, 2011, **133**, 14090.
65. J. Yamazaki, E. Minami and S. Saka, *J. Wood Sci.*, 2006, **52**, 527.
66. K. Barta, G. R. Warner, E. S. Beach and P. T. Anastas, *Green Chem.*, 2014, **16**, 191.

67. C. N. Satterfield, *Heterogeneous Catalysis in Industrial Practice*, McGraw-Hill, New York, 1991.
68. A. Corma and B. W. Wojciechowski, *Catal. Rev. Sci. Eng.*, 1985, **27**, 29.
69. G. W. Huber and A. Corma, *Angew. Chem., Int. Ed.*, 2007, **46**, 7184.
70. D. C. McCulloch, *Catalytic Hydrotreating in Petroleum Refining*, Academic Press, New York, 1983.
71. Y.-H. E. Sheu, R. G. Anthony and E. J. Soltes, *Fuel Process. Technol.*, 1988, **19**, 31.
72. S. Ramanathan and S. T. Oyama, *J. Phys. Chem.*, 1995, **99**, 16365.
73. A. Centeno, R. Maggi and B. Delmon, *Stud. Surf. Sci. Catal.*, 1999, **127**, 77.
74. S. R. A. Kersten, W. P. M. van Swaaij, L. Lefferts and K. Seshan, in *Catalysis for Renewables: from Feedstock to Energy Production*, ed. G. Centi and R. A. van Santen, Wiley-VCH, Weinheim, 2007.
75. M. Stöcker, *Angew. Chem., Int. Ed.*, 2008, **47**, 9200.
76. R. B. Gordon, M. Bertram and T. E. Graedel, *Proc. Natl. Acad. Sci. U.S.A.*, 2006, **103**, 1209.
77. D. Xiao, I. Warnke, J. Bedford and V. S. Bastista, *RSC Specialist Periodical Report – Chemical Modelling*, 2014, **10**, 1.
78. D. Xiao, L. A. Martini, R. C. Snoeberger III, R. H. Crabtree and V. S. Batista, *J. Am. Chem. Soc.*, 2011, **133**, 9014.

CHAPTER 8

Holistic Green Chemistry Metrics for Use in Teaching Laboratories

ADÉLIO A. S. C. MACHADO*[a]

[a]Chemistry and Biochemistry Department, Faculty of Science, Oporto University, R. Campo Alegre, 687, Porto 4169-007, Portugal
*E-mail: amachado@fc.up.pt

8.1 Introduction: The Rational Basis of Holistic Green Chemistry Metrics

Green chemistry (GC) aims at reshaping the current practice of chemistry to decrease its deleterious impacts on the human health and the environment. These impacts are numerous and globally very complex, owing to several factors: the large number and variety of chemicals prepared by industrial chemistry, as well as of the manufacturing processes used to obtain them;[1] the large number of natural resources, some of them in huge amounts, required by the industry to provide its feedstocks, predominantly non-renewable (including energy, of which the chemical industry is notably voracious);[2] the diversity of the ways chemicals are used in the industrial activity of the technosphere, including dissipative use that, for instance, is promoted trivially in industrialized agriculture, as well as by humans, in their personal and professional lives, for a innumerable number of purposes; the countless

Worldwide Trends in Green Chemistry Education
Edited by Vânia Gomes Zuin and Liliana Mammino
© The Royal Society of Chemistry 2015
Published by the Royal Society of Chemistry, www.rsc.org

types of physical, chemical and biological effects chemicals can yield, causing the above-mentioned negative impacts, *etc*. Such effects involve still more complexity than the production and use of chemicals; for instance, toxic effects on human health and the biosphere involve the complexity of biology, impacts can occur immediately or be delayed for a few decades. In summary, the complexity pervades the industrial manufacture, use, impacts, *etc*., of chemicals and a reasonable understanding of the situation requires systems thinking; chemistry, and especially GC, should be considered a systems science.[3] Indeed, chemicals play an important role in several complex systems, which are themselves linked by numerous connections of different types, as suggested by Figure 8.1. All these systems and therefore the global "chemistry super-system" have also connections to their surrounds, in the present case the earth environment, as shown in the figure. This includes also the intermediate frontiers of other two global systems supported by the environment, the human society and the economy, the other components of Sustainable Development beyond the environment. These are introduced in the figure to remind the important role chemistry plays in achieving sustainability, which alone justifies the urgency of GC development.

The ideal of GC would be a full alteration of the practice of the prevalent chemistry to provide the complete elimination of its negative impacts, for obtaining fully benign chemistry. However, the multiple complexities of the chemical endeavour, as shown by the several systems it involves, make this impossible except in rare situations. In practice, the benignity ideal serves as

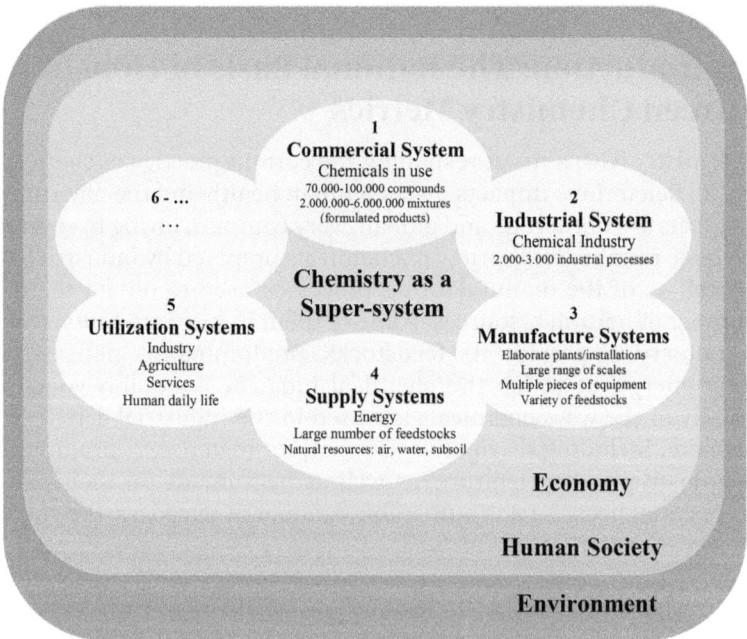

Figure 8.1   The complexity of chemistry.

a goal that, although unattainable, is very useful for defining lines of action for changes that decrease the deleterious impacts of chemistry. In colloquial language this decrease corresponds to an increase of (*chemical*) *greenness*, the very difficult to define but desirable characteristic of the chemistry of not causing harmful consequences; this is the GC global intention. Greenness is a complex feature owing to its broad scope and to the large number of different features it involves, as exemplified in Figure 8.2. Indeed, with reference to the scope of the concept, greenness may be referred to compounds, reactions, synthetic routes, chemical process, unit operations, feedstocks, ways of utilizing chemicals, *etc*. On the other hand, it is composed of a large number of components, depending on the characteristics of benignity to be improved, illustrated in Figure 8.2 (the lists included are far from

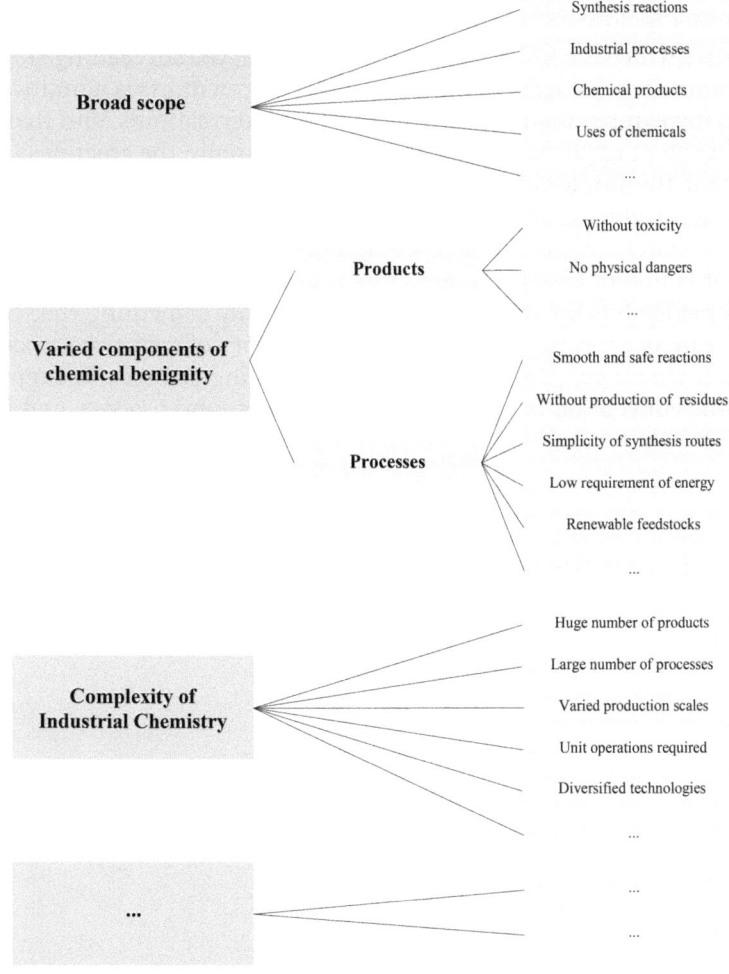

Figure 8.2   Scope and examples of components of chemical greenness.

being exhaustive). For instance, the greenness of synthetic reactions, the case mainly considered in this text, depends on several properties both of the compounds involved in them and of the reactions themselves (thermo-dynamic and kinetic parameters), as well as of the conditions for their exe-cution. Moreover, the complexity of the chemical greenness is compounded by the huge number of different situations found in the practice of industrial chemistry.

This is suggested by Figure 8.3, which represents the long chain of events between the conception of a green new molecule and its green synthesis route in the laboratory by GC, along the scale-up and process development by green chemical engineering, green manufacture and formulation to bring the corresponding chemical product to the market, and finally its green use by society. Starting from a green molecule, the greenness has to be kept along all the steps of the course until the use of the chemical; this *greenness chain* cannot be broken if a sustainable product is to be reached. However, the features which are involved in the greenness in the succeeding steps are of different nature, at least in part. In summary, greenness is a multivariate concept, involving numerous dimensions and interrelations, and therefore its management requires systems thinking. Essentially, the greenness chain means that the implementation of greenness involves lifecycle thinking: Figure 8.3 introduces the *greenness lifecycle*.

The assessment of greenness is very important for the accomplishment of GC, but it is not an easy task. Benignity is an unattainable theoretical goal that in practice has no absolute meaning; it is only something that can be improved by decreasing the number and intensity of negative impacts of chemicals, reactions, processes, *etc.*, by introducing changes in chemistry, both in the conception of new molecules, products and process, and in the

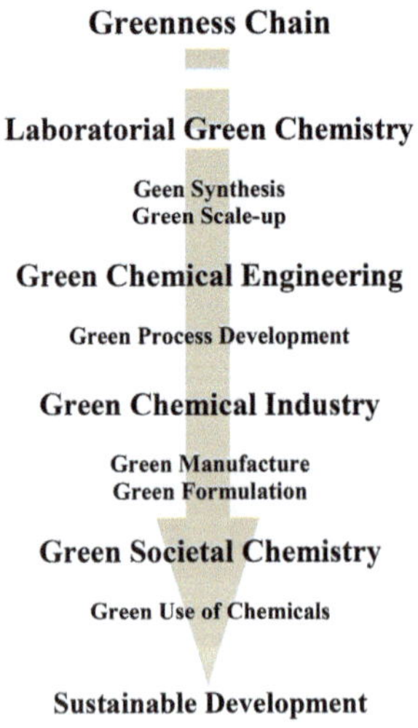

Greenness Chain

Laboratorial Green Chemistry

Geen Synthesis
Green Scale-up

Green Chemical Engineering

Green Process Development

Green Chemical Industry

Green Manufacture
Green Formulation

Green Societal Chemistry

Green Use of Chemicals

Sustainable Development

Figure 8.3   The chemical greenness chain.

improvement of those in use nowadays. The assessment of the improvement involves a comparison of two situations (after *vs.* before a change introduced to pursue benignity); greenness is therefore a relative quantity. For the purpose of evaluation of its more obvious components, metrics have been introduced since the beginning of the 1990s, when the atom economy (AE, percentage mass of atoms in the reagents that are incorporated in the product, assuming that reagents are used in stoichiometric proportions and 100% yield)[4] and the E-factor (ratio of total mass of residues to mass of product)[5] were developed as mass metrics of synthetic reactions, to assess the degree of materialization they require, with the purpose of supporting their dematerialization. These metrics evaluate, respectively, the efficiency of use of atoms from stoichiometric reagents (incorporation in the product) and the total loss, as residues, of matter proceeding from reagents and other additional materials required to perform the reaction (solvents, catalyst, *etc.*); this shows that the assessment involves two components, being made along two dimensions (atoms placed in the product molecule and matter lost in residues). Since then, more mass metrics[6] as well as metrics addressed to other dimensions of greenness have been devised.[7,8] These numerous metrics were conceived under the nowadays prevailing reductionist mindset used for the teaching and research of academic chemistry, which increased in importance along the 20th century, and stimulated excessive compartmentalization of subjects and the specialization of chemists. As a consequence, each of the GC metrics described in the literature is almost invariably addressed to a separate aspect of greenness; they are reductionist mono-dimensional metrics. As the variables to be considered in greenness are numerous, a large number of metrics have been proposed to cover its different aspects. Table 8.1 lists the main types of greenness metrics and their direct relations to the CG Twelve Principles[9] when the relations are well defined. The table shows that the coverage of the Twelve Principles by mono-dimensional metrics is neither homogenous nor complete.

For instance, in the above case of mass metrics, very important to assess the materialization of synthetic reactions to pursue the dematerialization of chemistry, the metrics AE and E-factor were found to be insufficient to account for all aspects of material greenness a decade after their conception,[10] and since then a lot of alternative reductionist metrics have been proposed.[6–8] This is not unexpected, as the calculation of metrics should be easy

Table 8.1   Main types of metrics of greenness.

Type	Green chemistry principles
Reaction efficiency/atomic productivity/ mass losses in residues	1 and 2
Energy	6
Environment/human health	Several
Safety	12
Economics	None

and use data already available or easy to obtain, which depend on the case at hand (*p. ex.* laboratory *vs.* industry), but above all because the metrics were conceived one by one in a reductionist framework, without attending to the systemic nature of the problem. To assess greenness, which is intrinsically multi-dimensional, with mono-dimensional metrics, a set of metrics with connections among themselves (a system of metrics, commonly called a *battery*) must be chosen to capture the relevant information on the dimensions of the chemical system that determine its behaviour in the situation under study, as prescribed by metrification studies in systems science.[11] For mass metrification of chemical reactions, the reaction mass efficiency (RME), introduced by Curzons *et al.*,[10] is a third metric suitable to constitute a battery to assess the two separate dimensions of dematerialization mentioned above. Indeed AE is a theoretical metric, determined by the stoichiometry of the reaction, that defines the maximum percentage of mass incorporation in the product of atoms of the reagents, when no excess of reagents are used and the yield is 100%. However, when the reaction is implemented, an excess of a reagent is often used or/and the yield obtained is less than 100%, that maximum is not fulfilled and another metric is required to evaluate the level of atom utilization attained. There is a complex relationship between AE and RME that depends on the yield and reagent excess[12] which means that the two metrics are not independent – although it is always RME ≤ AE, for instance, RME = (AE × Yield) when the reaction is performed with reagents in stoichiometric amounts. The three metrics, AE, RME and E-factor (or the mass intensity, MI,[13] the ratio of the sum of masses of all reagents and materials to the mass of product, used as an alternative to the E-factor, as E-factor = MI − 1) capture different features of the chemical reaction (see Table 8.2) and constitute a battery suitable to assess its material greenness along its two dimensions.[14]

These metrics are of two types, because they cover the already mentioned two different components that materialization/dematerialization involves (see left scheme in Figure 8.4): first, the parcel of stoichiometric reagents

Table 8.2 Components of a synthesis reaction captured by the mass metrics of a battery suitable to assess its material greenness.[a]

Component	AE	RME	E-factor[b]
Reagents (stoichiometric proportions)	+	+	+
Reagents (stoichiometric proportions, excess)	−	+	+
Reagents (auxiliary, *etc.*)	−	−	+
Residues	−	−	+
Solvents	−	−	+[c]
Yield	−	+	+

[a]Symbols: +, captured; −, not captured.
[b]The metric mass intensity (MI) may be used instead.
[c]Solvents are often ignored, especially water.

essential to perform the chemical transformation that lead to the product; second, the parcel of other materials required to promote the transformation (solvents, catalysts, auxiliary reagents, *etc.*). The dematerialization of the first type is achieved by placing atoms from the reagents in the product as much as possible, instead of losing them in co-products or residues; this is evaluated by metrics addressed to express what may be called the *atomic greenness*, AE and RME. The dematerialization of the second type is more complex because generally a large number of substances must be used to implement the reaction and is accomplished by decreasing as far as possible the amounts of these auxiliary materials, especially those used in larger quantities. However, generally, this second type of dematerialization is not measured; instead, the E-factor (or, alternatively, MI) measures the total materialization (that may be called *material* or *mass greenness*).

Two further points deserve comment. First, in practice, the two types of materialization (reagents *vs.* auxiliary materials) are not independent, for instance, decreasing the volume of the solvent used as reaction mean, to decrease the E-factor, may affect the yield of the reaction, and change the value of RME. Second, while the evaluation of the atomic greenness involves two metrics, the materialization *via* auxiliary materials is assessed by one only, the E-factor (or MI), because there is no theoretical upper limit for the amount of matter used in them (for the atomic greenness, the stoichiometric equation defines AE as such a limit). This situation makes difficult the comparison of auxiliary materials responsible for materialization in different reactions and, consequently, of their global materialization, because it is impossible to define a normalized scale for the E-factor (or MI). This is a

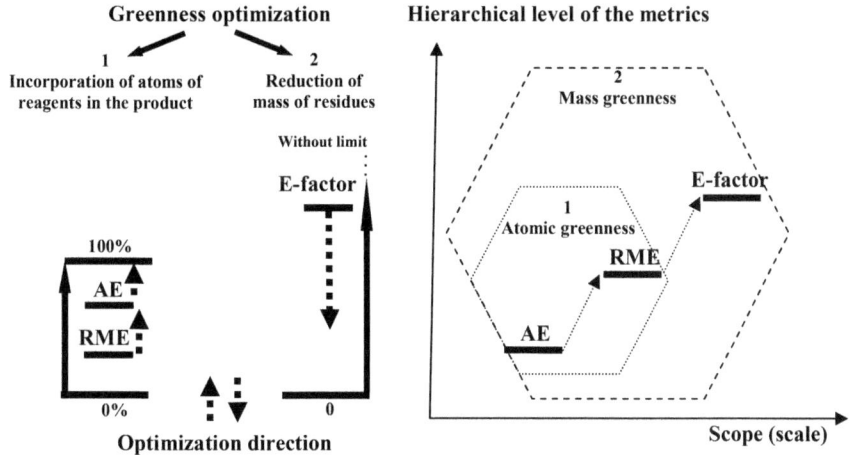

Figure 8.4 Battery of metrics for assessment of the material greenness of synthesis reactions. Left: The assessment (and optimization) involves two dimensions that require different metrics. Right: Scope and hierarchy of the metrics.

consequence of the complex nature of the chemical reaction, which involves several components with many interactions (see the left part of Figure 8.5) and therefore requires a system of metrics to provide information on the reaction system, for instance, the set AE, RME and E-factor. By definition, the scope of each of these metrics and the models used to define the frontiers for their calculation are different (see the right scheme of Figure 8.5), and there is a hierarchy in the set, as shown in the right scheme of Figure 8.4 (in which the scales of both axis are qualitative). This hierarchy is well defined for AE and RME through their relationship, as discussed above, but rather vaguer with respect to the E-factor because this assesses a different component of greenness, the total mass (more precisely, its fraction lost as residues). However, as the total mass includes the mass of reagents, the hierarchical level of E-factor is superior. In summary, the battery provides a holistic evaluation of the material greenness in synthesis reactions and, as the metrics are hierarchized, its use is efficient for the experimental optimization of the greenness of synthesis. Arrows in the right scheme of Figure 8.4 define the order of the use of the three metrics, but in practice this is not 'sacred' (the scheme is reductionist and does not show the closed loops used in systems' thinking). In conclusion, the greenness optimization of a reaction requires the improvement of two metrics in parallel (it is a bivariate optimization!), RME and E-factor (the arrows in the left part of Figure 8.4 show the directions of changes required for increasing greenness for both). This is a consequence of the systemic nature of the battery (and of the reaction itself) and denotes how systems' thinking is required by green chemistry.[3]

This example illustrates, for the restrict case of mass metrics, the need to tackle the greenness metrification of chemistry in a systemic mindset,[15]

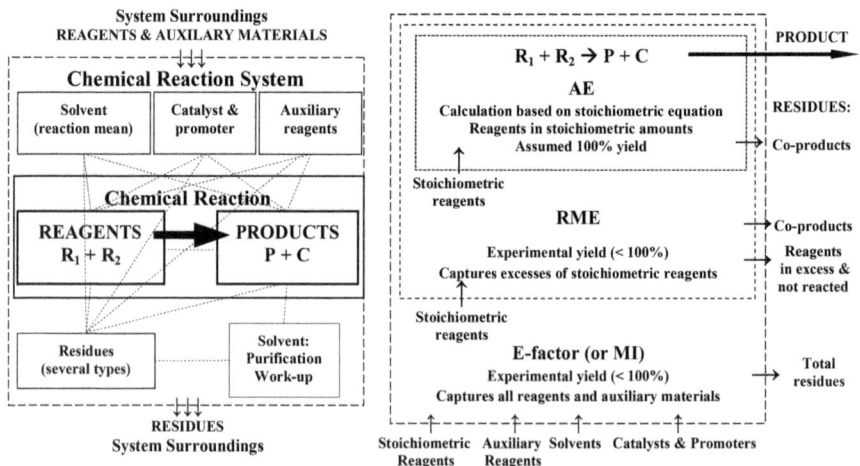

Figure 8.5   The synthesis reaction view as a system. Left: Components of the system and examples inter-relations between them. Right: The frontiers of the models of the system to show flows of matter captured by the different metrics.

which is still more important when the other types of metrics are considered (see Table 8.2) and the multi-dimensionality of the evaluation increases. However, for historical reasons, a large number of different types of alternative mono-dimensional metrics have been introduced casuistically, giving origin to a complicated situation. This results, for instance, from ignoring the importance of a clear fixation of the frontier for the definition and easy use of a metric, from forgetting that metrics should be intuitive and informative and not involve complicated calculations, *etc.*[11] Indeed, when different alternatives have been proposed, often some of them have not been adopted by green chemists at large because they are not instinctive and their calculation is not easy. On the other hand, the choice of the more suitable for the case under evaluation may be subjective and debatable, *etc.* All these problems result in a huge complication of the field of green metrics, which possibly has hindered the broadening of their use.

To overcome this limitation, a different approach can be used: to start the metrification with the assumption of a systems thinking mindset and pursuing the conception of holistic metrics that can provide a global view of the chemical greenness for each case to be studied. This view should include all, or at least a large number, of the relevant greenness dimensions, along which the evaluation is made in parallel.

This chapter reviews work developed in the Chemistry Department of the Faculty of Science of Oporto along this line of action, where metrics of this type for educational purposes have been developed since the teaching of GC was launched almost ten years ago.[16–20] The text is structured as follows: the following section describes briefly the metrics and Section 8.3 their construction; their use in teaching activities is presented in Section 8.4; finally, Section 8.5 discusses the metrics in the context of systems metrics, and their advantages and limitations.

8.2 Holistic Metrics Based on the Twelve Principles of Green Chemistry

8.2.1 The Basic Idea that Inspired the Metrics

The Twelve Principles of GC are qualitative prescriptions very useful for its teaching and that cover many of the aspects that should be considered to improve the benignity of chemistry, especially with respect to the reduction of wastes, the elimination of toxic substances (both as synthesis reagents and products), saving energy, *etc.* Indeed, as shown in Figure 8.6, the principles deal, although unevenly, with the greenness components captured by almost all types of metrics (however, economic metrics are an exception!).

For supporting the development of GC, as well as the reliable metrification of greenness, the principles should be kept in view together, because alterations of the conditions when seeking improvement of the greenness of, for instance, a synthesis reaction, may have unlike consequences with respect to different principles; the greenness may improve with reference to some of them but unintentionally worsen with reference to others.[21] From this requirement of a global vision of the principles, the inspiration for construction of the holistic metrics emerged: in these the principles are used together for evaluating simultaneously the dimensions of greenness they capture in the situation under assessment. For this evaluation, pre-defined criteria that provide values in three or two level scales for each principle were established. These different criteria yielded different metrics.

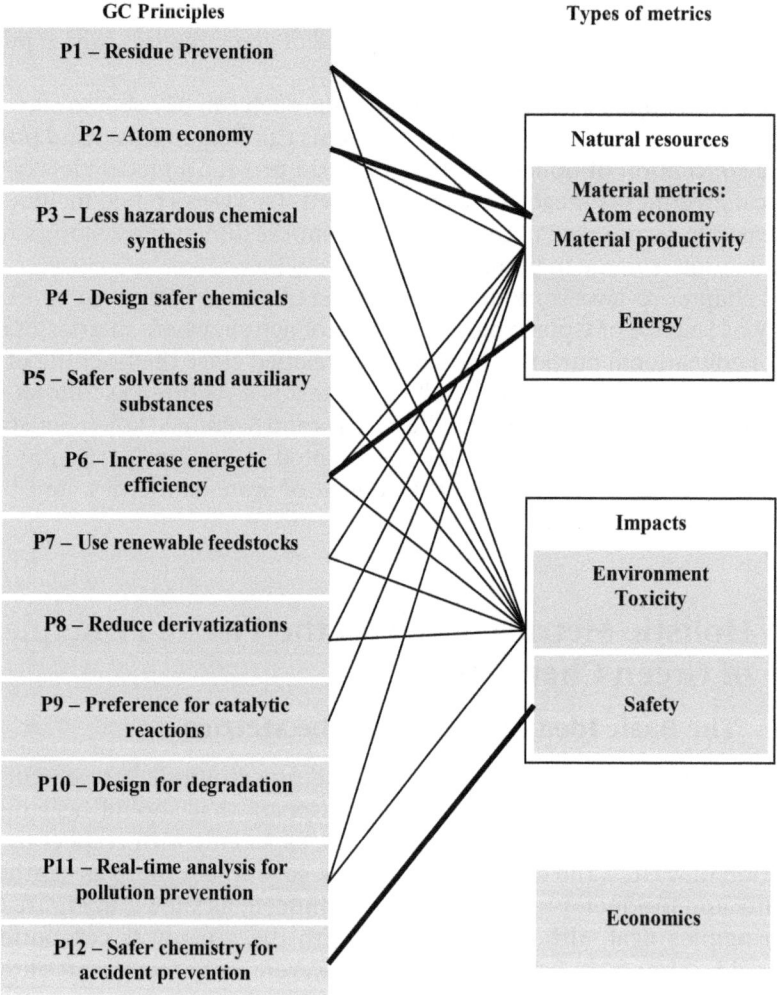

Figure 8.6   The connections between the Twelve Principles of green chemistry and the types of greenness metrics. The connections represented by thick lines are those specified in Table 8.2.

8.2.2 The Metrics: Green Star, Green Circle and Green Matrix

When a three-level scale was chosen, the results of the evaluation were used to devise a star as a graphic metric. The star was constructed with a number of corners equal to the number of principles used for the evaluation, all the twelve or only some if the remaining are not applicable (see Section 8.3.1), each corner with length proportional to the degree of accomplishment of the corresponding principle (see Section 8.3.2); this star, filled in green, obtained as a radar chart of an Excel spreadsheet, provides a semi-quantitative view of the global greenness of the reaction, that can then be acquired simply by looking at the star and appreciating its green area: the larger the area, the greener is the reaction. Therefore the star functions as a visual metric, the 'green star',[16–18] as shown in Figure 8.7 (centre).

If the evaluation of each principle is made with a simpler binary criteria (yes/no, *i.e.*, accomplishment/non-accomplishment of the principle), results can be represented in a circle divided in coloured sectors in number equal to the number of principles evaluated,[22] filled in green or red according the positive or negative result (see Figure 8.7, left). A look at the circle provides again a semi-quantitative impression of the greenness depending on the degree of predominance of the green over the red. The circle, obtained as a pie chart in

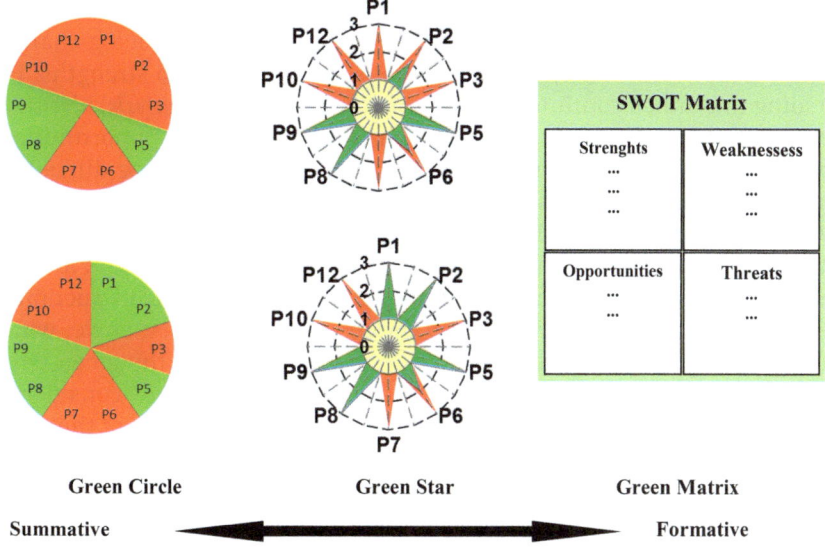

Green Circle   **Green Star**   **Green Matrix**

Summative ⟷ **Formative**

Figure 8.7   Holistic metrics based on the principles of green chemistry. The graphic metrics refer to an optimization experiment of the syntheses of tetramminecopper(II) sulfate monohydrate 16 (only ten of the twelve principles were used in the evaluation, principles 4 and 11 being excluded). When the excess of ammonia was deceased from 300% (top metrics) to 7% (bottom metrics), the green area in the graphics increased and the indexes API (green circle) and GSAI (green star) increased respectively from 30 to 50 and from 40 to 55.

an Excel, serves as a second graphic metric, the 'green circle'.[19,20] Although the green circle is rather less precise than the green star, its construction is easier.

The construction of a third metric, the 'green matrix', was motivated by exploring the purpose of having a tool with better characteristics for making students more well-informed about finding opportunities for greenness improvements. The evaluation procedure uses the same criteria for accomplishment of the principles as the green circle, but the results are expressed as tables constructed using a strengths, weaknesses, opportunities and threats (SWOT) analysis.[23] This allows not only a more profound appreciation of the features of greenness in each case, but also provides hints about how to improve it (and limitations for such an improvement). SWOT is more commonly used as a framework for brainstorming on complex problems,[22,24] rather than a tool for multi-dimensional evaluation like greenness assessment. For construction of an evaluative tool, the objective of the task must be well-defined at the start of the procedure; in the present case, it is to assess the greenness of a synthetic reaction and look for suitable changes of improving it, as defined by the accomplishment of the GC Twelve Principles. For this purpose, the dimensions of the *internal analysis* of the problem, *i.e.,* the aspects that support (strengths) or hinder (weaknesses) each principle are defined; the number of strengths/weaknesses is equal to the number of fulfilled/failed principles, and a greater difference between the numbers means more greenness. To improve the greenness by introducing changes in the reaction procedure, weaknesses must be converted to strengths. For pursuing this aim, a continuation of the analysis (*external analysis*) identifies the aspects that can change weaknesses into strengths (the oppportunities), as well as the external circumstances that may prevent or bring difficulties to realize the opportunities (these are the threats). The green matrix is the SWOT matrix obtained by this procedure (see Figure 8.7, right). As the evaluation criteria of the principles are the same for the green matrix and the green circle, the two metrics are complementary and upon construction of the first it is straightforward to draw the results as the green circle, which is then a compact graphical result of the matrix. However, the green circle can be obtained by simple application of the criteria without considering the full analysis required by SWOT. Indeed, the great limitation of the green matrix is that it requires previous acquaintance with SWOT, which university students in general do not have, and therefore it can only be used if there is enough time to teach its basics previously. If this is possible, the green matrix allows a deep analysis of the greenness as defined by the principles, including the aspects that contributed to their accomplishment and cues to improve the greenness when this was not achieved. In summary, the green matrix requires a great deal of thinking about the nature of greenness and the Twelve Principles that allows the user to acquire an enhanced learning of GC; this metric is quite formative while the green circle is eminently summative, providing an incisive presentation of the results, while the green star falls in an intermediate rank (see arrows at the bottom of Figure 8.7).

8.3   Construction of the Metrics

8.3.1   Basic Aspects

Two general items will be discussed in this section: first, selection of the Twelve Principles of GC to use in different greenness assessments; second, collection of data to be used in the metrics.

8.3.1.1   Selection of the Principles

This item refers to defining which principles are relevant for the greenness evaluation in each situation, the selection depending on the nature of the problem. In Figure 8.8 three typical situations common in chemistry are contemplated: while in a full process development for manufacture of a new chemical the complete set of principles should in principle be considered, laboratory activities on teaching chemistry deal with simpler situations where some of the principles are not involved. In preparative experiments prescribed in organic/inorganic teaching laboratories, usually no activities fall in the scope of principles 4 (design of safer chemicals) and 11 (real-time analysis for pollution prevention) and the greenness evaluation of the synthesis reactions requires only the remainder ten principles. For laboratory activities that do not involve chemical reactions, but are often made to train

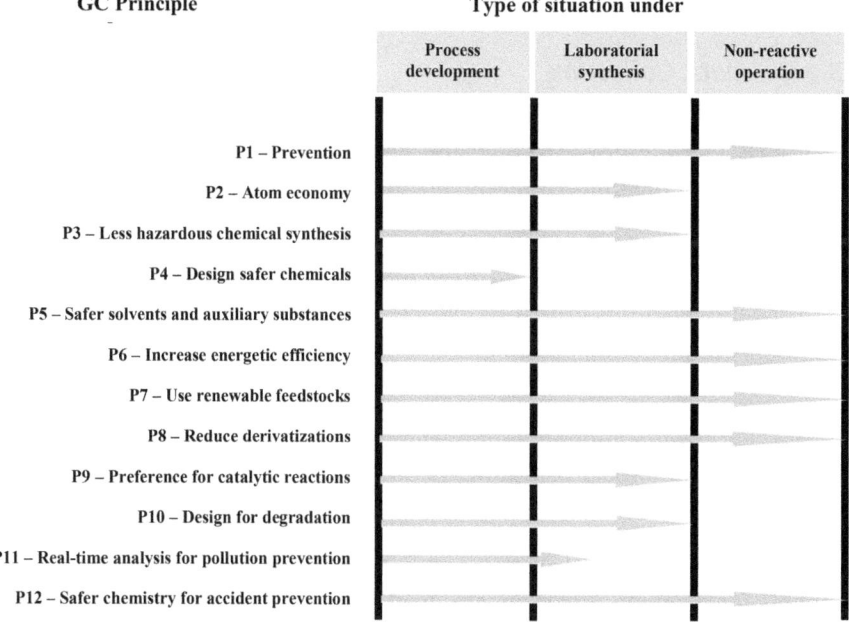

Figure 8.8   The principles of green chemistry used in the construction of holistic metrics in different situations.

students in laboratory techniques, like distillations, recrystallizations, *etc.*, the range of principles for the assessment is reduced to six, because in the absence of reaction principles 2 (atom economy), 3 (less hazardous chemical synthesis), 8 (reduce derivatizations) and 9 (preference for catalytic reactions) are not relevant. This last situation may be found also in synthesis experiments, if a greenness evaluation of the work-up operations is made separately from the reaction, to show their relative contributions to the overall greenness or lack of it (see below). These three situations are common, but there are many other that occur in chemistry, therefore each new case should be carefully analysed to identify the components of greenness with practical relevance and chose accordingly the GC principles to be used.

More precisely, for the case of the assessment of a synthesis reaction greenness, the protocol of the experiment is examined to obtain information about: (1) stoichiometric reagents in excess (this information is used to assess the accomplishment of principle 2); (2) the conditions of pressure and temperature (principle 6); (3) hazards to human health and the environment (principles 1, 3, 5 and 9) and of potential chemical accidents (principle 12), of all the substances involved (raw materials/feedstocks, products, by-products, solvents and other auxiliary substances such as catalytic reagents, solvents, separation agents, *etc.*) and wastes; (4) renewability of raw materials/feedstocks and tendency to break down into innocuous degradation products (principles 7 and 10); and (5) the use of derivatizations or similar operation (principle 8). This information allows the evaluation of the summative greenness of the synthesis from judgment of the levels of accomplishment of the individual principles.

8.3.1.2   Data Collection

Collecting the full data required to implement the construction of the metric may not be a straightforward task and here the case of synthesis reactions used in teaching laboratories, which normally follow protocols described in the literature, will be considered. These protocols should be scrutinized in detail to obtain a complete list of all reagents and products involved, including solvents, catalysts, auxiliary materials, *etc.* Conditions (temperature, pressure, *etc.*) prescribed to execute the reaction and work-up operations should also be recorded, as well as details on excess of stoichiometric reagents, amounts of solvents, concentrations of solutions, residues, *etc.* However, such details are not always provided in protocols, mainly because they did not deserve interest before the emergence of GC.[25] The inclusion of environmental impact components in the greenness evaluation requires much more data than for the material greenness metrics, on safety, dangers, toxicity of reagents and residues, *etc.* Any information of this type should also be annotated, but it is rarely provided. In summary, even if well worked protocols for teaching purposes are used, a lot of information of this type must be found in other sources.

In practice, after listing all compounds and materials, it is necessary to collect information for each substance on risks for human health and the environment, physical risks (inflammability, explosivity, reactivity, *etc.*), its origin (renewable/non-renewable resources) and end-of-life (innocuous/dangerous degradation products). Although these data can be collected from a variety of sources, the use by the students of safety data sheets (SDS) provided by the suppliers of reagents is a practical way to obtain such data (and simultaneously to make the students more aware of the safety problems of chemistry). SDS have been improving since the REACH (Registration, Evaluation, Authorization and Restriction of Chemicals) legislation made them the main instrument for transmission on hazards and risks of chemicals along the commercial chain down to the users[26] and prescribed a standardized format.[27] More recently, the increasing adoption of the GHS,[28] now in progress, endorsed this format, and pressed further the use of SDS. Although the original purpose of SDS was to facilitate the transmission of information on hazards of chemicals, they eventually started to be used in risk assessments for the manufacture of safer products and even for the assembly of safer manufacture processes,[29] which incentivized their use for the construction of holistic metrics.

Although the first versions of these metrics[16-20] were developed from SDS that used risk phrases (R-phrases) and safety phrases (S-phrases),[30] the metrics have been revised to adapt them to GHS, which required a new definition of the criteria to score the hazards and assess the accomplishment of the GC principles.[31] As in GHS the systemization of the hazards in classes and categories, to which hazard statements are assigned (hazard codes or H codes), is more detailed than in the system of the R- and S-phrases, the change provided a finer evaluation of hazards and made easier the construction of the metrics. The revision included also an improvement of the graphic look of the green star to increase the facility of reading the individual scores of the principles. These new versions of the metrics are those discussed in this text.

The SDS provide information on hazards of substances, renewability and degradability that is useful for construction of the metrics in several of sections of their standard format, as shown in Table 8.3. The information is used

Table 8.3　Sections of SMS with information for construction of holistic metrics.

Section		Type of information provided
2	Hazards identification	Generic information
8	Exposure controls/ personal protection	Values of exposure limits
9	Physical and chemical properties	Physical risks
10	Stability and reactivity	Accident risks (incompatibilities, forbidden conditions, *etc.*)
11	Toxicological information	Human health: toxicity parameters for different exposure types (inhalation, cutaneous, oral, *etc.*); information on carcinogenicity and mutagenicity
12	Ecological information	Animal health: toxicity for selected species; persistence, bioaccumulation and degradability; mobility (in soils)

Table 8.4   Data provided by SMS for evaluation of the greenness using the green
                chemistry principles.

Principle		Evaluation[a]
1	Residue prevention	Risks of residues: H, E, A
3	Less hazardous synthesis	Risks of all substances: H, E, A
5	Safer solvents and auxiliary substances	Risks of solvents: H, E, A
7	Use renewable feedstocks	Renewability of feedstocks
9	Preference for catalytic reactions	Risks of catalysts: H, E, A
10	Design for degradation	Degradability of all substances
12	Safer chemistry for accident prevention	Risks: A

[a]H, human health; E, environment; A, accident.

for greenness assessment in the GC principles listed in Table 8.4. It is practically impossible to suggest a general procedure for extracting data from SDS for metrics construction, owing to the singularities of the different cases to be evaluated, reagents and materials involved, quality of SDS, *etc.*, but these two tables provide some guidance to perform the task. If up-to-date SDS are used, the data on hazards are generally sufficient, but data about degradability are often still not provided. Moreover, data on renewability may be dubious (*p. ex.*, ethanol may be obtained by fermentation of natural sugar or distillation of oil). In conclusion, although the task may be laborious, after some practice the SDS provided nowadays by the chemical sellers are a good source for obtaining the data required to construct the metrics.

8.3.2   Construction

The strategy for construction of the metrics will be exemplified here very briefly for the case of the green star when used to evaluate a synthesis reaction. A tool for its implementation can be accessed in the net, in a specific site.[32] The tool calculates the metric automatically upon introduction of the data, by applying the criteria defined in two tables mentioned below (table 'H codes/score of hazards' and table 'score of hazards/scores of principles'), that are included in it. The site provides also an example of the use of green star that shows the results along the construction of the metric.

The strategy includes the following steps:

1. Detailed examination of the protocol of syntheses to obtain the reaction conditions and organize a inventory of all the substances involved: feedstock, products, by-products, solvents, catalytic reagents, *etc.*
2. Collection of data on the hazards to human health, the environment and the potential chemical accident for each substance, as well as information to assess whether the substances are renewable and break down to innocuous degradation products (from SDS and other sources).
3. Attribution of scores to the hazards of the substances using their H codes following systematic criteria (defined in 'table H codes/scores

of hazards', where all relevant information from GHS[28] is collected and systematized), a three numeric level scale being used (1/3 or low/moderate/high, the lowest level being attributed when no hazards are known); the renewability and degradability of the substances are similarly scored by pre-established criteria.

4. Attribution of a greenness score to each principle (scale 1/3, from minimum to the maximum value of greenness) following criteria defined from the hazard scores of substances and the reaction conditions (table 'score of hazards/scores of principles').

5. Construction of the graphic result of the evaluation as an Excel radar chart from the scores of the principles: the higher the scores are, the fuller is the star, therefore a visual inspection provides an indication of the global greenness.

6. If necessary for comparison of different green stars, calculation of a green star area index (GSAI) as a percentage (GSAI = 100 × green area of the green star/area of the green star of maximum greenness), to be used together with the graphic metric; this index varies between 100 (maximum greenness) and 0 (minimum greenness).

The green circle is obtained by a similar procedure, the only difference being a simpler binary criterion (accomplishment/no accomplishment) for evaluation of the principles, where the intermediate level in the 1/3 scale of green star is included in the lower level (no accomplishment). The metric is presented as an Excel pie chart, where the colours green/red in the sectors are used to identify the accomplishment/no accomplishment of the principles. For comparison of different green circles, an accomplishment of principles index (API) may be calculated as the percentage of principles accomplished (API = 100 × number of principles accomplished/total number of principles that apply).

8.4   Use of Holistic Metrics in Teaching Activities

The metrics have been used for a variety of activities related to GC teaching based on the assessment of greenness.

Green star was conceived to support GG laboratory teaching by a new procedure that involves as an intentional purpose the pursuing of the improvement of greenness in preparative experiments of syntheses.[16,17] Synthesis protocols for compounds commonly included in organic and inorganic teaching laboratories, obtained from the literature used for this purpose (book, laboratory manuals, *etc.*), were prescribed to students to be performed in the laboratory with special attention to their greenness (or, rather, the lack of it), to be evaluated by construction of the green star (and, in parallel, mass metrics). Then, the students were challenged to improve the procedure to replace steps with deleterious effects on greenness, and after the discussion of their effort, to write a revised protocol and execute it in the laboratory, with evaluation of the intended greenness progress by drawing a revised green star. The results were discussed again to assess the increase of greenness obtained with the revised protocol (or its failure) and to identify further modifications to be explored in a second iteration of the exercise (see Figure 8.9), or to conclude that this was practically impossible or did not deserve the effort. Independently of the results, the analysis, discussion, implementation and evaluation of the successive protocols press the students to feel what GC involves and that it has to be pursued with a purposeful determination. Moreover, the students have to consider multiple solutions and make decisions on the choice of best alternatives, which is a desirable departure from the usual laboratory experiences where their role is often quite passive; the procedure incentivizes pro-active thinking. On the other hand, this type of assignment provides the instructor with further ways for evaluating the students. However, it should be remarked that, as students are not acquainted with type of work, much more support by the instructor is required along the preparation and discussion of the alteration of protocols. Therefore this new teaching strategy requires classes with a limited number of students.

The results of an exercise of this type are exemplified in Figure 8.7 for the case of the synthesis of the complex tetramminecopper(II) sulfate monohydrate from copper(II) sulfate pentahydrate and ammonia.[16] For this very simple syntheses, the excess of ammonia was decreased from 300% (top graphs) to 7% (bottom graphs) without diminution of the yield and the green area in the

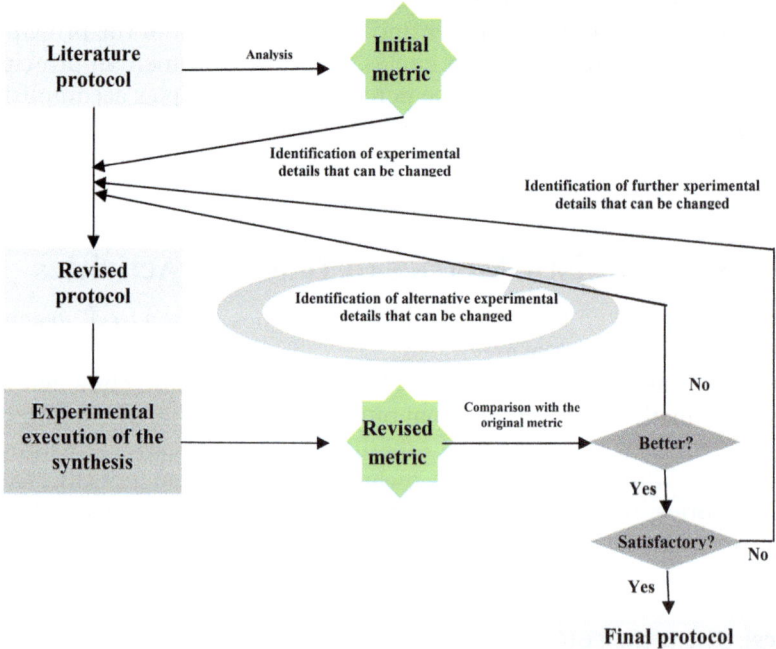

Figure 8.9   Use of holistic metrics in optimization of synthesis greenness.

green star increased, showing better greenness. However, the GSAI increased only to 55, a moderate value (from the initial value of 40). Further improvement proved impossible, mainly because of the toxicity of ammonia and the copper(II) compounds involved in the preparation, which obviously could not be changed, as they were essential to obtain the product. In Figure 8.7 the results of the assessment of the optimization with the green circle are also included, showing the increase of greenness with a rise in the API from 30 to 50. The comparison of the plots for the two metrics shows also that the evaluation with green star is more detailed than for the green circle, as mentioned above.

The results of the study of further cases of synthesis experiments already published are summarized in Table 8.5. These studies showed that often the protocols in the teaching literature prescribe a large excess of a reagent which is not necessary to reach the yield reported: upon optimization of the synthesis, the same yield is obtained with a lesser excess, frequently only a slight excess, meaning an increase of greenness. The use of a large excess of reagents in teaching experiments is probably a legacy of the importance given to chemical equilibrium and its displacement by the law of mass action, when no attention was paid to the production of wastes, the only objective being the obtention of the product.[25] The levels of greenness reached after optimization presented in Table 8.5 seem limited (the GSAI are lesser than 60), which is not unexpected because until recently greenness was ignored when choosing synthesis reactions for teaching laboratories. These results confirmed Andraos' remarks on the large amounts of residues produced in Canadian university teaching laboratories[33] and suggested that there is plenty of room to improve greenness by revising laboratory experiments used for teaching and conceiving new ones addressed proactively to GC.

In more recent activities, the green star is being used in a broader study of the synthesis experiments used in the 1st and 2nd years organic and inorganic laboratories of chemistry BSc courses of Portuguese universities.[34] This study showed that the metric is adequate to evaluate separately the three stages of the synthesis process, the reaction step (with ten corner stars), and the two post-reaction steps, isolation and purification (six corner stars). This assessment of the *micro-greenness* of the steps shows that the post-reaction

Table 8.5 Results of the optimization of syntheses assessed with the green star.[a]

Compound	Ligand/metal	Improvement of greenness[b]	Ref.
$[Cu(NH_3)_4](SO_4)\cdot H_2O$	Ammonia	$27.5 \rightarrow \mathbf{40.00}$	16
$Fe(II)ox_3\cdot 2H_2O$	Oxalate	$20.00 \rightarrow 36.25 \rightarrow 41.25 \rightarrow \mathbf{46.25}$	17
$Fe(III)acac_3$	Acetylacetonates[c]	$32.50 \rightarrow 40.0$	18
		$41.25 \rightarrow \mathbf{51.25}$	
		30.0	
$M(II)acac_2$	Mn or Mg	$22.50 \rightarrow 30.0$	18
	Ca	$46.25 \rightarrow \mathbf{57.5}$	

[a]Results are given as the per cent of maximum greenness.
[b]Values in bold type are the maxima.
[c]Different lines refer to different protocols.

steps are often more problematic for obtaining high levels of greenness than the reaction itself, owing to the liberal use of organic solvents, some of them toxic and physically dangerous and therefore unsuitable to be kept in GC.[34] Indeed, the limited greenness showed by the green star for the global synthesis process is often a consequence of the poor greenness of the six corner stars of post-reaction steps. On the other hand, when alternative protocols are obtained from the literature (and nowadays very easily also from the internet), and subjected to micro-greenness evaluation with the green star, greenness optimization is sometimes possible by combining the greenest steps of different protocols available to raise the greenness. The work in progress aims to define the scope of this optimization procedure, as well as to investigate whether there are rules that govern the relationship between the micro-greenness step stars and the global star of the synthesis process.

Green star was also used for a thorough study of the experiments used in chemistry programs of the 10th and 11th grades of Portuguese secondary schools.[35] Most of the experiments were simple and involved no synthesis reactions, being evaluated by six corner stars. The study showed that on the whole the greenness level was low, suggesting again that programmes require an extensive revision to support GC teaching.

At the moment, the metrics are also being used in a project under development (2013–2014),[36] with the aim of introducing GC to secondary schools in Portugal (in the Great Oporto region). The project is being developed in the schools by two educational PhD students and will reach more than a thousand of students of the 10th, 11th and 12th grades. The green circle, being easier, is used in the 10th and 11th grades and seems to be well accepted by students in a first introduction to GC. The more elaborate green star is used in the 12th grade for support of discussions on the greenness of synthesis experiments included in the program. However, a complete assessment of the results of this pedagogical experiment addressed to the real use of the holistic metrics in schools, based on questionnaires, discussions with students and school teachers, *etc.*, must wait until the end of the field activities.

A large number of the greenness evaluations in these activities are accessible in a site dedicated to GC teaching in the Chemistry Department of the Faculty of Science of Oporto,[37] although the evaluations were made with the first version of the green star. However, a bilingual site (English/Portuguese) to include the recent and future evaluations with the revised GHS metrics is under construction.[38]

8.5   Discussion

8.5.1   Comparison of the Holistic Metrics

The set of tools reviewed in this chapter provides diverse opportunities to introduce the evaluation of greenness in the teaching of GC at several levels. The green circle is useful for a quick assessment of the greenness of a chemical reaction or process without performing the experiment from a protocol, if this is detailed enough, or upon experimental execution in the laboratory. It is simple enough to be used in the terminal years of secondary schools in the exemplification of the Twelve Principles in laboratory experiments based on simple synthetic reactions. The green matrix is more suitable for university teaching, but the SWOT analysis, being probably a novelty for most first/second year chemistry students, requires time for a preliminary presentation. However, the construction of the SWOT matrix involves a detailed analysis of different aspects of the greenness, including those that may bring improvements to it, and this extra effort by the students provides an opportunity to enhance their knowledge about GC. If used together, the green circle and the green matrix allow a detailed description of the results of the greenness evaluation: the green circle allows a quick visual identification of the principles accomplished, or not, and the green matrix provides details about the aspects that support the assessment, evidencing those to be given prior attention to increase greenness.

With reference to rating of the greenness, these two metrics are both easier to construct than the green star, as they are based in criteria with only two levels, accomplishment/no accomplishment. However, they do not hierarchize the suggestions for improvements to increase greenness: owing to the limitation of using only two levels of discrimination, they do not consider partial accomplishment of the principles like the green star does. Therefore this last tool is superior; in particular it allows refined greenness and micro-greenness evaluation in a variety of situations.

8.5.2 Advantages of the Holistic Metrics

Holistic metrics allow a large scope in the evaluation of greenness, which captures both material components (material greenness), important to dematerialize chemistry, and components referring to the impacts on the environmental and human health (environmental greenness). The metrics are suitable to be used operationally to improve the greenness of laboratory experiments, as they are easily constructed for this type of situation, requiring only limited effort and time; for most cases data can nowadays be obtained from easily accessible SDS. Moreover, the graphic presentations of green circle and green star provide evidence of the principles that limit the greenness and require improvement and allow finding the characteristics of reactions and physical operations, reagents, solvents, *etc.*, involved in the synthesis process that should be revised. The metrics are easily recalculated to assess the increase of greenness achievable by alterations in the procedure, as only part of them needs reconstruction, to confirm whether the intended improvement was indeed achieved after testing the changed procedure in the laboratory.

The use of holistic metrics presents several further advantages for GC teaching that transcend the information they provide; for instance, their construction: (1) facilitates a better understanding of the GC Twelve Principles and allows training in their use for pursuing chemical greenness; (2) provides a way of feeling the importance of a holistic approach to the practice of GC, including the development of the capacity of dealing with multi-dimensional analysis of situations (and multi-criteria decisions); and (3) helps in assuming a proactive attitude toward the transformation of chemistry in GC.

8.5.3   Limitations of the Holistic Metrics

The main limitation of these holistic metrics based on the GC principles is that some of the risks are captured simultaneously by several principles, as exemplified in Figure 8.10, and consequently the evaluations made by the

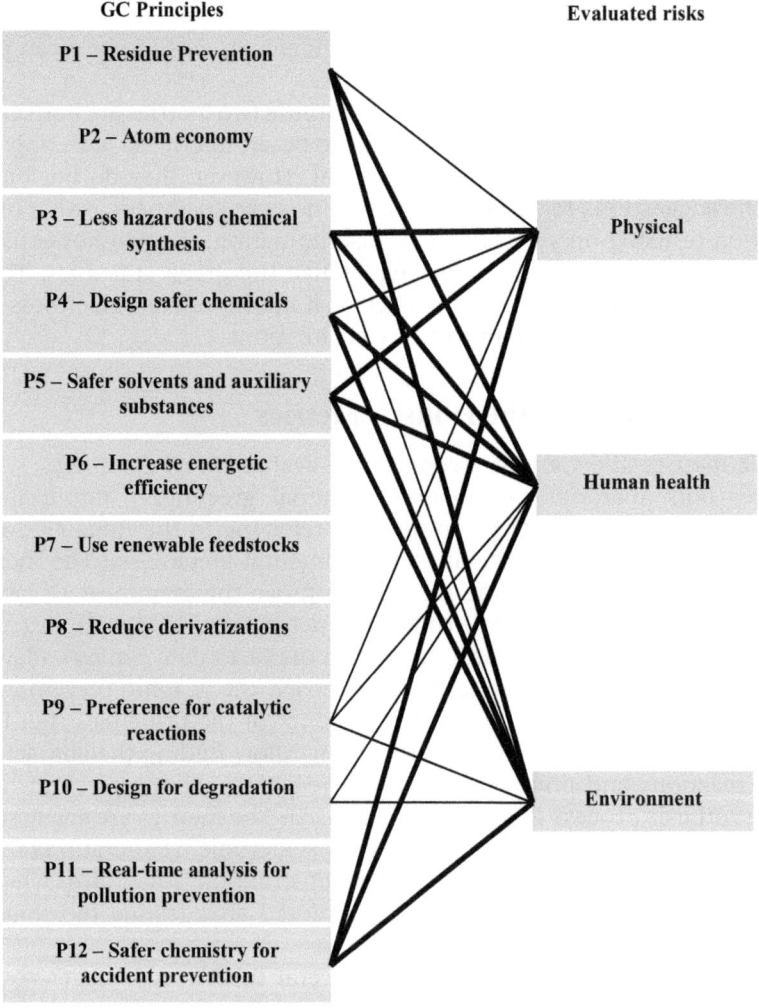

Figure 8.10   Several risks are evaluated in parallel by different principles and therefore the individual assessments by principle are not orthogonal.

different principles are not orthogonal; they may be correlated. This lack of independency is undesirable in batteries of metrics for assessment of systems as it may originate inadvertent effects with perverse consequences on the evaluation and improvement of the system operation.[11] However, so far, no problem of this type has been found with these holistic greenness metrics.

On the other hand, when the holistic metrics are used in parallel with the mass metrics referred to above, the results and conclusions about greenness may show differences. This is a consequence of the nature of the evaluation—systemic *vs.* reductionist—which implies that different components of the greenness may be captured in the assessment of the system by different metrics, as exemplified in Table 8.6. This table shows that some material components, like masses, yields, *etc.*, are caught only by mass metrics, as these provide quantitative values. This lack of completeness is a limitation of the holistic metrics (as well as of the mass metrics), but is inherent to their systemic nature. This situation suggests that in practice the use of mass metrics in parallel with the holistic metrics, as often made in the cases studied,[16–18] provides a fuller account of greenness.

Indeed, it should be remarked that the huge number of situations found in chemistry that require greenness evaluation implies that the level of difficulty in the use of the evaluation criteria for the GC Twelve Principles in holistic metrics varies from case to case and therefore their results can be more or less precise. This uncertainty must be taken into account when comparing alternatives. Metrics of systems are tools for helping in the decisions about them and their results should not be taken as an ultimate truth. Indeed metrics often do not capture the whole set of dimensions of the system and therefore the chemist in search of greenness should have a profound chemical knowledge about the case under study and consider it when making decisions; metrics should be used as guides to enlighten that knowledge, not to replace it.

Table 8.6 Comparison of components of greenness captured by holistic metrics *versus* mass metrics.[a]

Component	Holistic metrics	Mass metrics
Yield	N	E-factor, MI, RME
Mass of residues	N	E-factor, MI
Mass of solvents	N	E-factor, MI
Nature of residues	Y	N
Nature of solvents	Y	N
Energy efficiency	Y	N
Use of renewable feedstocks	Y	N
Reduction of derivatizations	Y	N
Nature of catalysts	Y	N
Degradability of substances	Y	N
Risks (human health, environment, accident)	Y	N

[a]N, no; Y, yes.

8.6   Conclusions

The term 'greenness' is unavoidably ambiguous and amenable to a variety of interpretations, because the concept it tries to convey in GC is extremely complex; the greenness synergistically compounds the complexities of the chemistry, the environment and their systemic connections, especially the impacts of chemicals and their manufacture on the environment. However, the evaluation of greenness, although only relative, is important for the prosecution of GC as noted in several recent articles on its development and future.[39]

Holistic metrics, which are based on systems thinking, present several benefits for teaching GC; for instance, familiarization with the Twelve Principles and with their implications as a whole; multi-dimensional evaluation of the greenness of chemical reactions and processes by simultaneous use of the set of principles; identification of aspects that could be optimized to improve greenness; further assessment of the effects on the greenness of the changes implemented; separate evaluation of the greenness of the chemical reaction and the physical operations of the synthesis process that constitute the work-up (micro-greenness assessment), *etc.*

In conclusion, these holistic tools enhance the understanding of GC and of greenness evaluation problems, as well as of the practice of the Twelve Principles, being also useful to facilitate their incorporation in teaching activities. Therefore they seem to be a contribution suitable for the development of practices in chemistry that must be proactively changed to reach sustainable chemistry, which is the ultimate purpose of green chemistry.

References

1. J. A. Kent (ed.), *Kent & Riegel's Handbook of Industrial Chemistry*, 11th edn, vol. 1, Springer, 2007.
2. (a) US Department of Energy, *Energy and Environmental Profile of the US Chemical Industry*, 2000, accessible from: http://eelndom1.ee.doe.gov/oit/oitcatalog2.nsf/ShowData?OpenAgent&Top=Program&Mid=Chemicals&Low=Energy+and+Environmental+Profile (accessed February, 2013); (b) B. S. Kim and M. Overcash, *J. Chem. Technol. Biotechnol.*, 2003, **78**, 995; (c) G. Wernet, C. Mutel, S. Hellweg and K. Hungerbuhler, *J. Ind. Ecol.*, 2011, **15**, 96.
3. T. E. Graedel, *Pure Appl. Chem.*, 2001, **73**, 1243.
4. B. M. Trost, *Science*, 1991, **254**, 1471; idem, *Angew. Chem., Int. Ed. Engl.*, 1995, **34**, 259.
5. R. A. Sheldon, *Chem. Ind. (London)*, 1992, w/ vol., 903–906; idem, *CHEMTEC*, 1994, **24**(3), 38.
6. F. G. Calvo-Flores, *ChemSusChem*, 2009, **2**, 905.
7. A. Lapkin and D. J. C. Constable, *Green Chemistry Metrics: Measuring and Monitoring Sustainable Processes*, Wiley, 2009.
8. C. Jiménez-González, D. J. C. Constable and C. S. Ponder, *Chem. Soc. Rev.*, 2012, **41**, 1485.

9. P. T. Anastas and J. C. Warner, *Green Chemistry – Theory and Practice*, Oxford UP, 1998, p. 30.

10. (a) A. D. Curzons, D. J. C. Constable, D. N. Mortimer and V. L. Cunningham, *Green Chem.*, 2001, **3**, 1; (b) D. J. C. Constable, A. D. Curzons, L. M. F. Santos, G. Green, R. E. Hannah, J. D. Hayler, J. Kitteringham, M. A. Mcguire, J. E. Richardson, P. Smith, R. L. Webb and M. Yu, *Green Chem.*, 2001, **3**, 7; C. Constable, A. D. Curzons and V. L. Cunningham, *Green Chem.*, 2002, **4**, 521.

11. (a) Committee on Industrial Environmental Performance Metrics (CIEPM), US Academy of Engineering, *Industrial Environmental Performance Metrics – Challenges and Opportunities*, US National Academy Press, 1999, accessible from: http://www.nap.edu/catalog.php?record_id=9458; (b) Committee on Metrics for Global Change Research (CMGCR), US National Research Council, Thinking Strategically – The Appropriate Use of Metrics for the Climate Change Science Program, US National Academy Press, 2005, accessible from: http://www.nap.edu/catalog.php?record_id=11292.

12. M. G. T. C. Ribeiro and A. A. S. C. Machado, *Green Chem. Lett. Rev.*, 2013, **6**, 1.

13. C. Jimenez-Gonzalez, C. S. Ponder, Q. R. Broxterman and J. B. Manley, *Org. Process Res. Dev.*, 2011, **15**, 912.

14. A. A. S. C. Machado, *Quim. Nova*, 2014, **37**, 1094 (in portuguese).

15. A. A. S. C. Machado, *Introdução às Métricas de Química Verde – Uma Visão Sistémica*, Editora da Universidade de Santa Catarina, Florianópolis (Brazil), 2014, in press (in portuguese).

16. M. G. T. C. Ribeiro, D. A. Costa and A. A. S. C. Machado., *Quim. Nova*, 2010, **33**, 759 (in portuguese).

17. M. G. T. C. Ribeiro, D. A. Costa and A. A. S. C. Machado, *Green Chem. Lett. Rev.*, 2010, **3**, 149.

18. M. G. T. C. Ribeiro and A. A. S. C. Machado, *J. Chem. Educ.*, 2011, **88**, 947.

19. M. G. T. C. Ribeiro and A. A. S. C. Machado, *Quim. Nova*, 2012, **35**, 1879 (in portuguese).

20. M. G. T. C. Ribeiro and A. A. S. C. Machado, *J. Chem. Educ.*, 2013, **90**, 432.

21. D. G Blackmond, A. Armstrong, V. Coombe and A. Wells., *Angew. Chem., Int. Ed.*, 2007, **46**, 3798.

22. M. Deetlefts and K. R. Seddon, *Green Chem.*, 2010, **12**, 17.

23. S. E. Jackson, A. Joshi and N. L. Erhardt, *J. Manage.*, 2003, **29**, 801.

24. H. Keller and J. R. Cox, *J. Chem. Educ.*, 2004, **81**, 520; M. E. Brown, R. C. Cosser, M. T. Davies- Coleman, P. T. Kaye, R. Klein, E. Lamprecht, K. Lobb, T. Nyokong, J. D. Sewry, Z. R. Tshentu, T. Zeyde and G. M. Watkins, *J. Chem. Educ.*, 2010, **87**, 500.

25. N. Winterton, *Green Chem.*, 2001, **3**, G73.

26. European Union (EU), *Regulation (EC) No 1907/2006, Registration, Evaluation, Authorization and Restriction of Chemicals - REACH*, 2006, accessible from: *http://ec.europa.eu/environment/chemicals/reach/legislation_en.htm* (accessed 2014.03.01).

27. European Chemicals Agency (ECHA), *Document ECHA-2011-g-08-en, Guidance on the Compilation of Safety Data Sheets – Version 1.1*, 2011, accessible from: *http://echa.europa.eu/documents/10162/13643/sds_en.pdf* (accessed 2014.03.01).

28. United Nations Economic Commission for Europe (UNECE), *Globally Harmonized System of Classification and Labelling of Chemicals (GHS)*, 5th revised ed., 2013, accessible from: *http://www.unece.org/trans/danger/publi/ghs/ghs_rev05/05files_e.html* (accessed 2014.03.01).

29. R. J. Willey, *Procedia Engineering*, 2012, **45**, 857.

30. The European Parliament and The Council, *Regulation (Ec) No 1272/2008 on Classification, Labelling and Packaging of Substances and Mixtures, Amending and Repealing Directives 67/548/EEC and 1999/45/EC, and Amending Regulation (EC) No 1907/2006, 2008*, accessible from: *http://eurlex. europa.eu/LexUriServ/LexUriServ.do?uri=OJ:L:2008:353:0001:1355:EN: PDF* (accessed 2014.03.01).

31. M. G. T. C. Ribeiro, S. F. Yunes and A. A. S. C. Machado, 2014, **91**, 1901.

32. *Metrics Calculations for Accessing the Greenness of Chemical Reactions*, accessible from: http://educa.fc.up.pt/avaliacaoverdura/.

33. J. Andraos and M. Sayed, *J. Chem. Educ.*, 2007, **87**, 1004.

34. R. C. C. Duarte, M. G. T. C. Ribeiro and A. A. S. C. Machado, *Quim. Nova*, 2014, **37**, 1085 (in portuguese).

35. (a) D. A. Costa, M. G. T. C. Ribeiro and A. A. S. C. Machado, *Química – Bol. SPQ*, 2009, **115**, 4; (b) *Idem*, 2011, **123**, 63 (both in portuguese); (c) D. A. Costa, *Métricas de Avaliação da Química Verde – Aplicação no Ensino Secundário*, PhD thesis, Faculdade de Ciências do Porto, 2011, accessible from: *http://educa.fc.up.pt/ficheiros/investigacao/61/VER%20 TESE%20%20de%20%20Dominique%20A.%20Costa%20.pdf*).

36. M. G. T. C. Ribeiro (coordinator), *Introdução do Ensino da Química Verde, como Suporte da Sustentabilidade, no Ensino Secundário*, Programa Escolher Ciência–Ciência Viva, Project PEC 123, 2013–2014, accessible from (in portuguese): *http://educa.fc.up.pt/projeto_encontros.php?id_projecto=18*.

37. M. G. T. C. Ribeiro and A. A. S. C. Machado, *Pedagogia da Química Verde – Educação para a Sustentabilidade*, accessible from (in portuguese): http:// pedagogiadaquimicaverde.fc.up.pt/.

38. M. G. T. C. Ribeiro (coordinator), *Catálogo Digital de Verdura de Atividades Laboratoriais para o Ensino da Química Verde*, accessible (in english/portuguese) from: http://educa.fc.up.pt/catalogo/en.

39. M. Epicoco, V. Oltra and M. S. Jean, *Technol. Forecasting Soc. Change.*, 2014, **81**, 388; J. Clark, R. Sheldon, C. Raston, M. Poliakoff and W. Leitner, *Green Chem.*, 2014, **16**, 18; W. J. W. Watson, *Green Chem.*, 2012, **14**, 251.

CHAPTER 9

Embedding Toxicology into the Chemistry Curriculum

NICHOLAS D. ANASTAS*[a]

[a]Poseidon's Trident, LLC, 83 Sassamon Avenue, Milton, MA 02186, USA
*E-mail: nanastas@poseidonstrident.net

9.1 Introduction

Designing safer and healthier chemicals is essential to promoting and supporting a sustainable future. Chemists are trained to craft molecules in myriad innovative ways to achieve functionality and to satisfy materials requirements. These advancements have contributed to increasing prosperity and a higher quality of life of the past several decades. However, not all of the products and materials made by chemists have been sustainable and have actually contributed to detrimental outcomes on public health and the environment. The way chemists are trained to characterize and design chemicals is an emerging philosophical change.

Green chemistry is the utilization of a set of principles that reduces or eliminates the use or generation of hazardous substances in the design, manufacture and applications of chemical products.[1] More university and college chemistry faculty are incorporating the principles of green chemistry into their curriculum in either the form of targeted topics or as wholly independent courses in toxicology. The flexibility to incorporate this new material into routine practice is critical to its widespread acceptance and adoption.

Worldwide Trends in Green Chemistry Education
Edited by Vânia Gomes Zuin and Liliana Mammino
© The Royal Society of Chemistry 2015
Published by the Royal Society of Chemistry, www.rsc.org

Over the past several years, significant advancements in hazard reduction have been made in hazard reduction using the Twelve Principles of green chemistry, primarily in the area of solvent reduction, atom economy and alternative synthetic methods. One focus area that has lagged behind the other focus areas of green chemistry is designing products and processes for reduced toxicity and, as and larger topic, hazard. Describing and predicting the toxicity of chemicals is a complex and demanding undertaking for a number of reasons, even for trained toxicologists and risk assessors. This challenge is even greater for chemists who require guidance on crafting safer chemicals through appropriate molecular design. The availability of design rules for hazard reduction is at this time inadequate and must expand to satisfy the demand by chemists for new, safer synthetic strategies.

Toxicology is an essential component of sustainable molecular design. The current American Chemical Society (ACS) curriculum requirements[2] do not incorporate the principles of toxicology as a required teaching element, therefore most practising chemists and current students have not encountered the principles of toxicology as part of their training. The intent of augmenting the current chemistry curriculum is not to produce graduates that are trained at the level of a professional toxicologist, but rather to develop scientists that can recognize potential hazard in a molecule, and then take affirmative steps to design safer, healthier and sustainable substances.

This chapter presents an overview of selected examples and a potential approach for embedding the structure–activity relationship, *i.e.*, toxicology, into the existing chemistry curriculum in an effort to produce a holistically trained modern chemist. Educators can adapt the framework as they identify the links between the fundamental principles of chemistry with the principles of toxicology. This is a paradigm shift in current thinking that will require significant commitment to alter the status quo. This knowledge gap needs to be filled by scientists comprehensively trained in both disciplines, a hybrid chemist perhaps. Melding of these two disciplines has been termed 'green toxicology'.[3]

9.1.1 The Role of Medicinal Chemistry in Safer Chemical Design

Medicinal chemistry fills a critical role by acting as a bridge between chemistry and developing new medicines, cosmetics and consumer products. The innovative approaches developed by pharmaceutical companies through fundamental and applied medicinal chemistry research have provided treatments for intractable disease, improved the quality of life and decreased mortality for a number of diseases. The pharmaceutical companies were early adopters of the philosophy of green chemistry focusing on minimizing hazards to human health and have realized tremendous achievements. However, green chemistry involves a broader definition of hazard reduction that includes physical, global, and ecological endpoints, as well as organisms other than humans. The need for expanding the focus of medicinal chemistry is met by including the principles of toxicology into all chemistry curricula.

9.1.2 Toxicology and Sustainable Molecular Design

Toxicology is central to designing products with minimal hazard to human health and the environment. Determining whether a substance is toxic or non-toxic requires a set of metrics representing both assessment and measurement endpoints for clearly defined adverse outcomes. A cornerstone of toxicology, articulated by the medieval physician Paracelsus, states that every compound is toxic at sufficient dose; in other words, 'the dose makes the poison'.[4] This central message has been expanded and appropriately refined to include time as a core component of the manifestation of toxicity. The idea that a chemical can be 'non-toxic' is actually a misnomer because all chemicals are toxic at some defined dose. There is also an inherent assumption that there is a threshold dose below which adverse effects do not occur and that all chemicals are therefore non-toxic at some dose.

9.1.3 Principles of Toxicology

Toxicology is the study of adverse effects of chemical, biological and physical agents on organisms. In other words, it is the study of poisons. This chapter focuses on the principles and practices of chemical toxicology intended as a baseline of knowledge for teachers, students of chemistry and practising chemists. A single chapter is clearly inadequate to present any area of toxicology in the detail necessary to become a skilled practitioner, especially a topic as complex as toxicology. Therefore, only an overview of the most critical aspects of toxicology that are necessary to inform safer chemical design are presented here. Several exceptional textbooks are available that serve as outstanding resources for those who want to investigate this fascinating field further.[5–8]

The choice of toxicology topics to emphasize and include in current chemistry curricula presented in this chapter, is not comprehensive but illustrates several opportunities to embed toxicology into the education of 21st century chemists. Many more examples of the natural nexus between chemistry and toxicology exist and provide a treasure trove of possibilities for ambitious educators to incorporate into teaching the next generation of chemists and allied scientists. Examples of the application of toxicology data in designing safer chemicals are provided throughout this chapter as concept demonstration exercises. Real-world examples that illustrate the influence of structure on toxicity can replace general—often abstract—examples regularly used as part of chemistry pedagogy that may be difficult for a student to grasp. Actual structures can replace the traditional pedagogy of using 'R-group' notation to represent a general reaction.

Before we continue, a few definitions of terms used throughout this chapter are necessary to maintain subtle but important differences among potentially toxic compounds. *Toxicity* is a relative property of a molecule's potential to cause harm. A *toxicant* is any agent capable of producing adverse responses in an organism. A *toxin* is a toxicant of natural origin, for example a natural product from a plant or a toxin from a venomous animal. A *xenobiotic* is a compound that is foreign to the organism. Often, toxicant, toxin and xenobiotic are used interchangeably but incorrectly.

9.1.3.1   Dose–Response Relationship

The central maxim of toxicology is that there is a quantitative relationship between the dose of a toxicant, toxin or xenobiotic, and therefore in the biological response it produces. This fundamental association is called the *dose–response relationship* and is essential to both toxicology and to pharmacology. It is customary to plot the dose as the independent variable on the *x*-axis and the response as the independent variable on the *y*-axis. When the dose is plotted arithmetically, a hyperbolic curve is generated showing the increased response with increased dose (Figure 9.1). If the dose is log-transformed and plotted against response, a line segment is obtained making the statistical manipulation easier to evaluate. The dose–response relationship describes the correlation between an increase in the dose of a chemical and the resulting increased response, which can be either beneficial or adverse. Though the relationship is elegant in its simplicity, it remains a formidable assignment to fully characterize the complex and often subtle nature of toxic responses. Toxicity is a function of dose, exposure and time.[9] Consequences of the interaction of a molecule with a biological target will propagate through molecular, biochemical, cellular and organism levels of organization ultimately resulting in a biological consequence. This consequence can be detrimental in the case of toxicity, or beneficial in the case of therapeutic compounds.

An advantage to the log dose–response plot is a much more straightforward interpretation of differences among potency among a group of toxicants acting through similar modes or mechanisms of action. Chemicals producing the same maximal effect but at a lower dose will occupy a position farther to the left on the plot of the dose–response curve, indicating greater potency.

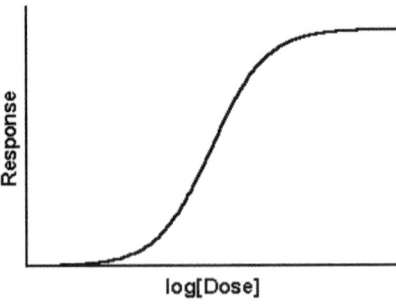

Figure 9.1   A representative dose–response curve.

The manifestation of adverse effects in biological systems is the result of perturbations of normal function or homeostasis, which involves a complex network of biochemical reactions that have evolved to maximize efficiency at the cellular, organ, tissue and whole organism level. These reactions are governed by the fundamental chemical principles that are part of typical chemistry curricula. Therefore, it is a natural extension to disclose opportunities to demonstrate the relationship between structure and toxicity.

9.2 Opportunities to Embed Toxicology into the Chemistry Curriculum

Every student and practitioner of chemistry has an obligation to understand and apply the principles of toxicology to their area of the chemical enterprise. This does not mean that every chemist must have the training equivalent to that of a practising toxicologist; however, a certain level of competence in toxicological sciences must be a core component of a modern chemist. Designing sustainable products and processes that are functional, economical and safe is challenging but necessary as society advances towards sustainable commerce. Understanding the toxicity of a substance and its dependence upon fundamental physicochemical properties, as well as adopting tools to characterize hazard, must be part of the training of all chemists. Intrinsic hazard is another physicochemical molecular property and cannot be treated in the curriculum as just a special topic in the education of scientists. Synthetic chemists are uniquely positioned to design less hazardous molecules. There is a growing demand to incorporate the intentional act of informed molecular design into the chemical enterprise. This can only happen when chemists are trained in the principles of toxicology. The same chemical principles apply to designing for functionality as for designing for reduced hazard. Instead of functionality, the spectrum of toxic effects must be considered in the design phase and must also be considered design flaws.[10]

9.2.1 Fundamental Molecular Forces Affect Toxicity

Teaching students that the fundamental forces that determine interactions and consequential reactions of toxicants at their sites of action, are the same as those associated with any other chemical reaction, (*e.g.*, covalent, ionic, electrostatic, ion–dipole, dipole–dipole, hydrogen bonding, van der Waals, hydrophobic, *etc.*) is a fundamental fact that offers a superb opportunity to show the relationships between molecular structure and toxicity. Weak forces occur when reactants are within close proximity and are complementary. Weaker than covalent bonds, they are significant because there are often many within a molecule of among complexes. One example is ionic bonds that rely on the attractive forces of oppositely charged species. At a physiological pH of approximately 7.4, many amino acids in proteins are positively charged, for example, arginine and lysine. These positively charged groups will react with negatively charged sites on carboxylic acids, aldehydes, ketones and other potential electrophiles. One example that can be used to illustrate this fundamental concept of ionic bonding influence on toxicity is the binding with receptors.

As background, receptors are generally macromolecules that are often embedded in cell membranes or can be located in the cytosol. They are proteins that may have additional structural features such as sugars (glycoproteins) or lipids (lipoproteins). An interaction of a substance (*i.e.*, ligand) with a receptor may lead to a biological response that can be beneficial (pharmacological) or detrimental (toxic). The initial interaction gives rise to a chain of events leading to a measurable, quantifiable outcome. An agonist is a compound that interacts with a receptor and causes a predictable response. An antagonist may bind with a receptor and inhibit the ability of an agonist to cause as response. A partial agonist acts at the same level as an agonist but does not result in a response as great as a full agonist.

Hydrophobic interactions play an important role in the manifestation of toxicity. Narcosis is non-specific toxicity that is related to the hydrophobic nature of a substance to interact with non-polar substances such as lipid membranes and storage fat. Anaesthesia is a type of narcosis that is thought to act through the general disruption of biological membranes.

Covalent binding to biological macromolecules is irreversible and often results in permanent damage. Many of these covalent reactions involve electrophiles (permanent or partial positive charge) with nucleophiles that are negatively charged. Molecular attributes such as charge density, degree of polarization, frontier molecular orbitals (HOMO–LUMO gap), *etc.*, dictate the course of these reactions.

A biological alkylating agent is a compound that can replace a hydrogen atom with an alkyl group under physiological conditions. These are general substitution reactions taught as part of undergraduate and graduate organic chemistry courses. Most of the covalent reactions involved with toxicity involve the heteroatoms nitrogen, sulfur and oxygen, acting as nucleophiles reacting with environmental and biological electrophiles in Sn1 and Sn2 type reactions.

In general, the relative rates of nucleophilic substitution depend on the nature of the alkylating agent and its access to the site of action which is dictated by steric, electrostatic and hydrogen bonding characteristics. Thiols are more reactive than amino groups which are more reactive than phosphates that are more reactive than carboxyl groups. Knowledge of these reaction rates and linking them to their fundamental quantum mechanical roots is crucial for predicting the toxicity profile of new and existing untested compounds. Computational toxicology depends on these fundamental relationships.

9.2.2   The Influence of pH on Toxicity

The behaviour of the hydrogen ion represents a fundamental chemical principle and is critical to the toxicity profile of many ionizable substances. Corrosivity is a dramatic, destructive and obvious result of very acidic or very basic substances as in the process of tissue necrosis. However, most compounds are weak acids or bases and their toxicity profiles are often governed by their ionization potential.

The definitive determinant of toxicity is a function of the concentration of the ultimate toxicant at the target site for a long enough period of time, which is governed by the time course of action (kinetics) and the response to the interaction at the target site (dynamics).

Toxic responses cannot occur unless an organism is exposed to a chemical and the ultimate toxicant reaches its site of action. The route by which the toxicant enters an organism can profoundly influence its ultimate fate. The major routes of exposure are inhalation, ingestion, dermal contact, and uptake by other species-specific organs, for examples gills in fish.

An example of the influence of pH on toxicity is the influence of ionization on toxicity. A chemical must reach its biological site of action to manifest toxicity. Often this process requires a substance to cross several biological membranes and also dissolve in various fluids. Absorption is the process of chemical, biological and physical agents crossing biological membranes. In most organisms chemical absorption is governed by the reactions occurring at biological membranes that are composed generally of lipid bilayers with polar head groups reflecting their amphipathic nature. The major sites of absorption are the gastrointestinal tract, the lungs, the skin (dermal absorption) and the gills of aquatic organisms.

Absorption across each of these anatomical structures is dictated by the properties of the compound, *e.g.,* molecular size, weight and chares, lipophilicity and ionization, and the properties of the membrane itself. Most compounds that transverse biological membranes through passive diffusion are non-polar, un-ionized organic chemicals because of the lipid nature of biological membranes. Absorption is primarily dependent upon the lipophilicity and charge of the compound and, in some cases, the presence of specific transporter systems. Specific membrane transporters include those for small molecules and certain amino acids. Generally, ionized molecules will cross membranes more slowly or not at all, therefore, the ionization constant is an essential property for estimating the likelihood of a substance crossing a membrane and reaching the site of action.

For toxicants that are weak acids and bases, the pH partition theory can be used to determine the extent of ionization of a toxicant that in turn can help characterize the extent of absorption. The Henderson–Hasselbach equation relates pH to the percent of compound ionized, as follows:

- for acids, pH = pK_a - log{[ionized]/[non-ionized]}
- for bases, pH = pK_a - log{[non-ionized]/[ionized]}

Non-ionized compounds are absorbed more efficiently than compounds that are ionized. The pH is an obvious influence in the ratio of ionized to un-ionized species and therefore the pH changes along the gastrointestinal tract, which profoundly influences the extent of absorption at a particular anatomical location. At physiological pH, most weak organic acids and bases will exist in various proportions based on their pK_a.

Curare is an arrow poison that has been used for thousands of years for hunting. It is a natural product, isolated from a plant. The molecular structure is complex and contains a quaternary nitrogen atom which is charged at elevated pH, for example, at approximately pH of 1.5 in the human stomach. This means that curare does not cross the gut very effectively. However, the exposure route predominantly used for hunting prey is intramuscular where the pH is higher, approximately 7.4. At this pH, absorption into the general circulation is higher and the molecule can reach the target site at a sufficiently effective concentration to cause adverse effects. This example effectively illustrates the importance of ionization and pH on a toxic outcome.

9.2.3   Applying Thermodynamics and Kinetics to Toxicology

The study of kinetics and dynamics are central parts of the chemistry curriculum providing an excellent opportunity to illustrate a natural link with toxicology. Two divisions of reactions describe the chemical journey from exposure to its final destination at its target: *toxicodynamics* and *toxicokinetics*. Toxicodynamics is the study of the interactions and subsequent responses of an organism from exposure to a toxicant. Potency and efficacy are two attributes associated with the toxicodynamic phase of the dose–response relationship. These reactions include the entire available chemical bonding schemes including covalent bonding, hydrogen bonding, ionic, non-covalent interactions, *etc*. It can be thought of as what the body does to the chemical. Potency is defined as the dose of a chemical required to achieve a maximal response.

The driving force for the toxicity cascade following binding of a ligand with a receptor follows the laws of thermodynamics and mass action. The affinity of a ligand or xenobiotic (X) for a receptor (R) is dependent upon the binding strength reflected in the equilibrium constant, Kd as depicted below:

$$X + R \in [X][R]$$

and

$$K_d = \frac{[X][R]}{[XR]}$$

A smaller value for K_d reflects a greater concentration of the complex and therefore a greater affinity of the xenobiotic for the receptor. A larger K_d reflects a lower affinity of the xenobiotic for the receptor. The toxicological terminology to reflect this fundamental behaviour is potency and efficacy. A more potent compound will show an effect at a lower concentration at the site of action reflecting a smaller K_d. Efficacy is a measure of the ability of a substance to elicit a full response compared to a standard.

At this point, the concept of effective dose can be introduced. An effective dose 50 (ED_{50}) is the dose of a compound required to reach 50% of a maximal response. For quantal responses, this value is the point where half the population responds. If the apical adverse effect is lethality (*i.e.*, death), the ED_{50} is referred to as a lethal dose 50 or LD_{50}. A lower value corresponds to a more potent substance, whether the response is therapeutic or toxic.

Efficacy, or intrinsic activity, is related to the affinity that at toxicant has for a particular receptor and with the resulting biological response.[11] The affinity of a toxicant for a ligand is related to the tendency to form a stable complex resulting in a biological response. This concept explains the differences between full agonists, partial agonists and antagonists. An agonist binds with a target site with a resulting complete response. A partial agonist binds to a target site with a predictable but a diminished response. An antagonist binds to a receptor with no resulting response. A xenobiotic that achieves the same maximal response at a lower dose than required for another compound to reach the same maximal response is considered to be more potent.

Toxicokinetics describes the processes associated with the time course of a xenobiotic along its pathway to its receptor site or sites. Generally, kinetics is the study of the time course of movement and the time course of chemical reactions including those processes associated with toxicity. In familiar terms, toxicokinetics describes the processes that the body performs on the xenobiotic.

How do toxicants access their sites of biological action? Unless they act directly at the exposure site, then they must be transported to the site of action through absorption, distribution, metabolism and excretion, commonly referred to by the acronym ADME. All of these factors have a role in determining the amount of toxicant reaching the target site as well as the length of time the xenobiotic remains in the organism.

Metabolism, or biotransformation, is the process of chemical transformation of a toxicant to different structures, called metabolites, which may possess a different toxicity profile than the parent compound. Biotransformation affects both endogenous chemicals exogenous (xenobiotic) entities. Metabolism can result in a transformation product that is less toxic, more toxic or equitoxic but in general more water soluble and more easily excreted. Chemical modification can alter biological effects through toxication of a substance, also called bioactivation, which refers to the situation where the metabolic process results in a metabolite that is more toxic than the parent. If the metabolite demonstrates lower toxicity than the parent compound, the metabolic process is termed detoxication. These processes can involve both enzymatic and non-enzymatic processes, all of which should be familiar to undergraduate and graduate chemists.

9.2.4 Redox Potential and Toxicity

Many examples exist that show the profound influence of oxidation–reduction reactions on the manifestation of toxicity. Examining several examples here will illustrate the effect of redox state of a molecule.

9.2.4.1 Nitrate and Nitrite Toxicity

Nitrate and nitrite are regulated in drinking water as pollutants that can cause methaemoglobinaemia in children at elevated concentrations. However, the toxicity is dependent directly to the nitrite ion, which is the oxidized form of the nitrate anion that binds to haemoglobin and inhibits oxygen from binding.

9.2.4.2 Chromium Toxicity

The toxicity of chromium is profoundly related to its oxidation state, whether the metal is in the +3 or +6 oxidation state. Chromium is a naturally occurring metal that is widely used for industrial purposes including plating, leather tanning, as a dye and as a wood preservative. Trivalent chromium (Cr^{3+}) is an essential trace nutrient required for proper glucose metabolism and other biological functions.

Hexavalent chromium (Cr^{6+}) is toxic to human and other organisms exhibited through a spectrum of adverse effects *via* inhalation, ingestion and dermal exposures. Inhalation of Cr^{6+} is associated with lung damage and it is carcinogenic through all routes of exposure through its interaction with DNA. Once it is inside the cell, it is reduced by intracellular electrophiles to Cr^{3+}. The process of reduction is believed to be responsible for its toxicity.

9.2.4.3 Cytochrome P450

The cytochrome P450 (CYP) family of enzymes are the major catalysts that are responsible for a variety of oxidation reactions associated with the biotransformation of many xenobiotics that can be used to illustrate several concepts of toxication and detoxication in a chemistry course.[12] The main function of this group of isozymes is to insert one atom of oxygen into a substrate thereby increasing hydrophilicity or water solubility. These enzymes are haem-containing proteins containing a reduced iron species essential for transferring electrons and work in concert with coenzymes NADPH and NADPH reductase.

The biotransformation of xenobiotics is catalysed by a number of enzymes that be divided into classes of reactions.[13] These are: hydrolysis, reduction, oxidation, conjugation. These reactions are part of the current curriculum and therefore provide an opportunity for connecting chemistry and toxicology.

A more complicated yet extremely enlightening example of the consequence of redox conditions on the ultimate expression of toxic outcomes is the mechanism of biotransformation through the oxidative family of enzymes know as the cytochrome P450 (CYP) superfamily. This enzyme system is responsible for an enormous array of biotransformations for substances that represent an extremely diverse range of molecular classes.

The primary function of the CYP enzyme is to insert one oxygen atom from molecular oxygen, into a molecule thereby imparting functionality, most often increasing hydrophilicity that leads to an enhanced potential for excretion in all organisms. The functionality imparted into the molecule can result in toxication or detoxication depending on the nature of the molecule and the potential biological targets. The fact that all biotranformation do not result in less toxic molecules is an important learning point. Polyaromatic hydrocarbons can react with CYP forming two epoxides that act as electrophiles capable of interacting with the nucleophilic sites on DNA, specifically the N^7 of guanine. This reaction forms a 'DNA adduct' that is a crucial step in carcinogenesis. The parent compound is called a 'pro-carcinogen', the diol metabolites are 'proximate carcinogens' and the epoxide is the 'ultimate' carcinogen. Many other examples of substances that follow this type of mechanism can be used as illustrative examples in the classroom.

9.2.5 Metals

Metals provide an illuminating example of how atomic and chemical structure influences toxicity. Metals are generally positively charged and therefore act as electrophiles allowing reactions with many biological nucleophiles especially those containing sulfhydryl groups. The tissues most vulnerable to adverse effects are the kidneys and lungs although metals have a wide spectrum of adverse outcomes on other organs as well as being implicated in many types of cancers. Educators can provide students with the opportunity to research the toxicity profile of a particular metal as an independent study project.

9.2.6 Influence of Isomerism on Developmental Toxicity: Thalidomide

The processes associated with reproduction and development in all organisms and plants are extremely complex, involving awe-inspiring events requiring flawless execution of elegant combinations of timing and of process fidelity. Any errors at critical stages in the process can lead to devastating consequences including physical malformations, increased reproductive failures, physiological deficits and death.

Exposure to certain xenobiotics is associated with adverse endocrine effects. These chemicals are commonly referred to as endocrine disrupting chemicals. Endocrine disruption can interfere at all levels of physiological organization and with all endocrine organs, for example mimicking natural hormones, blocking endogenous receptors, or directly affect the system itself.

Developmental toxicology is the study of adverse effects in a developing organism. Teratology is a sub-discipline of developmental toxicology that focuses on the specific time period between conception and birth. Many teratological effects are quite obvious, for example, cleft palate and missing or radically malformed limbs. The timing of exposure of an organism to a potential teratogen is of paramount concern especially for developmental toxicity. Critical exposure periods of susceptibility correlate with the timing of organ development and are quite precise. Any perturbations of the normal timing significantly increase the likelihood of adverse developmental effects including teratogenesis. Guidelines for evaluating developmental toxicants have been established to assess developmental risks.[14]

Thalidomide is among the most notorious examples of how a slight change in molecular structure can influence biological effects and can serve as a specific example when discussing isomerism as part of a general chemistry course. Thalidomide was a drug prescribed to pregnant women to mitigate morning sickness. Within a year of its introduction onto the market, reports of severe limb malformations in newborns were disclosed. The drug was removed from the market soon after the incidents of these adverse effects were made available.

Thalidomide exerts its teratogenic effects by interfering with organogenesis between days 24 through 33 of development.[15] The exact mechanism is not known; however, intercalation into DNA is one of the leading hypotheses among the more than thirty proposed. Thalidomide exists in two isomeric forms; the (S)- and the (R)-enantiomers (Figure 9.2). Research has indicated that only the (S)-enantiomer is capable of intercalating into DNA, resulting in toxicity. Understanding the relationship between the structural requirements and exposure limitations necessary for developmental toxicity has enabled pharmacologists to identify new clinical uses for thalidomide, including against leprosy, treatment of AIDS and against some aggressive forms of cancers.[16]

9.2.7 Linking Chemical Reaction Mechanisms with Mechanistic Toxicology

Organic chemists, especially those engaged in synthesis, are acutely aware of reactions and their mechanisms, including substitutions, eliminations, additions to double bond, rearrangements and oxidation-reductions.[17] These reactions are often classified by functional group for convenience and to illustrate patterns of chemical behaviour. The effect of structure on reactivity is crucial to understand mechanisms in both chemistry and toxicology.

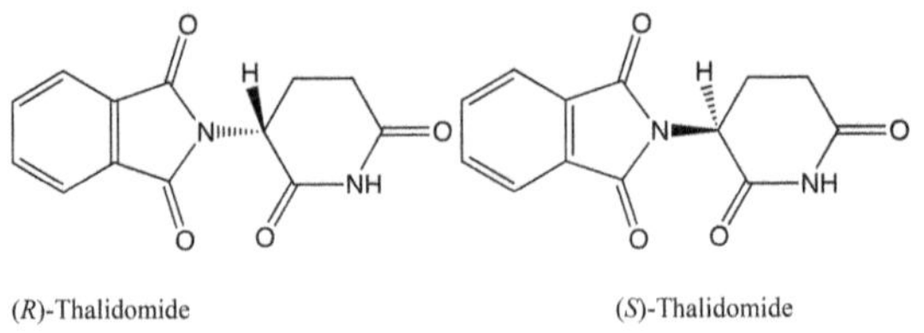

(R)-Thalidomide   (S)-Thalidomide

Figure 9.2 *R* and *S* isomers of thalidomide.

Mechanistic toxicology describes the process of how chemicals exert their toxicological effects on living systems.[18] These mechanisms employ the same fundamental reactions as those applied to any chemical reaction, therefore, the link between chemistry and toxicology becomes clear. Mechanisms of toxicity can be studied at a number of levels including whole organism, tissue, cellular and molecular level. Investigation at the molecular level forms the nexus between chemistry and mechanistic toxicology.

Mechanistic toxicology focuses on elucidating and describing the molecular events from exposure to the events that lead to the disruption of biological targets and describes the resulting adverse outcomes on living systems. A *mechanism* of action is defined as a detailed description of the key molecular events associated with a toxic response. A *mode of action* is a generic description of the key events and processes, starting with the interaction of an agent with a cell, through functional and anatomical changes, resulting in toxicity.[19] Advancements in elucidating mechanisms and modes of action at the molecular level have informed decisions regarding the relationship between molecular structure and adverse outcomes. Ideally all of the steps in the pathway of toxic response are identified and are connected to the manifestation of toxicity. This connection is extremely challenging and few mechanisms have been described in detail. Significant progress has been made in the past several decades identifying mechanisms and modes of action. However, there is much work to be done. Chemists are familiar with reaction mechanisms as they pertain to synthetic reactions, but are not often familiar with those same fundamental mechanisms applied to biological situations.[20] Determining individual steps involved with the manifestation of an adverse response provides a starting point for documenting opportunities to control the structure–toxicity relationship which is a critical step in the process for designing safer chemicals.

One example that can be used to illustrate the role of substitution reactions in adverse outcome pathways is the cancer process. Carcinogenesis is a multistage process associated with the induction of neoplasms that lead to a family of disease states commonly termed cancers. Cancer is not a single disease but is made up of multiple conditions that share common traits. Three main steps that comprise carcinogenesis are initiation, promotion and progression. Initiation involves a permanent change in the fundamental nature of a cell. A cell can remain in the initiated state indefinitely until acted upon by a promotor. Mutagens are compounds that act directly on DNA causing mutations.

A promoter is the trigger for an initiated cell to multiply into larger groups of neoplastic cells. Promoters can act in a number of ways, for example acting on oncogenes, killing normal cells that surround initiated cells, inhibiting the action of suppressor genes thereby resulting in the loss of cell cycle control and unchecked cell proliferation. Promotion is thought to follow a dose–response relationship unlike the initiation phase of carcinogenesis. A complete carcinogen acts as both an initiator and a promotor. Progression is the final step in the carcinogenic triad. Cells that have reached this stage tend to demonstrate the familiar attributes of malignant tumours, specifically invasion of nearby tissues, metastasis and loss of differentiation. Many carcinogenic molecules are electrophiles or undergo bioactivation to electrophiles that can bind to cellular nucleophiles like DNA, proteins causing myriad adverse effects including covalent binding to DNA, disruption of critical enzyme pathways and destruction of structural cellular components.[21]

9.2.8 Quantitative Structure–Activity Relationships (QSAR)

The relationship between chemical structure and biological activity has been a source of investigation since the late 19th century. Myer and Overton established the first relationship between lipid solubility of a chemical and narcosis in tadpoles. More quantitative treatment of structure–activity relationships using n-octanol as the preferred lipophilic solvent was done by Hansch and others.[22,23] These studies established the octanol–water partition coefficient, K_{ow}, as the standard for characterizing lipophilicity in biological systems.[24] QSAR models as they relate to adverse biological outcomes can be used for establishing quantitative relationships between structure and activity or structure and property, to predict the potential activity for compounds of unknown toxicity and for designing safer chemicals. QSARs can fit seamlessly into discussions of structure–property relationships, for example when discussing the Hammett or Taft equations.

Molecular descriptors are any parameter used in the development of either a SAR or QSAR to model any type of property under investigation (*e.g.*, molecular structure, $\log P$ with some type of biological attribute or response (*e.g.*, toxicity, LC_{50}, carcinogenicity, *etc*.). A plethora of potential property response combinations exist and many have been developed with a high degree of repeatability.[25] Some of the more common properties are listed in Table 9.1.

The partition coefficient between water and n-octanol occupies a central role for predicting the behaviour of non-polar organic molecules and is the most

Table 9.1 Examples of molecular descriptors used in QSARs.

Type of parameter	Descriptor
Physicochemical	Lipophilicity (as $\log P$)
	Water solubility
	Henry's law constant
	Rates of reaction
Topological	Molecular connectivity
	Molecular volume
	Structural fragments
Electronic	Hammett constants
	pK_a
	Dipole moment
	Frontier molecular orbitals
Steric	Taft constants
	Molecular weight
	Molecular volume and surface area

important of the molecular descriptors used to date. In a biphasic system containing immiscible liquids, a partition coefficient can be determined by measuring the chemical under investigation in each of the two phases at equilibrium.

When the organic phase is *n*-octanol, the partition coefficient is represented by the K_{ow}, or the octanol–water partition coefficient. The relationship is parabolic when the log of the inverse of the concentration is plotted against the log of the partition coefficient and reflects the influence that the degree of lipophilicity has on the movement of compounds across biological membranes. At low lipid solubilities, compounds do not pass as readily through membranes as do compounds that are more lipid soluble.

The partition coefficient values can span five or six orders of magnitude, therefore the log base 10 values are often used. Most xenobiotics that have a $\log K_{ow}$ between 2 and 6 will cross membranes easily and effectively. Above values of $\log K_{ow} = 6$, the compounds are said to be 'super-lipophilic' and their passage across membranes diminishes because the compounds tend to dissolve in the membrane lipids very strongly and never move or the time to equilibrium is too great to be measured on biologically important time frames.

QSARs have been used to develop mathematical relationships between structure and biological effect using regression analysis. An early regression equation was derived using data generated by Overton for the ability of several alcohols to induce anaesthesia in tadpoles. The regression analysis produced the QSAR model is $\log(1/C) = 0.909 \log P + 0.727$, where C is the concentration of the chemical needed to produce anaesthesia in tadpoles and $\log P$ is the log of the partition coefficient. One must understand the underlying mechanisms and modes of action as completely as possible as well as all other factors associated with modifying toxicity for the chemical under study. Modifying factors among and between species include anatomical and physiological differences, referred to as interspecies and then there are the intra-species differences accounting for variability in the populations associated with genetic polymorphism, age, gender and disease state.

9.2.9 Steric Hinderance and Radical Stability: Toxicity of Nitriles

Designing safer nitriles provides an example of how an understanding of the mechanistic nature and QSAR of a class of toxic chemicals can lead to a statement of design rules that inform the design of safer chemicals. DeVito and co-workers have developed a mechanistic-based model for predicting the acute toxicity of nitriles based on the rate of hydrogen atom abstraction by CYP 450 and hydrogen radical stability as the primary variables by using a combination and mechanistic QSAR approach.[26,27] A comprehensive evaluation of this type for the numerous other substances that have an elucidated mechanism of action can be tailored to demonstrate the relationship between a specific molecular property and an adverse outcome. Of course this same approach can be used to demonstrate beneficial outcomes as well.

Nitriles are a class of chemicals widely used for a variety of applications including as a solvent, in medicines and in other industrial application. Nitriles occur naturally in both plants and animals and are also synthesized. Their ubiquitous nature and volume of use mean that the number of individual potentially exposed to nitriles is significant, therefore evaluating and reducing the risk associated with exposure to this class of compounds is warranted. All nitriles contain the cyano functional group (CN). The toxicity of nitriles is similar to the toxicity of cyanide intoxication implying that that the cyanide moiety from the molecule is the ultimate toxicant.

An evaluation of the toxicity, as measured by the LD_{50}, compared with the structural characteristics of selected nitriles revealed that the critical mechanistic step is the rate of α-hydrogen abstraction. A higher rate led to a greater acute toxicity. From this evaluation, structural modifications for reduced hazard, or design rules, were derived. Among the molecular attributes that are associated with lower acute toxicity were: (1) steric hindrance around the α-hydrogen to restrict cytochrome P450 enzyme access, (2) add groups that reduce the stability of the α radical, and (3) avoid hetero-containing groups on the α carbon (Scheme 9.1).

9.2.10   Environmental Toxicology

The principles of environmental chemodynamics and environmental or ecological toxicology are another area of investigation that can be incorporated into existing chemistry curricula to show how fundamental chemical principles are applied to evaluate the behaviour and toxicology of chemicals in the environment. Specialized structure (*i.e.*, morphology), specialized biochemical mechanisms, persistence and bioaccumulation can all modify toxicity. Insects have a protective outer layer called chitin made of polysaccharides that provides a barrier to penetration by most organic compounds. Aquatic toxicology has developed sufficiently over the past several decades to emerge as a separate, focused and independent discipline examining the adverse effects of xenobiotics on aquatic organisms taking into consideration their unique exposure conditions and morphology.

Highly persistent and bioaccumulative chemicals are generally recognized as environmental hazards prompting national and international regulatory agencies to develop standards and guidelines to characterize these chemical attributes.[28] Persistent chemicals resist natural environmental breakdown and may be present in the environment for many decades. Persistence

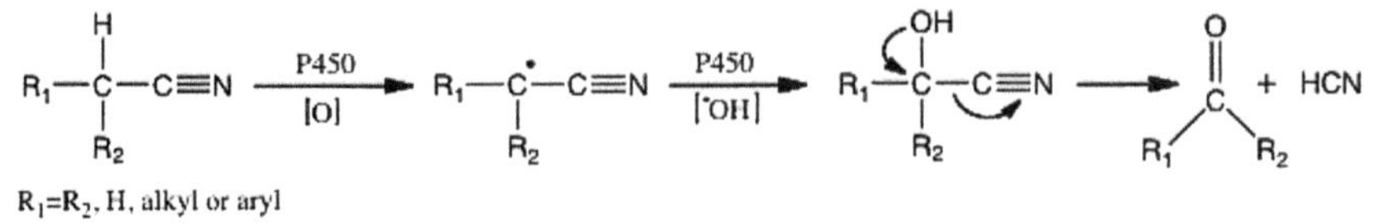

$$R_1 \text{—}\underset{\underset{R_2}{|}}{\overset{\overset{H}{|}}{C}}\text{—}C\equiv N \xrightarrow[\text{[O]}]{P450} R_1\text{—}\underset{\underset{R_2}{|}}{\overset{\bullet}{C}}\text{—}C\equiv N \xrightarrow[\text{['OH]}]{P450} R_1\text{—}\underset{\underset{R_2}{|}}{\overset{\overset{OH}{|}}{C}}\text{—}C\equiv N \longrightarrow \underset{R_1}{\overset{O}{\underset{}{\|}}}\underset{R_2}{C} + HCN$$

$R_1 = R_2$, H, alkyl or aryl

Scheme 9.1   The mechanism of cyanide release from nitriles.

is determined by evaluating chemical half-lives in various environmental media. The inherent stability of a molecule to the available degradation pathways in each environmental medium (*i.e.*, air, water, soil/sediment, organisms) will determine its half-life and consequently its persistence and bioaccumulation potential.

Chemicals that resist degradation in the environment because they are slow to transform are termed recalcitrant or refractory pollutants. Persistent is a relative distinction dictated by the choice of measurement parameters used to determine degradation under standardized conditions, thus persistence is established by the circumstances of use. Environmental persistence is therefore operationally defined, often guided by regulatory requirements by comparing the media-specific half-lives of the chemical in question with a series of predetermined measurement criteria. Chemical transformations can occur biotically (*e.g.*, enzymatic metabolism), abiotically (*e.g.*, hydrolysis, photolysis, oxidation) or through a combination of both pathways. An approach to designing chemicals has been developed that outline structural features that will favour efficient degradation in the environmental to innocuous substances.

Chemicals can be designed to degrade quickly to innocuous substances in the environment. Chemicals that resist environmental degradation persist in the environment resulting in longer exposure time. Designing chemicals for increased rates of biodegradability to innocuous products will reduce toxicity by limiting exposure times for hazardous agents and ideally build in functionality that promotes metabolism to less hazardous molecules. Incorporating molecular chemical characteristics that promote biodegradation is a prudent risk reduction strategy. The challenge is to present these principles in context.

The influence of structure on biodegradability has been investigated recently and certain generalizations for designing biodegradable molecules have been proposed.[29] Certain molecular properties can be correlated to the rate of biodegradability just as with the other types of environmental behaviour. These attributes include molecular weight, whether a molecule is branched or is a straight chain, the position and identity of substituents in a compound, structural group stability and lipophilicity. Examples are provided in Table 9.2 of molecular features that either increase or decrease biodegradation rates.

Table 9.2 Molecular features influencing biodegradation.

Increase biodegradation rates	Decrease biodegradation rates
Presence of straight chains	Chain branching
Ester linkages and other functional groups susceptible to hydrolysis	Presence of halogens or other electron with-drawing groups
The presence of oxygen atoms	Presence of heterocycles
Molecular weights greater than 1000 g mol^{-1}	Greater number of rings in polyaromatic hydrocarbons

9.3 Conclusions

The importance of integrating toxicology into the chemistry curriculum cannot be understated. Incorporating the principles of toxicology into the chemistry curriculum is an essential paradigm shift in the training of the next generation of scientists if we are dedicated to provide scientists that are trained in a trans-disciplinary manner. The desire for designing safer chemicals has been articulated for a number of years and has been incorporated into pharmaceutical and industrial chemistry research strategies.[30] The first step in making this change to the fundamental education of chemists is to demonstrate to the current chemistry educators the relationships between molecular structure, functionality and adverse biological (*i.e.*, toxicological) outcomes using examples that they currently recognize. Chemists are familiar with the properties that are required for functionality, for example in the preparation of dyes, solvents, surfactants, pharmaceuticals and their commercially important products. These chemists may not be as familiar with the process of evaluating the structure–hazard relationship for potential toxicity.

The line between toxicology and chemistry is becoming less apparent, which can be attributed in no small part to the advancements in computational chemistry and an ever increasing understanding of the mechanisms of toxicity. This intersection provides an opportunity to advance the discussion of computational chemistry and mechanistic toxicology using a common language to inform safer chemistry.

Most organic chemical reactions involve breaking and forming bonds between electron deficient centers (*i.e.*, electrophiles) and electron rich centers (*i.e.*, nucleophiles). Thousands of specific examples of substitution reaction exist providing detailed mechanisms (see Carey and Sunberg[20]). The challenge for the chemical education community is to find examples of these reaction mechanisms in the toxicology literature where the reaction mechanism involves an adverse outcome pathway. Many examples currently exist, however, not as abundantly nor as organized as in synthetic organic reactions.

This chapter has only scratched the surface of the opportunities and possibilities to demonstrate the natural relationship between chemistry and toxicology. Some of the topics that were not discussed include oxidation and oxidative stress, the influence of electrolytes on calcium homeostasis, immunotoxicology, focusing on biochemistry and applying frontier molecular orbital theory to predicting adverse outcomes. A discussion of catalysis may include the role of enzymes in toxication and detoxication, for example the mechanisms of action of organophosphate pesticides on the function of acetylcholinesterase.

The exciting journey has just begun and will undoubtedly grow steadily as the current knowledge gap between toxicologists and chemists narrows through the enlightened and indefatigable efforts of dedicated chemistry educators.

References

1. P. T. Anastas and J. C. Warner, *Green Chemistry Theory and Practice*, Oxford Press, New York, 1998, p. 135.
2. American Chemical Society (ACS), *Undergraduate Professional Education in Chemistry, ACS Evaluation Procedures for Bachelor's Degree Programs*, American Chemical Society Committee on Professional Training, Washington, DC, 2008.
3. N. Anastas, *Green Techniques for Organic Synthesis and Medicinal Chemistry*, ed. Wei Zhang and Berkeley Cue, John Wiley, New York, 2012, vol. 1.
4. J. F. Borzelleca, Paracelsus. *Toxicol. Sci.*, 2000, **53**, 2.
5. C. D. Klaassen, *Cassarett and Doull's Toxicology: The Basic Science of Poisons*, ed. Curtis D. Klaassen, McGraw-Hill, New York, 2008, vol. 7, p. 1280.
6. A. W. Hayes, in *Principles and Methods of Toxicology*, ed. A. Wallace Hayes, CRC Press, New York, 2007, vol. 5, pp. 2296.
7. G. Rand, in *Fundamentals of Aquatic Toxicology: Effects, Environmental Fate and Risk Assessment*, ed. G. Rand, Taylor and Francis, 1996, vol. 2, p. 1148.
8. E. A. Hodgson, *A Textbook of Modern Toxicology*, John Wiley and Sons, Hoboken, NJ, 4th edn, 2010.
9. R. Krieger, in *Handbook of Pesticide Toxicology: Principles*, ed. R. I. Krieger and W. C. Krieger, Academic Press, San Diego, CA, 2001, vol. 2, p. 1908.
10. N. D. Anastas and J. C. Warner, Linking hazard reduction to molecular design: teaching green chemistry, in *Green Chemistry Education: Changing the Course of Chemistry*, ed. P. T. Anastas, I. Levy and K. E. Parent, ACS Symposium Series, 2009, vol. 1011, p. 117.
11. A. L. Goldstein, L. Aronow and S. M. Kalman, *Principles of Drug Action, The Basis of Pharmacology*, John Wiley, New York, 1974, vol. 1.
12. F. P. Guengerich and J. S. Macdonald, *AAAP J.*, 2007, **8**(1).
13. A. Parkinson and B. W. Ogilvie, in *Cassarett and Doull's Toxicology: The Basic Science of Poisons*, ed. Curtis D. Klaassen, McGraw-Hill, New York, 2008, vol. 7, p. 3.
14. USEPA, *Guidelines for Developmental Risk Assessment*, US Environmental Protection Agency, Risk Assessment Forum, Washington DC, EPA/600/FR-91/001, 1991.
15. T. D. Stephans, C. J. W. Bunde and B. J. Filmore, Mechanism of Action of Thalidomide Teratogenesis. *Biochem. Pharmacol.*, 2000, **59**, 1489–1499.
16. H. A. Fine, W. D. Figg, K. Jaeckle, *et al.*, *J. Clin. Oncol.*, 2000, **18**, 708.
17. M. B. Smith and J. March, in *March's Advanced Organic Chemistry: Reactions, Mechanisms and Structure*, ed. M.B. Smith, Wiley, New York, 5th edn, 2001.
18. U. A. Boelsteri, *Mechanistic Toxicology: The Molecular Basis of How Chemicals Disrupt Biological Targets*, CRC Press Boca Raton, FL, 2nd edn, 2007.
19. USEPA (U.S. Environmental Protection Agency), *United States Environmental Protection Agency. Guidelines for Carcinogen Risk Rssessment. Risk Assessment Forum. SAB Review Draft*, U.S. Environmental Protection Agency, Washington, D.C., 1999.

20. F. A. Carey and R. J. Sundberg, *Advanced Organic Chemistry: Part B: Reactions and Synthesis*, Springer, 5th edn, 2007.
21. G. Zoltan, in *Cassarett and Doull's Toxicology: The Basic Science of Poisons*, ed. Curtis D. Klassen, Mc Graw-Hill, New York, 7th edn, 2008, p. 3.
22. C. Hansch, P. P. Maloney, T. Fujita and Q. M. Muir, *Nature*, 1962, **194**, 178.
23. C. Hansch and T. Fujita, Rho-sigma-pi analysis. *J. Am. Chem. Soc.*, 1964, 16161.
24. C. A. Hansch, R. Steward, S. M. Anderson and D. Bentley, *J. Med. Chem.*, 1968, **11**.
25. C. D. Selassie, S. B. Mekapati and R. P. Verma, *Curr. Top. Med. Chem.*, 2002, **2**, 1357.
26. S. C. DeVito, in *Designing Safer Chemicals,* ed. S. C. DeVito and R. Garrett, American Chemical Society, Washington DC, vol. 640, 1996, pp. 194.
27. J. Grogan, S. C. Devito, R. S. Pearlman and K. R. Korzekwa, *Chem. Res. Toxicol.*, 1992, **5**, 548.
28. U. Nations, *Globally Harmonized System of Classification and Labelling of Chemicals (GHS)*, United Nations, New York and Geneva, 2nd edn, 2007.
29. R. S. Boethling, E. Sommer and D. D. Fiore, *Chem. Rev.*, 2007, **107**, 2207.
30. W. G. Flamm, *Science*, 1994, September 9, **265**(1678), 1519.

CHAPTER 10

Green Chemistry and Sustainable Industrial Technology – Over 10 Years of an MSc Programme

JAMES CLARK*[a], LEONIE JONES[a], AND LOUISE SUMMERTON[a]

[a]Green Chemistry Centre of Excellence, Department of Chemistry, University of York, Heslington, York, YO10 5DD, UK
*E-mail: james.clark@york.ac.uk

10.1 Introduction

In 2001, the Master's degree course in green chemistry at the University of York was established by Prof. James Clark and colleagues, funded by a large grant from the UK research councils and generous support from two major companies. The course was very much part of the establishment of green chemistry activities at the university. Following on from the creation of the Green Chemistry Network[1] and the journal *Green Chemistry*,[2] staff at the University of York saw an opportunity to initiate a Master's course in the field of green chemistry. Originally entitled the 'MRes in Clean Chemical Technology', the course was renamed in 2007 as the 'MSc in Green Chemistry and Sustainable Industrial technology' to greater reflect the significant input from industry and applied chemistry. First of its kind in Europe, the course

Worldwide Trends in Green Chemistry Education
Edited by Vânia Gomes Zuin and Liliana Mammino
© The Royal Society of Chemistry 2015
Published by the Royal Society of Chemistry, www.rsc.org

has been a benchmark for a number of graduate level courses internationally. Green chemistry programmes are now available around the world, for example in Spain, Greece, Australia, Brazil and the USA.[3]

This reflects the growing recognition of the need to promote uptake of green and sustainable methodologies amongst the next generation of scientists and equip them with the requisite tools, knowledge and experience, in order to achieve a step-change towards sustainability in the chemical and chemical-using industries.

The course has evolved over at least the last decade to become self-sufficient and incorporate changes to accommodate new developments such as new legislation and regulatory pressures; changing perceptions of chemical stakeholders from retailers and consumers to government and NGOs, and new technical and socio-economic challenges. The course has also adjusted in synch with changing student needs, such as the desire to prepare students for a diverse range of careers and the increasing international dimension of the student intake.

Herein we describe the innovative composition of the course and the significant adaptations that have been made since its inception that reflect its ability to 'move with the times' and keep up to date with the requirements of not only the students themselves, but also those of chemical-dependant industries which rely upon new graduates with the requisite skills to make a positive impact to their business in sustainability terms, giving them a competitive edge over their rivals.

10.2   Course Content

The twelve-month programme consists of a blend of taught material and research. The knowledge, understanding and skills are delivered and assessed *via* a range of different methods (see Figure 10.1). In broad terms, the course covers the principles of green chemistry, its application and commercialization and additional transferrable skills. One of the most noteworthy advancements in the chemical sciences of late has without a doubt been the emergence of green chemistry and is recognized worldwide to describe the development of more sustainable chemical products and processes.[4] Despite the moniker, green chemistry is much more than a field of chemistry; rather it is a multi-disciplinary subject that brings together a combination of expertise including chemists, biologists, environmental scientists, engineers, economists and legislators, amongst others. In the past, green chemistry courses tended to have been heavily chemistry biased, but York among others, have broadened their syllabuses to encompass these other aspects. This is reflected not only in the content of the course at York but also in the expertise of its tutors, and is highly favoured by the students on the course:

> The course covered a really broad range of subjects from organic chemistry and chemical engineering through to intellectual property and how to set up a business. It was great to have that overview of lots of different areas that are really important but aren't given much attention in most science degrees. (Will Soutter, 2010–2011 MSc cohort)[5]

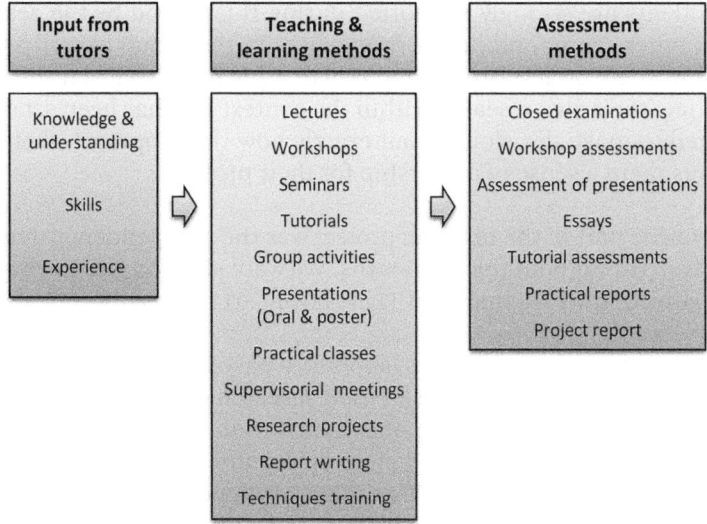

Figure 10.1 Blend of teaching, learning and assessment methods utilized on the Master's course.

10.3   Research Projects

Approximately half of the course (just over 55% of total marks awarded) is dedicated to the *Independent Study Module* which consists primarily of a research project. This is of fundamental value to the students as it allows them to apply knowledge from the taught part of the course, in an area of specific interest to them. It also has the benefit of exposing them to a wide range of techniques, developing their investigative skills and providing direct experience of working in a research environment. Past projects have covered diverse areas relevant to green chemistry and sustainable industrial technology, examples of which are listed below:

- *From Food Waste to Bio-fuels and Beyond*
- *Antioxidant Properties of Phenolic-rich Extract from Microwaved Biomass*
- *Utilization of Waste Fatty Acids for Developing Hydrophobic Surfaces*
- *Supercritical Extraction and Fractionation of Renewable Feedstocks*
- *Catalysis for the Formation of Amide Bonds*
- *Bio-derived Platform Molecules*
- *Green Oxidation of Alcohols in Water*
- *Biocatalytic Routes to Esters in Supercritical Carbon Dioxide*
- *Starbons® as Adsorbents for Water Purification*
- *Generation of High Energy Chars from Biomass Utilizing Microwaves*
- *The Recovery of Pharmaceuticals from Waste Streams*
- *From Ash to Bio-boards*
- *Adhesion Promoters for Water-based Links*
- *Development of PVC Replacements*
- *Novel Bio-derived Lubricant Additives.*

The students are strongly supported during this period by the academic staff, and other senior and junior researchers in the group as well as other members of their project area group (PAG). They are also given the autonomy to direct their own research within the context that has been set and are encouraged to make decisions about exactly how they approach their work, which gives them a sense of ownership for their project:

> My favourite part of the research project was the independence that we were given throughout the six months. We were able to take the project in any direction we wanted. (Ray Hale, 2010–2011 MSc cohort).[5]

Research projects are not always laboratory based; several students in the past have conducted in depth desk-based research projects, for example in the areas of food waste mapping and valorization; greener substitutes for brominated flame retardants and new environmental legislation and its effects on the chemical industry. This offers students greater flexibility, and tailors their project to their own individual skills and interests.

Research projects are frequently undertaken in conjunction with industrial partners or other relevant organizations, from a wide range of sectors including chemical manufacturing, pharmaceutical, engineering, aerospace, retail, food, home and personal care, oil and polymers. This allows the students a unique chance to be involved in solving real-life industrial challenges, some of whom have been successful enough to go on to demonstrate their project at pilot-scale and beyond with their industrial partner. It also helps to prepare graduates of the course who go on to seek employment in industry by giving them an insight into and appreciation of the interests, priorities and constraints of a business. Collaborative projects can also open the door to direct employment opportunities for graduates, as has been the case for some of our previous students. For many students the opportunity to carry out original research is very rewarding:

> What was quite surprising was that I wasn't just applying the knowledge that I learnt from the first half of the course, I was really learning more all the time throughout the research project. By the end of it, I really felt like I'd built up a deep understanding of the area I was working in'. (Will Soutter, 2010–2011 MSc cohort).[5]

10.4   Course Delivery

10.4.1   Overview

The course is delivered by academic experts in the field, from York and other universities, in collaboration with a wide range of companies and organizations involved with chemical use, manufacture, management and policy. These make up the pool of course tutors with diverse but complementary expertise and experience. One of the key aspects that ensures the course is up to date and demonstrates the applicability and relevance of green chemistry, is the well-established contribution from industry and other externals in terms of provision of course content and current case studies. The students are also given the opportunity for site visits to relevant workplaces:

> The opportunity to hear from those in industry about the subject is really important, it can offer more real world perspectives. (Angus Swinscoe, Boots, external tutor).

This strength is recognized by the students taking the course:

> What was really great about the course was that a lot of the material was delivered by professionals from industries that are actually working in the subject they teach. (Sasha Borisova, 2010–2011 MSc cohort).[5]

In addition to industrial collaboration, the course also seeks to give the students an appreciation of technology transfer and intellectual property protection to further demonstrate the business case for the application of green chemistry. This can also assist students with making decisions about their future career paths:

> The industrial focused topics of the MSc, spanning over a number of areas, provided a good stepping stone between an undergraduate degree and a job in industry, helping me decide what area I wanted to work in. (Claire Saville, 2009–2010 MSc cohort).[6]

10.4.2   Perspectives of a Course Tutor

Two of the staff involved in the delivery of the course as course tutors, Dr Andrew Hunt and Dr James Comerford, are also former students of the course and hence both are in a rare position to have an in-depth understanding of the course from perspectives. In the course of writing this chapter, we interviewed both of them regarding their experience of the course.

10.4.2.1   Dr Andrew Hunt

Dr Andrew Hunt was part of the first cohort of students on the course in 2001, and went on to complete his PhD in the group on the extraction of high-value chemicals from British upland plants. As well as being heavily involved in teaching the MSc, he is also the Scientific Leader of the Alternative Solvents Technology Platform at the Green Chemistry Centre of Excellence (GCCE) at York.[7]

What made you do the Master's course?
I did a BSc in Chemistry and Computer Science as an undergraduate (Swansea), then I decided that I really wanted to focus on chemistry, do a PhD and go into research. The Master's course was really exciting, innovative

and at that time there were no other courses of its type in the world. I believe green issues are key to the long term success of both industry and also the world as a whole. So I thought it was really important to not just do chemistry but to do green chemistry. I very much wanted to go into research and I saw the MSc as a stepping-stone to a career in research.

To what extent has gaining this qualification impacted on your career path?
Getting the MRes has really impacted on my career massively. I wanted to do research and the course allowed me to take the next step. I'm now in the position that I am, leading research and helping guide other people to do research simply because I was given the opportunity to do the course.

What benefit do you see from having a multi-disciplinary cohort?
Having a wide variety of both the disciplines and nationalities of students means that the course is really interesting; everybody brings something different to the table and students gain a lot from interacting with their cohort.

How do you think the course has change over the years?
Since the early days green chemistry has really developed as a subject and the breadth of the course has increased massively to reflect that. For example, the movement towards renewable materials now has a much bigger emphasis.

In recent years the GCCE have introduced the Biorenewables Development Centre (BDC) in collaboration with the Centre for Novel Agricultural Products (CNAP) in the biology department which has allowed us to think about scaling up processes for things like supercritical fluids and microwave process. Students have gained a hands-on approach in terms of working on a lab scale and then scaling up to a demonstrator scale so we are able to give insight into that.

What does the future hold for the course?
The course is still very vibrant and exciting; I believe that international collaborations will become an important part of the course.

10.4.2.2   Dr James Comerford

Dr James Comerford was part of the 2005–2006 MSc cohort and was recently employed by the GCCE to act as MSc tutor. His role involves acting as the first point of contact for students on the Green Chemistry and Sustainable Industrial Technology MSc course, as well as mentoring and meeting with the MSc cohort on a regular basis.

James is also an active postdoctoral researcher. His research interests centre on clean synthesis with focus on asymmetric induction in enantioselective catalysis, utilization of CO and CO_2 as a sustainable source of chemicals, heterogeneous catalysis and novel reaction technologies including cold plasma reactors and continuous flow synthesis.

What are your memories of the MSc courses as a student? What was the highlight of the course for you?

Personally, I found the opportunity to research a real industrial problem a highpoint. Few MSc courses offer such long project periods with freedom to explore chemistry in a top research laboratory and plenty of academic support. Collaboration with academics and colleagues gave me the sense of inclusion within the Green Chemistry Group and made the MSc a very enjoyable experience. The diversity of external lecturers teaching on the course allowed exposure to different points of view and broadened of understanding in Green Chemistry and its application in industry, policy making and University research.

To what extent has gaining this qualification impacted on your career path?

The MSc course prepared me thoroughly for a PhD. Having already experienced the process of researching chemical problems, I found that I was capable of making an impact in PhD research early on. The research performed in the MSc allowed me to develop core skills in project management, written and oral communication, as well as critical thinking and problem solving. Having been awarded a PhD, this allowed me to pursue other career paths such as my part qualification as a patent attorney.

What makes the MSc course at York different from other green chemistry courses?

In contrast to typical MSc courses, students are able to put taught knowledge directly into practice by solving industrial research problems during an extensive 6 + month project period. They are able to develop a wide range of new practical techniques and skills, *i.e.*, use of energy–efficient reaction technologies including microwave, continuous flow and sonication reactors along with supercritical fluid reactors/extractors. Students gain the ability to consider alternatives to non-sustainable chemicals and 'batch style' chemistry, realizing first-hand how such advances in clean chemistry can significantly reduce risk in the lab. The wide range of research activities performed at the GCCE allows exposure to many different areas, such as catalyst design and clean synthesis development, renewable and sustainable chemical feedstocks, greener products, (including major work on designing and using greener solvents) and waste valorization.

Team work is fundamental and we have developed a system where all students work in teams (technology platforms and project area groups) especially in the laboratory where an interest in each other's projects is encouraged.

What benefits do you see there are to having a multi-disciplinary student intake to the course?

A multi-disciplinary intake allows different expertise and views to be expressed on the course. Ideally, this has the effect of broadening the scope of green chemistry principles which can be practised in a number of different sectors, making more of an impact.

10.4.3 Views from External Contributors to the Course

One of the great strengths of the course is the significant involvement of external contributors. We interviewed a cross-section of the external tutors to seek their perspectives on the course at York, the benefits they gain from contributing to the course, and their views on essential skills for future green chemistry graduates.

10.4.3.1 Fiona Dickson

Fiona Dickson is a solicitor and partner in Maynards, a legal practice that is authorized and regulated by the Solicitors' Regulation Authority. She is also director of an environment and sustainability consultancy, Légiste Ltd. She contributes to the *Commercialisation of Green Chemistry* module topic 'Intellectual Property and Environmental Legislation'.

What do you think is distinctive about the MSc course at York?

- The integration of green chemistry teaching with business, law, policy and professional skills training.
- Raising awareness of the environment and sustainability imperatives driving the evolution and application of green chemistry principles and practices.
- The strong emphasis on communication skills and team working.

I wish such a course had been available at the time I studied for my first degree!

What impact do you feel the contribution from external tutors has on the course?

External tutors can play a significant role in helping students to make a transition from an academic environment to careers in business, research, regulation *etc*. From reading past student profiles, students have clearly been encouraged to apply their skills and qualifications to a broader range of careers (including law and regulation), and to embrace a policy context in their post-graduate research.

External tutors can infuse their teaching with their practical experience to ensure principles and practices are presented to students in a 'real-world' context. We also endeavour to introduce students to some of the practical problems and resistance they can expect to encounter in careers when seeking to introduce new ideas, ways of working *etc*. and help them to learn how they can approach dealing with those situations.

What is your motivation for contributing to the MSc course at York?

I thoroughly enjoy sharing my experience and enthusiasm for the subject with others and hope it can make a difference to students' educational experiences and to their chosen careers. I particularly enjoy teaching students from around the world as it also encourages me to research materials from other jurisdictions in order to present an international and comparative perspective.

What, in your opinion, will be the most important skills for green chemistry graduates to be equipped with in the future?

- Communication skills because, as some of my coursework seeks to illustrate, some of the most pressing environmental problems have arisen from dislocations between scientists, policy-makers and the public.
- Resource efficiency will be an increasing global challenge and skills relevant to understanding and solving problems of that nature will be of growing importance.
- Skills that can contribute towards developing and implementing new technologies, policies, *etc.*, relevant to climate change mitigation, but perhaps more likely, adaptation.

10.4.3.2   Dr Paul Ravenscroft

Dr Paul Ravenscroft is the retired former Director of Synthetic Chemistry at GlaxoSmithKline plc. He contributes to the *Clean Synthesis* module regarding the selection of alternative routes.

What do you think is distinctive about the MSc course at York?

The fact that it gives a good opportunity for students at postgraduate level to actually work and interact with staff from industry. Within the MSc course we have staff from various industries supervising students and often an academic supervisor is involved too. The tripartite arrangement is great for feeding ideas off each other. The bottom line is industry using the output for the benefit of their commercial processes. Very little new chemistry is invented in industry so we have to work together for the benefit of the business as a whole.

What impact do you feel the contribution from external tutors has on the course?

I hope that the external tutors bring an attitude of realism to the course. Students typically see decision-making as simply 'black or white' when in reality most issues fall into the grey zone. For example, they often perceive 'green' chemistry as always good and anything else as bad. Life is not that easy! Like it or not, in a capitalistic society, economics usually rule and it is important that the students develop an understanding of the types of issues involved and get a grasp of why, in their mind, industry often seems to take bad decisions. The external tutors can also help bridge the academia–industry divide. They can bring an industrial perspective to the course and thus help students (and academics) to make their work more meaningful to business.

Conversely, the academics can bring a bit of more 'blue-sky' thinking to the table and help industry to think more widely about the green credentials of their respective businesses. When it works well it can be highly synergistic.

What is your motivation for contributing to the MSc course at York?

A desire to use the skills I have learnt and developed over the years for the benefit and education of students and thus make them better prepared for the working world, particularly in terms of bringing more efficient green chemistry to the chemical industry in all its various guises. I also very much enjoy working with young people and seeing the challenges of the world they live in.

What, in your opinion, will be the most important skills for green chemistry graduates to be equipped with in the future?

This may sound rather obvious but it is crucial that green chemistry graduates have a good understanding of all aspects of theoretical chemistry. This is a basic requirement and not a 'nice to have'. Although these days it is often seen as trendy to work in the 'green' area of science, if they do not have the necessary chemistry skills then they will not be able to have meaningful discussions or make sensible decisions regarding their own work and its relevance in the wider industry.

10.4.3.3   Dr Kamelia Boodhoo

Dr Kamelia Boodhoo is a senior lecturer in the School of Chemical Engineering and Advanced Materials at Newcastle University. She contributes to the *Application of Green Chemistry* module topic 'Chemical Engineering and Advanced Materials'.

What do you think is distinctive about the MSc course at York?

The green chemistry and sustainable technology focus of this MSc course provides a unique opportunity to enable a number of distinct areas to be covered, giving the students a broad exposure to a range of technologies. The inherently multi- and interdisciplinary nature that such a course opens itself up to can only bring benefits in shaping up graduates of this programme.

What impact do you feel the contribution from external tutors has on the course?

It is important for students to be taught by people who are research active in any given technology or scientific area; this is the only way for students to access up-to-date material. It is difficult to do this without getting external lecturers on board, as no single university department can claim to have all the required expertise in-house.

What is your motivation for contributing to the MSc course at York?

To share knowledge of my area of research in process intensification with as wide an audience as possible. The course gives me access to students with a science background rather than an engineering background and it is important to inform students with such academic backgrounds of engineering developments related to green chemistry and sustainability. In this way, when students graduate and hopefully get a job in a relevant industrial sector, they can spread their knowledge of this technology around. Process intensification is an area which would benefit from widespread dissemination if there is to be greater industrial uptake.

What, in your opinion, will be the most important skills for green chemistry graduates to be equipped with in the future?

- To be able to work and communicate effectively in multi-disciplinary environments alongside engineers to put ideas into practice. Greening of processes cannot be done by chemists, biologists or engineers on their own; it necessarily involves a team effort.
- To be aware of latest developments in technology and be equipped to understand and apply a range of green chemical technologies.

10.4.4   Course Delivery Summary

As can be seen from the interviews above, external tutors are able to bring valuable insights into real-world issues and the practical applications of green chemistry. Research-active tutors from a wide variety of fields play a vital role in keeping material up to date, providing cutting edge research and case studies. In addition to students having a good grounding in basic principles, tutors highlighted the importance of excellent communication skills and the ability to work in a multi-disciplinary environment. As the course has evolved, training in transferrable skills has been incorporated into the programme (see Section 10.6.6).

10.5   Students

10.5.1   Academic Background

The multi-disciplinary nature of the course content subsequently has an influence on the academic backgrounds of students admitted on the course at York, which are by no means limited to those who have taken chemistry and related courses. The variety of qualifications of former students ranges from degrees in environmental chemistry through to pharmaceutical science, biochemical engineering and material science and approx. 25% of students hold degrees in (non-chemistry) areas of specialisation (see Figure 10.2). The distribution of academic backgrounds of the students can be seen in Figure 10.2.

The subject of green chemistry is rarely taught in depth at undergraduate level. The students on the course build upon the skills and knowledge they have developed during their undergraduate degrees in a more applied

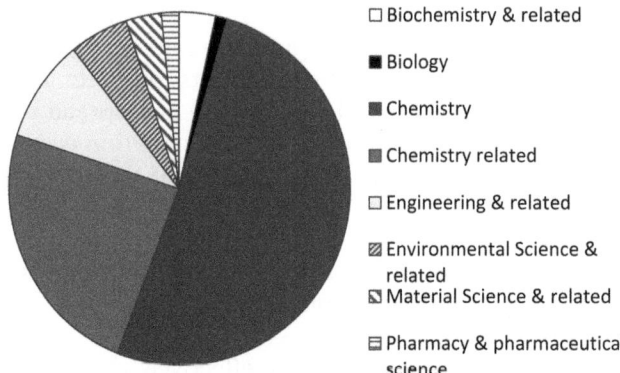

Legend:
□ Biochemistry & related
■ Biology
■ Chemistry
■ Chemistry related
□ Engineering & related
▨ Environmental Science & related
◩ Material Science & related
▤ Pharmacy & pharmaceutical science

Figure 10.2 Academic background of the student intake, by degree subject.

setting, with particular focus on a holistic approach to lowering environmental impact of products and processes. The breadth of course content and multi-disciplinary nature provides the students with the opportunity to change direction in their educational and career paths.

10.5.2 Internationalization of the Student Intake

In the very first year, student numbers were low as the course was completely new and there was nothing similar available around the world. From that time the course has grown considerably and, since 2001, 179 students have graduated from the course, equipped with a technical, scientific and global understanding of sustainability and environmental issues. The early student cohorts were predominately of UK origin, which could in part have been due to the funding opportunities available which were limited to UK and other European Union (EU) students. This began to significantly change from 2008 onwards, when the number of overseas students (from outside the EU) increased dramatically. Figure 10.3 shows how the blend of student nationalities has changed over the years.

This change coincided with the course administrator at the time being in regular contact with applicants to support them by providing advice and information through the entire process, from completing the application form and the interview process, through to applying for visas, organizing accommodation and making arrangements for individual requests for practical needs, cultural differences, administrative and financial reasons. This undoubtedly contributed towards the conversion of offers of a place on the course into acceptances. This pastoral support continued after the students arrived in York.

There has been a broad range of different nationalities of students being accepted onto the course and past students have come from Brunei, China, Cyprus, France, Ireland, Lithuania, Malaysia, Malta, Nigeria, Oman, Spain, Tanzania and Thailand amongst many others. The complete distribution can be seen in Figure 10.4.

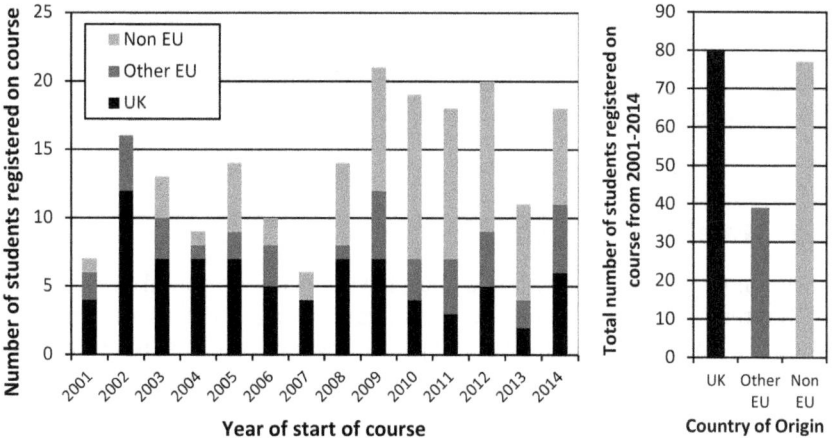

Figure 10.3 Distribution of students' countries of origin by year of registration (left) and total number of student from each region during the period 2001–2014 (right).

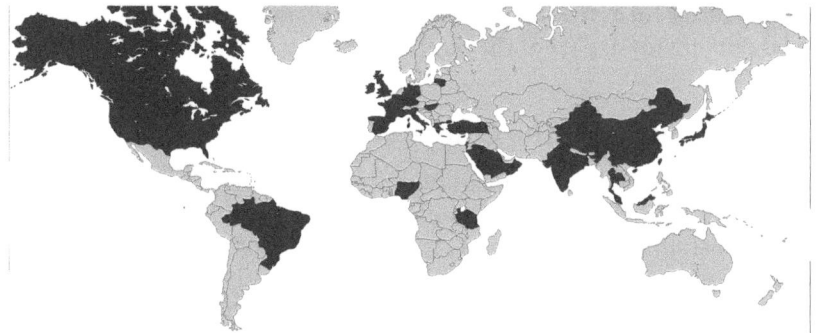

Figure 10.4 Map showing the range of nationalities of the student intake.

Due to the global relevance of the course content, there is an increasingly international blend of students that are eager to convey the skills and knowledge acquired from the course and wherever possible to implement them when returning to their home countries. Particularly in rapidly developing countries that possess a wealth of natural resources and waste that has the potential to be exploited to produce valuable chemicals, materials and fuels, this knowledge could open up a range of new opportunities. Past international students have tailored the content of their research projects to correlate with the situation in their home countries; one student from Oman looked at valorization of a native renewable feedstock and another from Tanzania specifically chose to work on surfactants due to the particular relevance to the industry there. An additional benefit of this increased international dimension is the enhanced cultural experience for the green chemistry group at York as a whole:

The Green Chemistry Centre attracts a lot of people from different countries and we can make friends with people from different cultures. It's a really good opportunity for me to improve my academic (skills) and my English. (Maggie Zhu, 2010–2011 MSc cohort).[5]

10.6 Evolution of the Course

Additional to the modifications to the MSc at York mentioned previously in this chapter, there have been other significant changes successfully made to the content, structure and delivery of the course over the past ten years plus. The drivers for these changes originate from student feedback (*via* questionnaires and student cohort meetings), external examiners annual review of the course and subsequent recommendations, and steering from the External Advisory Board and the course tutors. Some of the more fundamental changes are described below. Other more minor modifications have been trialled over the years, but were not always implemented long term. For example, optional modules/topics were piloted at one point, but these were found not to be successful as feedback from students showed that they did not want to 'miss out' on any aspect of the course.

10.6.1 Renaming of the Course

A particular noteworthy modification to the course was the change of name in 2007 from 'MRes in Clean Chemical Technology' to 'MSc in Green Chemistry and Sustainable Industrial Technology'. One of the main reasons for this change was to include the 'green chemistry' terminology. When the idea for the course was originally conceived, the university was advised not to include the term 'green' due to potential political connotations. However, as the term 'green chemistry' became more accepted, it became apparent that, particularly for marketing reasons, it was critical to make the shift as potential applicants searching for a course in green chemistry would not find the course at York, despite the fact that it was inherent in the course content! Changing from MRes to MSc was also perceived to be important to increase international recognition of the type of qualification.

10.6.2 RSC Accreditation

In 2012 the MSc in Green Chemistry and Sustainable Industrial Technology at York became the first discrete Master's course of its kind to be accredited by the Royal Society of Chemistry. To attain this, the course was subject to a rigorous evaluation process, which is respected both within the UK and internationally. This process covers all aspects of the course, from admissions/student selection criteria onwards.

10.6.3   Funding and Student Bursaries

As stated earlier, the course was originally funded *via* a research council MTP (Masters Training Package) which later became a CTA (Collaborative Training Account) in addition to generous industrial sponsorship. This covered the first five years of the course, after which in order to survive the course had to become financially self-supporting. The course now receives some support from the university proportionate to student numbers recruited, and additional finance must be secured from other sources. This has impacted not only on the course itself, but also on the bursaries that can be made available to prospective students. *Via* the CTA, the university was able to offer either fees and stipend (UK students) or fees only (other EU students) to a limited number of applicants. A few students were provided with a 'top-up' from industrial funding. When CTA funding ceased, attracting home (UK and EU) students became increasingly difficult and a limited amount of funding from the Chemistry Department is now currently available to encourage these students to apply and is awarded on the basis of academic merit and financial need. The GCCE is currently exploring ways to maximize this in future.

In 2012/2013 a number of Brazilian students attended the course *via* the 'Science Without Borders' scholarship programme, a large-scale student mobility initiative set up by the Brazilian government to allow students to study STEM (science, technology, engineering, maths) subjects and creative industries (focussing on technological and innovative development) at leading universities around the world.[8]

10.6.4   Modularization

In 2009 the structure of the course was changed for two main reasons:

- To improve the course in terms of structure and cohesion.
- To enable certificate and diploma options.

The existing course material was repackaged into 20 credit modules and 60 credit blocks, with some modification to the assessment but maintaining the course work to project balance. The students now have the option of either completing the full Master's course (180 credits: 80 credits from taught material, and 100 credits from the research project) or obtaining a Postgraduate Certificate or Postgraduate Diploma qualification (at 60 and 120 credits respectively). This new structure is shown in more detail in Table 10.1 and the modules required to complete the Certificate (C), Diploma (D) and Master's (M) are indicated.

The modularization of the course in 2009 also had the added advantage of simplifying what was a complex programme of individually assessed units, including the replacement of one of the formal examinations with assessed coursework assignments. Continual assessment throughout the

Table 10.1 Structure of the MSc course showing exit points for the post-graduate Certificate (C), Diploma (D) and Masters (M) options plus content and timing of modules (Autumn (A), Spring (Sp) and Summer (Su) terms).

Module	Topics covered	Credit	Teaching methods	Programme C	D	M	Term	Assessment
Principles of Green Chemistry	Introduction to Green Chemistry	20	Lectures & Workshops	■	■	■	A	Workshop
	Control of Environmental Impact		Lectures & Workshops				A	Assignment
	Alternative Reaction Media		Lectures & Workshops				A	Examination
	Catalysis for Green Chemistry		Lectures & Workshops				A	
Application of Green Chemistry	Clean Synthesis	20	Lectures & Practical	■	■	■	A	Workshop
	Renewable Resources		Lectures & Practical				Sp	Practical
	Energy Efficiency & Emerging Technologies		Lectures & Workshops				Sp	Assignment
	Chemical Engineering and Clean Technology		Lectures & Workshops				Sp	Presentation
Transferable Skills	Advanced IT Skills, CV & Interview Techniques	20	Workshops	■	■	■	A	Workshop
	Green Chemistry Presentations		Workshops Seminars				Sp	Presentation
							A	Assignment
	Literature Seminars		Workshops				A	
	Public Awareness		Lectures Workshops Presentation				Sp	

Attainment of 60 Credits – Completion of Certificate or continue to Diploma or MSc

Module	Topics covered	Credit	Teaching methods	Programme C	D	M	Term	Assessment
Commercialisation of Green Chemistry	Greener Products	20	Lectures & Workshops		■	■	Su	Workshop
	Intellectual Property & Impact of Environmental Legislation		Lectures Seminars & Workshops				Su	Presentation
	Commercialisation: Business Plan Development		Workshops				Su	Assignment
Green Chemistry Research Project	Diploma Research Project	40	Research Project		■		A	Report
							Sp	
							Su	

Attainment of 120 Credits – Completion of Diploma

Module	Topics covered	Credit	Teaching methods	Programme C	D	M	Term	Assessment
	Masters Research Project	100	Research Project			■	A	Report
							Sp	Presentation
							Su	

Attainment of 180 Credits – MSc Awarded

course requires the students to work to deadlines and balance their workload accordingly, which is another valuable learning outcome:

> This challenge taught me how to work for sustained periods under high pressure, which has been a useful skill in my new job, as has the ability to juggle and prioritise projects. (Katie Privett, 2011–2012 MSc cohort).[9]

However the change of examination style towards a greater proportion of in-depth, essay-style examination questions in 2009 was modified the following year to ensure ability to demonstrate in-depth knowledge of the subject was not restricted by language skills.

10.6.5 Project Area Groups

The concept of project area groups, or PAGs, was introduced to the course in 2011 in order to provide students with a wider range of experiences during the project period including significant additional opportunity for team working, as well as to accommodate the increasing number of students on the course with regards to laboratory space. As was the case previously, the students work on independent research projects, but are assigned into small groups (most often three per PAG) where their projects are linked *via* a common theme, for example, food waste valorization. The PAGs meet regularly with their PAG leader (a senior researcher within the GCCE with research interests in the project themes) to discuss their progress, as well as working together on a team assignment. Overall the result of this change was to create a more inclusive environment for the students during their research period, providing them with a peer group working in a similar area with whom they could share ideas and issues and provide support for each other, in addition to the support from academic staff.

10.6.6 Transferrable Skills, Including Science Communication

The course tutors at York recognized that the graduates from the course should not only be provided with up-to-date technical information and specific skills related to the subject area of green chemistry, but also that emphasis should be placed on the development of transferrable skills. This led to the incorporation of a dedicated module on these skills, also referred to as 'personal development planning', that complements the more academic side of the course. This module encompasses:

- Green chemistry posters: preparation and presentation
- Literature seminars
- Advanced IT, CV and interview techniques
- Public awareness of science and sustainability

Proficiency in these areas is extremely valuable to the modern scientist, and in fact these skills are of relevance for any future career, in either chemical or non-chemical related employment. All of these activities are highly successful in increasing the students' confidence and improving communication skills. Opportunities for the students to work together are also invaluable for team building and the development of collaborative ideas and solutions to problems:

> This Master's course develops you as a well rounded individual in terms of transferrable skills. (Ray Hale, 2010–2011 MSc cohort).[5]

One aspect of this that has been particularly popular with the students has been the successful incorporation of 'public awareness of science and sustainability' into the course in 2009. Direct communication with the public is indispensable as a means to influence and inspire the next generation of 'green chemists'. Engaging a wider audience in discussion on topical issues and challenges that affect sustainable production and consumption can raise awareness of the role of green chemistry in improving the quality of our everyday lives:[10]

> Participating in this green chemistry outreach activity is a great opportunity for a scientist who loves chemistry and respects the environment. [There is] an opportunity to communicate your enthusiasm and knowledge as well as to inspire people of different ages and interests. (Andri Constandinou, 2011–2012 MSc cohort).

Training and practical experience of designing, planning and taking part in a range of public facing events is provided by a combination of staff at York and external experts. The students work in small groups to come up with a concept for a new green chemistry outreach activity, as well as being given the opportunity to participate in the Green Chemistry Centre's annual agenda of outreach events and assist with the development of new educational material. For example this could involve hosting an exhibitions or workshops at science centres, shopping centres or festivals of science aimed at raising awareness of green chemistry in school children and the general public. Conveying their own enthusiasm for the subject, in particular to the younger generation, is not only beneficial but is also a rewarding experience for the students themselves:

> The activities were challenging to set up and fun to carry out, and taught me a lot on the pragmatic side of dissemination and public engagement. It was also a great opportunity to get involved with the local community, which was a precious experience for me being an international student. (Giulia Paggiola, 2011–2012 MSc cohort).

10.7   Destinations of Graduates

The MSc in Green Chemistry and Sustainable Industrial Technology at York is specifically designed to prepare its graduates for a wide range of future careers. It is rewarding to see our graduates fulfilling a varied set of roles across the world in research, education, environmental services, process development, retail, consultancy, government, finance and law amongst others. This diversity demonstrates the demand for green and sustainable practices in sectors outside traditional chemical and related industries. The wide-ranging opportunities available to graduates of the course are possibly one of its greatest selling points:

> What's really good about this course is that it doesn't limit graduates to one specific area but provides enough skills and knowledge to go and work in lots of different sectors. (Sasha Borisova, 2010–2011 cohort).[5]

For many, the time spent on their project awakens a passion for research and subsequently several of our students choose to go on to study for a PhD before embarking upon their careers.

In 2013, for the first time, a careers event was incorporated into the course, where representatives of the MSc alumni were invited to return to the department to speak about the career paths they took, how green chemistry influenced their choices and what impact it has on their current role. This opportunity to meet with former students and learn from real experiences is invaluable at a time when time when the students are considering the next stage of their career, and it has been decided to continue to hold this symposium annually.

10.8   Graduate-Level Courses in Green Chemistry around the World

To obtain alternative perspectives of running an MSc course in green chemistry we have sought the viewpoint of MSc course coordinators at Imperial University, UK, and the University of Zaragoza, Spain.

10.8.1   MSc in Sustainable Chemistry, University of Zaragoza, Spain

Dr Luis Salvatella is Coordinator of the Master's course in sustainable chemistry at the University of Zaragoza. This Master's course was established in 2006, though the modular organisation of the curriculum was modified in 2008. Typical student intake is about 10 students per year. The course units comprise of:

- *Biotransformations and Advances Processes in the Industry* (6 ECTS)
- *Catalysis* (6 ECTS)
- *Design and Control of Processes* (6 ECTS)
- *Non-conventional Solvents and Reaction Methods* (9 ECTS)
- *Fundamentals of Sustainability and Sustainable Chemistry* (6 ECTS)
- *Environmental Legislation and Toxicology* (6 ECTS)
- *Renewable Resources* (6 ECTS)
- *Master's Thesis* (15 ECTS)

The course provides a comprehensive overview of green chemistry and a number of related issues (toxicology, legislation, economics, engineering). Recognized lecturers (such as Luis A. Oro and Pedro Arrojo) have participated. High level chemistry training is offered (the University of Zaragoza is placed in the 51–75 range of the Jiao Tong ranking of universities in chemistry).

Unfortunately the future of this course is uncertain. Luis confirmed that:

> A number of high level Master's courses on chemistry-related issues will be launched or modified in 2014 or 2015 at the University of Zaragoza. Furthermore, the 2012 increase of tuition fees in Master's courses and the poor job perspectives for Master's graduates in Spain may discourage the intake of new students. The continuity of the Master's course in sustainable chemistry may be in danger unless a sufficient student intake is achieved.

In Luis' opinion, the most important skills for green chemistry graduates to be equipped with in the future are two-fold: (1) criticism and self-criticism: 'green chemistry literature is full of studies focusing on a single-issue for a reaction (or set of reactions), but disregarding a whole analysis of the problem'; and (2) multi-disciplinary training: 'green chemistry must be applied in teams including non-chemistry specialists (economists, lawyers, engineers, *etc.*)'.

Further details on this course may be found at http://ciencias.unizar.es/masterqs/index.html.

10.8.2 MRes in Green Chemistry: Energy and the Environment, Imperial College London, UK

Dr James Wilton-Ely is the Director of the MRes in Green Chemistry: Energy and the Environment at Imperial College London, UK. He is also a senior lecturer in inorganic chemistry and co-director of the MRes in Catalysis: Chemistry and Engineering.

The MRes course was established at Imperial College in 2007 by Prof. Tom Welton. Since then over 75 students have graduated, going on to PhD research, industry, environmental NGOs and government agencies in many countries. Over the existence of the course, the average intake has been 11 students but has risen to 20 in each of the last two years. The course is research led with 70% of credit coming from the proposal, dissertation, oral presentation and viva based on the research project. Only 15% credit comes from examined courses in chemistry and chemical engineering. The remainder comes from group work (*Renewable Energy* poster project) and a literature report on a different topic to the student's research (and supervised by another member of staff).

The MRes at Imperial is distinctive due to the fact that the course is heavily weighted towards the research component (70% of all credit) and the project, which lasts from November to August. Each student is supervised by two academics (or an academic and an industrial partner) with different but complementary expertise, encouraging a multi-disciplinary approach. The close cooperation and large number of projects with the Chemical Engineering Department also make the course stand out. Increasing interactions with the EU-funded (EIT) Climate-KIC initiative (www.climate-kic.org) has led to the MRes programme becoming a partner of the Climate-KIC, which has provided five PhD studentships for MRes green chemistry graduates.

When asked what the future holds for the course at Imperial, James reports that they:

> Plan to maintain the current student numbers without compromising on our aim of only accepting strong academic performers (2i degree or higher) who have a genuine interest in the aims of the course. We envisage an even greater degree of cooperation with engineering departments and, in particular, industrial partners.

In James' opinion the most important skills for green chemistry graduates to be equipped with in the future are:

- A rounded skill set and enough fundamental knowledge to be able to apply green and sustainable approaches to a broad range of areas
- Sufficient understanding of engineering to be able to see beyond the academic laboratory environment
- The ability to view a process or system in a holistic way rather than concentrating on a particular part in isolation
- The ability to recognize when claims of 'green-ness' are limited or non-existent!

Further details on this course may be found at: http://www3.imperial.ac.uk/chemistry/postgraduate/mres/greenchemistry.

10.9 Future Vision of the MSc in Green Chemistry and Sustainable Industrial Technology at York

In 2014 the GCCE moved into a new purpose-built building providing twice the footprint of its previous location. The new building has state-of-the-art laboratories where the MSc students carry out their work alongside more experienced PhD and postdoctoral researchers. The new centre also includes a unique industrial engagement facility where the students will be able to meet and talk to industrialists on a day-to-day basis.

In the pipeline, we have plans to take advantage of lecture capture software to record course lectures and other relevant material. We envisage that having the lectures recorded will provide an invaluable resource for the students, for example for revision purposes and/or to allow international students to listen again to lectures if they are having language difficulties.

Greater use of a virtual learning environment and recorded lectures could also facilitate the use of some student contact time for more interactive engagement for example in the form of so-called 'flipped lectures'. In addition, it also leads us to other opportunities regarding distance learning for continuing professional development purposes.

Each new student cohort brings with it new ideas for the green chemistry outreach as part of the *Transferrable Skills* module, and one of these ideas we are currently exploring is the creation of a green chemistry centre blog, to which the MSc students will contribute content. The annual influx of new students keeps the dissemination activities of the Green Chemistry Centre current and up to date with new methodologies for getting the message out there to a global community. The GCCE at York continually endeavours to make additional connections with other countries particularly those with emerging markets. The Centre has recently helped launch a Global Green Chemistry Centres initiative, which seeks to bring together leading green chemistry centres from around the world to work collaboratively on global challenges.[11]

As elaborated upon in this chapter, the MSc in Green Chemistry and Sustainable Industrial Technology at York is constantly evolving and improving in order to keep up with the changing demands of the chemical using community and the students themselves, as well as new technical advancements and additional legislatory drivers. This will continue indefinitely into the future.

Acknowledgements

The authors of this chapter would like to acknowledge Rachel Crooks (MSc course administrator 2010 to present) and Alison Edmonds (MSc course administrator 2005 to 2010) for their assistance in compiling data on the course.

References

1. Green Chemistry Network, http://www.greenchemistrynetwork.org/, (accessed 19/02/2012).
2. J. Clark, *Green Chem.*, 1999, **1**, G1–G2.
3. http://www.acs.org/content/acs/en/greenchemistry/education/academicprograms.html, (accessed 19/02/2012).
4. P. Anastas and J. Warner, *Green Chemistry: Theory and Practice*, Oxford University Press, New York, 1998.
5. http://www.york.ac.uk/chemistry/postgraduate/taught/taughtpostgraduatevideos/, (accessed 19/02/2014).
6. http://www.york.ac.uk/chemistry/postgraduate/studprofiles/uktaughtalumni/, (accessed 25/02/2014).
7. http://www.york.ac.uk/chemistry/research/green/contact/tpleaders/, (accessed 21/02/2014).
8. http://sciencewithoutborders.international.ac.uk/home, (accessed 21/02/2014).
9. http://www.york.ac.uk/chemistry/postgraduate/studprofiles/uktaughtprofile/, (accessed 25/02/2014).
10. L. Summerton, A. J. Hunt and J. H. Clark, *Educ. Quim.*, 2013, **24**, 150–155.
11. http://g2c2.greenchemistrynetwork.org/

CHAPTER 11

The State of Green Chemistry Instruction at Canadian Universities

JOHN ANDRAOS*[a,b] AND ANDREW P. DICKS[c]

[a]CareerChem, Don Mills, Ontario, M3B 2W4, Canada; [b]Department of Chemistry, York University, Toronto, Ontario, M3J 1P3, Canada; [c]Department of Chemistry, University of Toronto, Toronto, Ontario, M5S 3H6, Canada
*E-mail: c1000@careerchem.com

11.1 Introduction: Green Research and Teaching at Canadian Institutions

The development of green chemistry as a *bone fide* sub-discipline in Canadian university chemistry departments can be traced from both the applied research and teaching fronts. Here we present a brief historical account of both tracks, as it is important to underscore that the establishment of research groups at different institutions was a catalyst for the introduction of green chemistry content in several undergraduate courses. The emergence of green chemistry research activity in Canada began in 2000 at McGill University in Montreal, Quebec, where T.H. (Bill) Chan launched

Worldwide Trends in Green Chemistry Education
Edited by Vânia Gomes Zuin and Liliana Mammino
© The Royal Society of Chemistry 2015
Published by the Royal Society of Chemistry, www.rsc.org

the Canadian Chapter of the Green Chemistry Institute, an organization based in Washington, DC, under its first Director, Joseph Breen. This was soon followed by the launch of a nationwide Green Chemistry Network in 2002[1] that assembled the names and contacts of all researchers in Canada who were doing any kind of research that fit the aims of green chemistry. Under the leadership and vision of the then chairman, Bruce Lennox, a stream of new faculty members were recruited to McGill with the mandate to mount vigorous research programmemes in green chemistry. The first person to be hired in 2003 was Chao-Jun (C.J.) Li, a former graduate student of Bill Chan. After post-doctoral training at Stanford University under Barry M. Trost, Li took up his first academic position at Tulane University where his work on developing organic reactions in aqueous solvents was recognized with a US EPA Presidential Green Chemistry Challenge Award in 2001 on 'Quasi-Nature Catalysis: Developing Transition Metal Catalysis in Air and Water'.[2] This work developed new methodologies to carry out Grignard and Barbier reactions in water using indium.[3] These efforts in turn led to the development of a new multi-component reaction involving the linear coupling of aldehydes, alkynes and amines using various metal catalysts,[4] which was subsequently called the 'A3 reaction'.[5] Further work in aqueous organic synthesis methodology led to the cross-dehydrogenation coupling (CDC) reaction.[6] The next hires in green chemistry at McGill included Audrey Moores (2006), Ulf M. Lindstrom (2006, now at the University of Lund, Sweden), Tomislav Friscic (2011), and Jean-Philip Lumb (2012). Their investigations span the application of green chemistry in polymer science, organometallic catalysis, mechanochemistry, and organic synthesis methodology in the broadest sense. Together they currently form the largest cluster of green chemistry researchers at a Canadian university.

The second major academic institution offering a green chemistry research programme is Queen's University in Kingston, Ontario. Green Centre Canada[7] was launched there in 2010 at Queen's University Innovation Park under the leadership of Philip Jessop, a graduate of the University of British Columbia, who was recruited from UC Davis in 2003. His research work led to the concept of 'switchable solvents and surfactants' that use carbon dioxide and amines to form zwitterions as a means to reversibly change the polarities of organic solvents from hydrophobic to hydrophilic and *vice versa*.[8] This technology was applied to the problem of separating oily substances from water in emulsified solutions such as those encountered in chemical processing in various industries, most notably the plant-based oil and petroleum industries.[9] At Memorial University in Newfoundland, Francesca Kerton was hired in 2006 from the University of York, UK, to head a research programme in the areas of solvent replacement, catalysis, and renewable feedstocks.[10] With the launch of such vigorous research campaigns nationwide in this emerging field, both the Canadian and Ontario governments recognized important achievements with the following awards:

- Green chemistry related Canadian Research Chairs (CRC): Jason Clyburne (Simon Fraser University then St. Mary's University: Environmental Studies and Materials Tier 2); C.J. Li (McGill University: Green Chemistry Tier 1); Audrey Moores (McGill University: Green Chemistry Tier 2)
- Green chemistry related Ontario Research Chairs (ORC): Michael Cunningham (Queen's University: Green Chemistry and Engineering); Suresh Narine (Trent University: Green Chemistry and Engineering).

Furthermore, the Canadian Green Chemistry and Engineering (CGCE) Award and the Ontario Green Chemistry and Engineering (OGCE) Award are given annually to individuals and organizations that innovate and actualize green chemistry principles in pure and applied research with technical, human health and environmental benefits. Table 11.1–11.3 summarize the recipients of these awards since their inception.

Table 11.1 Winners and citations for the Canadian Green Chemistry and Engineering Award (Individual).

Year	Winner	Affilation	Citation/Award Lecture
2010	C.J. Li	McGill University	Exploration of synthetic chemistry for a sustainable future
2011	R. Tom Baker	University of Ottawa	Towards a renewable transportation fuel: Catalysed ammonia–borane dehydration for chemical hydrogen storage
2012	Philip Jessop	Queen's University	Carbon dioxide triggered switchable solvents and surfactants
2013	Flora T.T. Ng	University of Waterloo	Development of green processes for the production of chemicals and fuels from biomass; Catalytic distillation
2014	Douglas Stephan	University of Toronto	Frustrated Lewis acid Pairs: A new approach to carbon dioxide capture and utilization

Table 11.2 Winners and citations for the Ontario Green Chemistry and Engineering Award (Individual).

Year	Winner	Affilation	Citation/Award Lecture
2010	Leo W. M. Lau	University of Western Ontario	Green chemistry: Experiences and research
2011	Franco Berruti	University of Western Ontario	Turning waste into black gold
2012	Paul Charpentier	University of Waterloo	Supercritical fluids – A tool for sustainable nanotechnology and alternative energy

Table 11.3 Winners and citations for the Ontario Green Chemistry and Engineering Award (Organization).

Year	Winner	Citation/Award Lecture
2010	Woodbridge Foam Corporation	A world leader in green chemistry
2011	EcoSynthetix	The challenges and achievements of following the twelve principles of green chemistry
2012	Xerox Research Centre of Canada	Green chemistry and engineering
2013	No award presented	—
2014	Orbite Aluminae Ltd	A green and eco-friendly technology for producing alumina from clay and fly ash

In parallel with the emergence of green chemistry as an area of active research, teaching faculties at Canadian universities have largely spearheaded instruction of the subject in the chemistry undergraduate curriculum. The earliest green chemistry courses were offered at McGill University and York University, Toronto, in 2002. A detailed account of the course given at York between 2002 and 2008 has been given elsewhere.[11] In 2005, it was recognized as the most innovative Canadian contribution to green chemistry education by the Green Chemistry Education Roundtable held at the United States National Academy of Sciences in Washington, DC.[12] Currently, we are aware of at least eight Canadian chemistry departments out of over sixty universities in the whole country that offer formal green chemistry instruction in the form of one-semester courses. These offerings can be classified into two broad groups: (1) stand-alone courses and (2) hybrid courses. Most are electives and are therefore not mandatory for an honours bachelor degree in chemistry. All have a lecture component and at least two have a laboratory component. Although the number of courses offered is presently small it is gaining momentum as the field is still emerging.

11.2   Green Chemistry Courses: Content

In this section we present details of eight established green chemistry-related courses taught at Canadian universities according to the following criteria: institutions, course titles, undergraduate year, instructors, brief descriptions, prerequisites, textbooks/other reading materials, and course evaluations. The first group (four courses), designated as Type A, are stand-alone lecture-only courses and summarized below:

Institution: University of Calgary, Alberta
Title: CHEM 423: Green Chemistry Principles and Techniques
Year: Third-year undergraduate
Instructor: Dr Kal Mahadev

Description: The following is a tentative list of topics that will be covered in lectures:

1. Introduction to green chemistry
2. The Twelve Principles of green chemistry
3. Atom economy
4. Waste: production, problems and prevention
5. Environmental performance
6. Catalysis and green chemistry
7. Renewable resources
8. New technologies in green chemistry
9. Green metrics
10. Industrial chemistry (some examples)

Prerequisites: CHEM 333 (second-year Inorganic Chemistry: Transition Metals); CHEM 353 (second-year Organic Chemistry II) or CHEM 355 (second-year Organic Chemistry II for Chemists)
Textbooks/Reading Materials:

1. *Green Chemistry – An Introductory Text* by Mike Lancaster. RSC Publishing, 2010
   Various journal articles in *Green Chemistry* (Royal Society of Chemistry)
2. *Green Chemistry Letters & Reviews* (Taylor & Francis).

Institution: McGill University, Quebec.
Title: CHEM 462: Green Chemistry.
Year: Fourth-year undergraduate.
Instructors: Prof. C.J. Li; Prof. Audrey Moores
Description: New reactions and methods which can be used for the production of chemicals from renewable feedstocks; the use of new environmentally benign solvents, catalysts and reagents; organic reactions in aqueous media and in supercritical carbon dioxide; bio-catalysis and bio-processes. At a time when human beings realize that their actions have had a definite, irreversible and detrimental impact on their environment with consequences such as global warming or loss of biodiversity, we need, as chemists, to ask ourselves a simple question: 'What can we do to help the planet?' An active part of research in chemistry has pursued the goal of answering that question, leading to a domain of chemistry called 'green chemistry'. After an introduction to the problem and the definitions of green chemistry, we will investigate four major domains in which green chemistry finds application. Throughout the class detailed examples from the most recent literature will be used. We will present chemical reactions as they are: a complex set of many factors, like energy, solvent uptake, bio-compatibility of the products, sustainability, waste, *etc*. that have to be looked upon to evaluate a reaction. The goal of this course is three-fold: (1) the student will discover the most

recent advances in the domain and examples that show how reinventing old reactions could dramatically improve their eco-friendliness; (2) the student will deepen his or her knowledge of principles of chemistry itself; and (3) the student should learn throughout the course how to develop a critical look at chemistry and green chemistry and to identify where and how reactions could be improved. Three invited speakers from prestigious institutions will provide a more focused look on their practice of green chemistry.

Prerequisites: CHEM 302 (third-year Introductory Organic Chemistry 3); CHEM 381 (third-year Inorganic Chemistry 2)

Textbooks/Reading Materials: Recommended library readings from various books and journals.

Institution: Memorial University, Newfoundland.
Title: CHEM 4206/6206: Green Chemistry.
Year: Fourth-year undergraduate, cross-listed as a graduate course.
Instructor: Prof. Francesca Kerton
Description: This course is designed to equip students with the tools, techniques and general understanding of environmental, economic and social factors important in the implementation of clean technology and sustainable chemical development. It will include aspects of this field specifically related to inorganic chemistry such as the development of new catalytic processes and use of alternative reaction media. Once introductory lectures on Green Chemistry have been provided, the course will move on to using examples from the current chemical literature and assigned readings will be provided. Lectures will cover the principles of green chemistry, catalysis, alternative reaction media, alternative synthetic routes (*e.g.*, C–H bond activation), energy efficiency and emerging technologies, and renewable resources. One class will be used to discuss business opportunities (*i.e.*, entrepreneurship) and legislation. There will be several classes where students are asked to prepare for in-class discussions by reading recent articles from the chemical literature and chemical news magazines.

Prerequisites: CHEM 2401 (second-year Introductory Organic Chemistry II), CHEM 3211 (third-year Transition Metal Chemistry)

Textbooks/Reading Materials:

1. *Green Chemistry – An Introductory Text* by Mike Lancaster. RSC Publishing, 2010
2. *Green Chemistry: Theory and Practice* by Paul Anastas and John Warner. Oxford University Press, 2000
3. Recommended library readings from various books and journals.

Title: CHMD89H: Introduction to Green Chemistry.
Year: Fourth-year undergraduate.
Instructor: Dr Effiette Sauer
Description: The course will begin by introducing the Twelve Principles of green chemistry followed by the use of green chemistry metrics for quantifying 'greenness'. In this context, we will move on to explore major areas

of green chemistry research including alternative solvents, catalysis and renewable feedstocks. Examples from industry and from the current literature will be used to reinforce the material and highlight recent advancements. The topics covered will be multi-disciplinary in nature and will draw on aspects of organic, inorganic and polymer chemistry.

Prerequisites: CHMB31 (second-year Introduction to Inorganic Chemistry), CHMC41 (third-year Organic Reaction Mechanisms) or CHMC42 (third-year Organic Synthesis)

Textbooks/Reading Materials: Recommended library readings from various books and journals.

The second group (two courses), designated as Type B1, are hybrid green chemistry lecture-only courses and are summarized below.

Institution: Queens University, Ontario.
Title: CHEM 326: Environmental and Green Chemistry.
Instructors: Prof. R. Stephen Brown; Prof. Philip Jessop.
Year: Third-year undergraduate.
Description:

1. Overview: define environmental chemistry and green chemistry; 'spheres' of importance; outline of text; review of concentrations and calculations (one lecture)
2. Chemistry of the atmosphere: review of gas-phase reactions; radical reactions and thermodynamics; chlorine radicals and the ozone 'layer', CFCs and other ozone depleting contaminants, catalysis on condensed phases; hydroxyl radical, ozone production, proton abstraction, VOCs, NOx, and photochemical smog (four lectures)
3. Greenhouse effect and global warming: infrared absorbance spectra and greenhouse effect; major greenhouse gases – CO_2, H_2O, CH_4, N_2O, aerosols, others; predicted effects; energy sources and alternatives (two lectures)
4. Chemistry of contaminants: review of organic chemicals, classes and nomenclature; principles of toxicology, mechanism and dose–response; persistence, bioaccumulation and toxicity; pesticides – chlorinated, DDT and others; dioxins and furans; partition, fugacity and long-range transport (three lectures)
5. Chemical contaminants: PCBs, PBDEs and others; PAHs; oestrogenic contaminants; heavy metals; environmental and health effects (two lectures)
6. Water: natural waters – oxygen and redox chemistry, acid/base chemistry and carbonate system; drinking water – purification, disinfection, impact of chlorine; groundwater – contaminants and remediation; wastewater – phosphate, oxygen demand, fate of organic compounds, wastewater treatment (four lectures)
7. Soil and sediments: major contaminants – behaviour, fate and transport; chemical and biological remediation methods; heavy metals, lead, arsenic (two lectures)

8. Introduction to green chemistry: history, goals and principles, economic and legislative drivers (three lectures)
9. Measures and metrics: E-factors and related measures, multi-variant assessment of impact, energy consumption (three lectures)
10. Web resources (one lecture)
11. Solvents: solventless conditions, preferred organic solvents, water, supercritical fluids, expanded liquids, ionic liquids, and liquid polymers (three lectures)
12. Alternative feedstocks and reagents: biomass, waste polymers, CO_2 (two lectures)
13. Synthetic methods and strategies (two lectures)

Prerequisites: CHEM 223 (second year Organic Reactions) or CHEM 281 (second-year General Organic Chemistry I)
Textbooks/Reading Materials: *Environmental Chemistry*, 4th edition, by Colin Baird and Michael Cann. Freeman & Company, 2008.

Institution: York University, Ontario.
Title: CHEM 3070: Industrial and Green Chemistry (note: this course no longer has a green chemistry component as of 2009; however, there is a small component of environmental chemistry)
Year: Third-year undergraduate.
Instructors: Dr John Andraos (2002 to 2008); Prof. Pierre G. Potvin; Prof. Gino Lavoie; Prof. Gerald Audette (2009 to present)
Description (2002–2008): see Andraos[11]
Description (2009 to present): The chemical industry has become a huge complex of operations that range from large multi-national corporations to small, locally owned factories. Collectively, they manufacture materials and products that compose at least some part of almost every item used in our society today. This course serves as an introduction to industrial and environmental chemistry. Various aspects related to the production of chemicals on a large scale will be presented, ranging from a general introduction of the chemical industry to specific manufacturing and legal issues.
Credits: 3
Prerequisites: CHEM 2020 (second-year Introduction to Organic Chemistry)
Textbooks/Reading Materials: Course kit; recommended library readings from various books and journals.

The third group (two courses), designated as Type B2, are hybrid green chemistry combined lecture and laboratory courses and are summarized below.

Institution: University of Toronto (St. George campus), Ontario.
Title: CHM 343H: Organic Synthesis Techniques.
Year: Third-year undergraduate.
Instructors: Dr Andrew Dicks; Prof. Robert Batey; Prof. Mark Taylor.
Description: This course is designed to provide the opportunity of developing practical skills in (1) synthesizing organic compounds, primarily on a

micro-scale/semi-micro-scale, with a special particular emphasis on modern catalytic methodologies; (2) proving the structure of the compounds by application of modern physical methods; and (3) analysing the purity of the prepared compounds by different techniques. The CHM 343H lecture material has been designed to closely align with laboratory work and has heavy emphasis on spectroscopy at the beginning of the course. The laboratory additionally showcases the relevance and importance of organic synthesis in 'everyday life'. Highlights include: (1) preparation of a pharmaceutical currently prescribed as an anti-depressant, an anti-fungal analogue, and a sunscreen analogue (featuring phase-transfer catalysis and organo-catalysis); (2) 'green chemistry': using water to replace common organic solvents under conditions of palladium catalysis, reactivity using a recyclable glycerol-based solvent, and an atom-economical multi-component reaction featuring both Lewis acid and Lewis base catalysis; (3) one three-step synthesis where the product generated during one laboratory session is used as a starting material the following week; and (4) a laboratory practical examination that will assess understanding of green principles taught earlier in the course.

Prerequisites: CHM 247H (second year Introductory Organic Chemistry II) or CHM 249H (second-year Organic Chemistry)

Textbooks/Reading Materials: The following two practical textbooks are highly recommended:

1. *The Synthetic Organic Chemists Companion* by M. Pirrung. John Wiley & Sons, Inc.: Hoboken, NJ, 2007
2. *Advanced Practical Organic Chemistry* by J. Leonard, B. Lygo, G. Procter. Blackie Academic & Professional: 1995.

Institution: St. Mary's University, Nova Scotia
Title: ENVS 2100: Green Chemistry
Year: Second-year undergraduate
Instructor: Prof. Jason Clyburne
Description:

Lecture 1: introduction to green chemistry
Lecture 2: chemistry top 10 – industrial, fine, and pharmaceutical
Lecture 3: principles of green chemistry, sustainable development and green chemistry
Lecture 4: toxicity
Lectures 5 + 6: atom economy – rearrangement reactions, addition reactions, substitution reactions, elimination reactions
Lectures 7 + 8: design for degradation – the cost of waste, waste minimization, waste treatment, designed for degradation, and recycling and disposal
Lectures 9 + 10: real-time analysis for pollution prevention: the importance of measurement, lifecycle assessments, green process metrics, environmental management systems, eco-labels, legislation

Lectures 11 + 12: less hazardous chemical synthesis, safer chemistry for accident prevention
Lecture 13: solvents – general features
Lecture 14: solventless reactions
Lectures 15–18: solvents – water, liquid carbon dioxide, super-critical water, super-critical carbon dioxide, fluorous solvents, deep eutectics, ionic liquids
Lecture 19: renewable feedstocks
Lectures 20 + 21: catalysis, heterogeneous catalysis, homogeneous catalysis, phase transfer catalysis, biocatalysis
Lectures 22–24: energy efficiency, renewable energy, hydrogen economy, methanol economy, fuel cells

Prerequisites: Two first-year undergraduate chemistry courses
Textbooks/Reading Materials: Recommended library readings from various books and journals.

A third course in category B2 is currently being designed for implementation at the third-year undergraduate level at the University of Ottawa and is titled *Catalysis and Sustainable Synthesis*. Finally, Table 11.4 summarizes the evaluation breakdown for all eight courses presented. Courses offered at the University of Toronto Scarborough and at York University have significantly changed their grading schemes since 2011 and 2009, respectively.

11.3   Green Chemistry Courses: Similarities and Differences

In this section we compare and contrast various pedagogical aspects of the above eight green chemistry courses.

11.3.1   Similarities

Most of the courses are listed as electives and are typically offered in the third or fourth year of undergraduate study. Often one or two introductory organic or inorganic courses are required as prerequisites. Almost all of them have a large written assignment or project that accounts for a significant proportion of the overall course evaluation. The courses are largely designed to mimic actual research practice, and involve evaluation of documented chemical processes with respect to green chemistry principles. Reactions are critically reviewed according to some set of given criteria and a decision is made as to which of at least two processes is greener than another. Such in-depth assignments train science students to write significant reports, which are customarily not prepared in other lecture-only courses. Moreover, if the assigned chemical transformations have never been examined in the literature before by green chemistry principles, students quickly take ownership of their work and become aware that their critiques are new, important contributions and

Table 11.4 Summary of course evaluations among green chemistry type courses offered at various Canadian universities described in this chapter.

Category	University								
	McGill	Memorial	Calgary	UTSC-1	UTSC-2	Queen's	St. Mary's	York[a]	Toronto
Term tests	20	30	20	0	0	20	20	15	0
Quizzes	0	0	0	0	0	0	10	5	0
Problem sets	0	0	0	0	40	9	0	40	0
Short written assignments	0	0	10	0	0	26	0	0	0
Long written assignments	20	20	20	50	35	0	0	35	15
Laboratory	0	0	0	0	0	0	30	0	50[b]
Final exam	40	30	30	20	0	45	25	0	35
Presentations	20	15	15	15	10	0	15	5	0
Class discussions	0	5	5	15	15	0	0	0	0
TOTAL	100	100	100	100	100	100	100	100	100

[a]Breakdown shown applies for course offered between 2002 and 2008; from 2009 the course content shifted significantly from green chemistry to traditional industrial chemistry based on petrochemicals with the following breakdown: quizzes (2%), short paper (8%), long paper (20%), term tests (40%), and final exam (30%); see ref. 11 for details.
[b]50% of final course grade composed of (1) weekly laboratory work and reports (45%) and (2) a practical examination (5%).

not just another exercise they need to do to fulfil a course and degree requirement. Due to feasibility issues with respect to class sizes, this long format assignment may be given and evaluated as a team-based exercise rather than as one evaluated by the traditional method based on individual performance. In such a setting students learn to work in groups (a key skill to acquire and to master) as they will need it in their future careers in workplace environments, particularly when dealing with multi-disciplinary problems that demand working with people in other areas of expertise beyond chemistry. Students often learn more from each other and the instructor in a collaborative setting as they debate the application of green chemistry principles to a given chemical process they are examining. Instructors have used various innovative techniques in evaluating such assignments, including staging of an assignment throughout the course in order to allow students to make adjustments based on immediate feedback, rather than handing it in for evaluation in one large block at the end of term and waiting for the final verdict. Also, various techniques have been implemented to combat procrastination and plagiarism such as using Turnitin (http://www.turnitin.com) grading rough notes, breaking up the assignment into smaller chunks over the course of the term where each stage is evaluated separately and using student peer-review in addition to the instructor's evaluation as an integral part of assigning the final grade.

Decision-making *via* a comparative analysis based on a set of given criteria is a key skill that is developed among all eight courses in assessing the green attributes of a new process over a traditional one. Often students for the first time realize that such exercises do not lead to clear-cut unique 'right' answers as they previously experienced in traditional chemistry courses. Trade-offs are a hallmark of green chemistry subject territory that will likely impact each student's comfort zone. Students often discover exaggerated claims made in the literature, sometimes referred to as 'green washing', which can be disquieting given that they enter science believing that a high degree of integrity and veracity is practised among all scientists, and whatever is published must be true. Validation of references is also a key skill that students acquire, particularly when assessing resources found online.

Most instructors find that inspiration for showcasing green chemistry in action comes directly from the front-line scientific literature; hence they tend to heavily use journal articles rather than follow a textbook closely. Positive spin-offs of this approach are that students learn about the organization of the chemistry literature, and how to perform efficient searches. They also move away from the safety of having a textbook that is normally viewed as a failsafe crutch when preparing for texts and exams, though student feedback regarding this point has been mixed. Also, librarians play a key role in instructing students based on the availability of databases and resources that a library has access to.

All courses are heavily weighted to the application of green chemistry to organic reactions. There may be exceptions if the course is a hybrid that covers topics in industrial chemistry, particularly gas phase reactions that generate first, second, and third generation feedstocks from the petrochemical industry. This remains as an ongoing challenge to expand the range of chemistries that can be assessed by green chemistry principles as will be discussed in Section 11.4. Opportunities exist to achieve this if curricula are designed to be more cross-disciplinary such as at Memorial University and UTSC.

11.3.2 Differences

Given that green chemistry is a multi-disciplinary and multi-dimensional subject, it is not surprising that instructor preferences would manifest themselves based on their own level of expertise in various topics in green chemistry, and their own knowledge and practice of pedagogical techniques in delivering lectures.

For example, at UTSC there is a high percentage of course evaluation allotted for class participation and oral presentations. This may pose challenges for students who are intrinsically introverted, or who have under-developed public speaking skills but can demonstrate full understanding of material presented to them in other ways. The instructor also experimented with open book term tests. In comparison, team-based assignments are routinely given at Queen's University. The most comprehensive green metrics analysis is taught there, including several environmental impact factors beyond elementary metrics such as reaction yield and atom economy. The course is closely tied to topics in environmental chemistry, where the fate of chemicals released in the environment are traced and studied. A unique feature at Memorial University is that business and entrepreneurship aspects are included in the syllabus.

At McGill University, various expert guest lecturers for key topics have delivered talks throughout their course on the following topics: chemical waste and toxicity (Wayne Wood), the Montreal Protocol and pollution (Parisa Arya), establishing a full mass balance and atom economy (Barry Trost), biocatalysis (Peter Lau), mechano-chemistry (Tomislav Friscic), and drivers of clean innovation (Steve McGuire). At the University of Calgary, student peer-review of long written assignments according to specified criteria have been implemented as part of the evaluation process. The final grade is determined based on revisions and replies to feedback from three peer-review student reports simulating a real submission to a journal editor. Student participation in giving feedback to other students' reports is mandatory. Final oral presentations are evaluated based on feedback from the instructor, an external expert, and fellow students.

The course at York University offered between 2002 and 2008 was unique in the following respects: (1) it introduced students to the patent literature and the chemical industry enterprise; (2) problem sets were heavily used to train students in critical thinking and problem solving using recent literature examples in industrial process chemistry and organic synthesis; (3) metrics analysis was emphasized as the major skill set to master in deciding degrees of greenness of chemical reactions, synthesis plans, and chemical processes in industry; and (4) a special three-hour lecture was devoted to career development in the chemical sciences for academic and industrial positions regardless of the type of chemistry.

The courses taught at the University of Toronto and St. Mary's University have a significant laboratory component where students can implement green chemistry principles learned in lecture into their own practice. The St. Mary's course units closely follow the Twelve Principles of green chemistry. The Toronto course emphasizes the importance of catalysis in organic synthesis. The focus of undergraduate experiments at both institutions revolves around solvent reduction and/or replacement,[13] use of catalysts, catalyst and solvent recycling,[14] and microscale techniques.[15] Scheme 11.1 highlights some key example reactions taken from CHM 343H at the University of Toronto. All four reactions showcase the use of catalysts. The Heck[16] and Suzuki[17] couplings are carried out in aqueous solvents while the Biginelli three-component condensation[18] is done in a solvent-free reaction environment. In the geraniol oxidation experiment, N-methyl morpholine oxide (NMO) is used to recycle TPAP which is used in catalytic amounts, thereby reducing its toxicity impact.[15b] A full description of this course including experiments has been given elsewhere.[19] An excellent resource for various green chemistry undergraduate experiments called The Greener Organic Chemistry Reaction Index (TGOCRI) has also been compiled.[20] Another resource for 'greening up' high school experiments has additionally been published.[21]

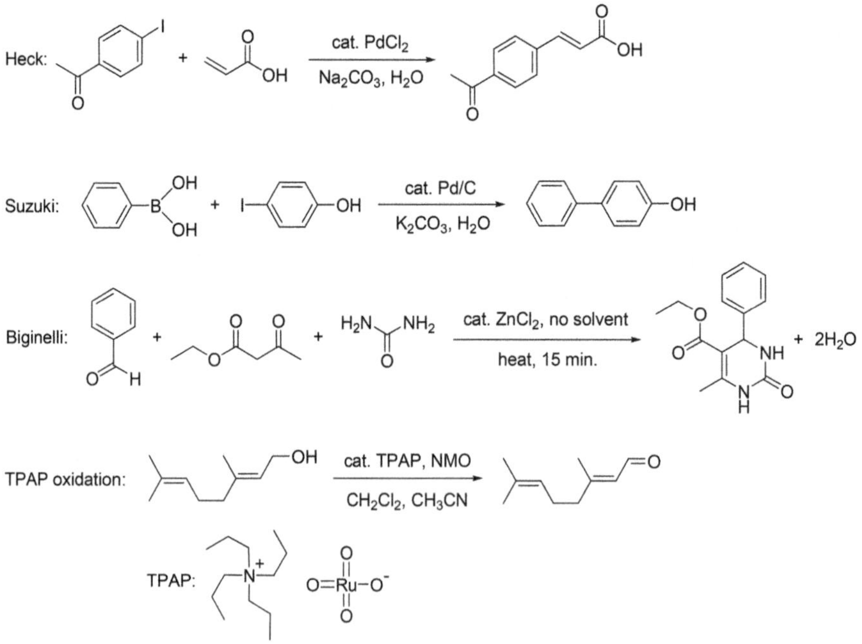

Scheme 11.1   Sample reactions used in CHM 343H at the University of Toronto to illustrate green chemistry principles in the undergraduate laboratory.

11.4 Topics Not Yet Covered in Green Chemistry Courses

We have previously reviewed the pros and cons of pedagogical practice in mounting specialized green chemistry lecture and laboratory courses, and in including this subject in traditional curricula.[22] Here we briefly highlight topics, typically not considered, that we believe would add significant value to any green chemistry course syllabus. Examples include the following:

- Reaction thermodynamic analysis applied to organic chemistry such as determination of heat of reaction, by calculation or experiment, for a chemical transformation in a given solvent under a given set of temperature and pressure conditions
- Energy consumption for chemical reactions (particularly for undergraduate laboratory reactions), since this is never measured in literature procedures, although guesses can be made based on reaction temperature and pressure conditions and operational procedures such as distillation; it should be noted that reactions are never carried out above 1 atmosphere in the undergraduate laboratory due to safety concerns and the use of expensive and specialized equipment such as autoclaves
- More attention should be paid to patent literature since this is essential to probe what the chemical industry has done in the area of green chemistry; currently it is weakly covered in course syllabi
- Green chemistry principles applied to analytical, physical, inorganic, and general chemistry are still relatively rare;[23-26] the main focus is on organic chemistry examples usually taken from the pharmaceutical industry, since references in this area are abundant in the literature where green chemistry principles have been successfully applied
- Recycling or reuse of catalysts in reaction procedures
- Cost analysis as part of a green assessment
- Inclusion of more examples using bio-catalytic, chemo-enzymatic or fermentation procedures so that their performances may be directly compared with petrochemical routes to the same target molecules; the main difficulties are the lack of experimental details disclosed in publications, and the inability of writing out balanced chemical equations since such transformations are usually unknown
- Reaction network analysis as a means to study the hierarchy and interconnectedness of industrial processes to commodity chemicals
- Greater emphasis on life cycle assessment beyond elementary discussions about atom economy; this is carried out in the Queen's University course and at UTSC in 2013 using an in-house automated Microsoft Excel spreadsheet based on a penalty-point algorithm developed by Andraos in collaboration with the Green Chemistry Initiative at the University of Toronto. This student group will be discussed in the next section.

11.5   Feedback

During 2012–2013 two surveys were circulated among the eight departments
offering green chemistry courses: one for students and the other for course
instructors. There were 58 responses from students and five responses
from instructors. Here we present typical responses to key questions from
both surveys reflecting experiences from both undergraduate and faculty
perspectives.

11.5.1   Student Voices

The following two questions were posed and the students' responses are
given in Section 11.8:

- Would you recommend a green chemistry lecture and/or lab to your fellow classmates?
- How has your overall perception of the subject of chemistry changed as a consequence of taking a green chemistry course?

From the comments received (see Section 11.8) we observe that most students view the formal study of green chemistry as a strong positive force in their education in the chemical sciences and beyond. These kinds of sentiments resulted in the founding of the Green Chemistry Initiative (GCI) at the University of Toronto in September 2012 by graduate students in the Department of Chemistry.[27] This is the first student-led initiative in Canada in the area of green chemistry coming nearly a decade after its birth at McGill University in 2002 as mentioned in the introduction. GCI members describe their motivation and philosophy as follows:

> In search of a way to decrease the environmental impact of their own chemistry research, while educating others in the department and the community on the principles and merits of green chemistry, the Green Chemistry Initiative was founded to promote sustainable practices in the lab and in everyday life. The group is proud to host monthly seminars by cutting-edge researchers and industry partners, waste reduction campaigns, social events and workshops, and to provide resources and literature to the department and the public at large.

In November 2013, the GCI also launched a journal club where selected research articles published in leading green chemistry journals are discussed and critiqued. Laura Hoch, one of the co-founders, who had prior research experience in green chemistry applied to material science at Pennsylvania State University and at Los Alamos National Laboratory, described her ideas as follows:

> When I arrived at U of T, I wanted to try to incorporate green chemistry into my research but found it very difficult to find the information I

needed to make my reactions greener. After some great discussions with other students and some very helpful faculty, I realized that I was not the only one who felt this way. We wanted to start the GCI so that other researchers in our chemistry community would have an easier time learning about and incorporating green chemistry principles into their work.

As part of their education mandate, in the spring of 2013 the group launched a green chemistry workshop taught by the authors of this chapter and Dr Effiette Sauer (UTSC). This opportunity was open to any graduate student in any science department on campus. John Warner, one of the founders of the Twelve Principles of green chemistry, gave a plenary lecture to the department and another to workshop participants. The event was capped by a challenge competition where participants were asked to come up with proposals that they could actualize green chemistry principles learned in the workshop in either their own doctoral research work or in undergraduate laboratory education. It is hoped that this kind of enthusiasm for the subject will continue to grow in other universities. It is clear that future generations of chemists will require the necessary training in green chemistry to tackle pressing issues surrounding the environmental impact of their actions. The good news is that modern students, like the GCI, are no longer content to wait for current faculty to respond and are prepared to take direct ownership of their education themselves.

For comparison purposes we also surveyed 413 undergraduate students at a separate Canadian university who had taken a traditional second-year organic chemistry course where one of the laboratory exercises was a green chemistry experiment, namely, an aldol condensation between benzil and 1,3-diphenylacetone in the absence of a reaction solvent (Scheme 11.2). This is an example of green chemistry being added as a component to an existing traditional course, rather than presenting the subject in a dedicated course. Selected responses to the same two questions give a flavour of this pedagogical strategy (see the responses in Section 11.8).

We can see that the responses are mixed (although very insightful) for students who are two years younger than their senior counterparts who took formal courses in the subject. Some students were unconvinced of the 'green' merits of the exercise, since it was only a single isolated example of green chemistry in action, while others could see the potential based on their own background knowledge of the subject and feelings of environmental issues,

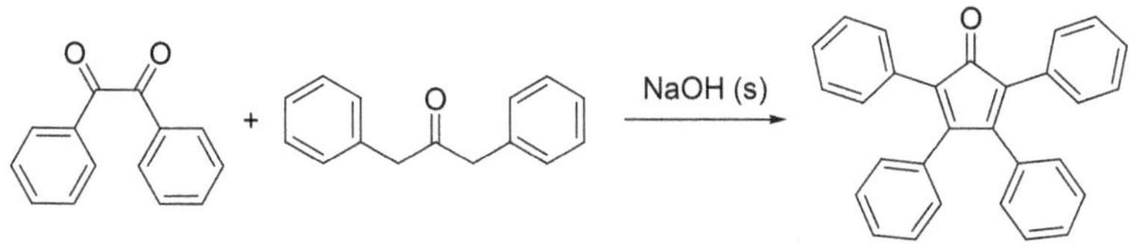

Scheme 11.2 Aldol condensation between benzil and 1,3-diphenylacetone performed in a second-year undergraduate organic chemistry course.

but were somewhat frustrated about why they did not have the opportunity to learn more about it in more depth in a structured way. Many recognized that the chemical industry needed to change its image and was taking steps to improve its waste performance, and that gaining green chemistry knowledge would be an asset for them in the marketplace. When comparing these responses with the ones presented earlier for students who took *bone fide* green chemistry courses, clearly the idea of 'parachuting in' one such experiment has a less positive pedagogical outcome. Some students rightly pointed out that lecture and lab material were not reinforcing one another to convince them of the attributes of the subject. This problem is a universal one, regardless of introducing green chemistry ideas in the curriculum, since subject matter in lectures and labs are generally not taught in phase with one another. It is customary that the two streams are run in parallel and are taught independently from one other by different teaching faculty. This is especially true in so-called 'service courses' containing large numbers of students who are not necessarily destined to become chemists.

11.5.2   Lecturer Voices

Instructors who responded to the survey were first asked about basic statistics concerning their course. Table 11.5 summarizes data regarding enrolments, lecture hours, and preparation times.

Instructors were then asked to give a scaled score response to two questions relating to challenges they may have experienced in mounting their course as shown below:

- In preparing a green chemistry course, were you challenged in your own understanding of fundamental concepts in organic chemistry?
- In preparing a green chemistry course, did you revise your own understanding of fundamental concepts in organic chemistry?

Instructors were requested to answer the first question on a scale of 1 to 5, where 1 = strongly agree, and 5 = strongly disagree. Four out of five

Table 11.5   Summary of data for green chemistry courses by survey instructor respondents.

School	Enrollment	Lecture hours per week	Prep time for green chem. course (hours per week)	Prep time for traditional org. chem. course (hours per week)
McGill	16	3	18	3
Memorial	14	3	4	4
UTSC	10	3	4	3
Calgary	18	2.5	3	Not available
York	30+	3	4	3

instructors gave an answer of '2' corresponding to 'agreement' with the statement, whereas only one gave a '4' corresponding to 'disagreement'.

The following comments were given in response to the second question, which sought to probe specifics about affirmative responses to the first question:

> I'm teaching more the inorganic section and materials, nonetheless I had to revise on some mechanisms, such as polymerization mechanisms. The need to balance chemical equations for the purposes of teaching the atom economy concept forced me to relearn and strengthen my understanding of the fundamentals, particularly for reduction and oxidation-type reactions, where by-products are not often declared in the literature and in textbooks.

Instructors gave the following comments concerning the main feedback comments they received from students regarding their 'green' lecture course at the conclusion of the semester:

- Respondent 1

The concepts associated with green chemistry are very broad and students often complain about the difficulties associated with exam preparation. Students like the fact they get the 'big picture' on many different chemistries. They enjoy the fact they can relate chemistry with things they know and understand such as pollution, consumer products, environment, and health.

- Respondent 2

In the past, students have said that they want a course book and that they want a more detailed book than *Green Chemistry: Theory and Practice* by Anastas and Warner, hence the adoption of the Lancaster text. As a professor, especially as this is a 4th year/graduate level course, I would be just as happy teaching from my own notes/literature. Most students enjoy the course and wish they were aware of green chemistry sooner. It seems to make them a little sad that they didn't find out the 'secret' sooner.

- Respondent 3

Students were very positive in their evaluation of this course. Comments included: 'most interesting course taken', 'most stimulating course taken', and 'favourite course in undergrad'. Both on course evaluations and in conversations with students, many remarked that the course strongly promoted critical thinking skills. In this respect, they found the course very applicable to the rest of their studies, both chemistry and non-chemistry courses.

- Respondent 4

Most students found that the staged written assignment was the main highlight of the course where they took ownership of their own work and learned about the resources available to them in the library that they were totally unaware of in the first two years of undergraduate study prior to taking the green chemistry course. They thought the workload for the course was not expected for one listed as an elective but they commented that their experiences were well worth the effort. Some students expressed concern that they may have been short changed in the past with respect to learning fundamental organic concepts incorrectly or incompletely in prerequisite courses as a consequence of realizing how broad and deep an evaluation of any given reaction is with respect to green chemistry principles. Many students liked the real-world examples and 'stories' behind the discoveries they read about in the course and others they knew of from past readings. Almost all students found the problem sets challenging but rewarding even though they may not have heeded the advice of starting on them early.

11.6   Green Chemistry Publications

As well as the experiments designed and discussed in Section 11.3.2, other work has been published in the context of teaching green chemistry in Canada. Andraos has introduced a visual representation of material consumption metrics calculations using radial pentagons and applied this technique to a survey of undergraduate experiments carried out at York University, the University of Toronto, and Malaspina University College (now the University of Vancouver Island).[28] The radial diagrams were used as an effective teaching tool to illustrate the extent of waste generation by various reaction types and to pinpoint the sources of that waste. Furthermore, knowledge gained from this exercise was then used in class discussions where students suggested synthesis strategies that could be used to reduce waste impacts for specific reactions they studied.[29] Consistent with a hybrid green and environmental chemistry course, the Jessop group at Queen's University has developed a multi-variate metrics exercise that included several environmental impact parameters in addition to materials efficiency metrics.[30] Students evaluate several plans to a given target molecule and then decide which plan is relatively greener based on the results of a broad metrics analysis. At St. Mary's University a solvent-free experiment has been developed for the synthesis of *N*-organophthalimides using the co-crystal controlled solid state technique.[31] Reactions were also evaluated according to atom economy and E-factor. At the University of Manitoba, a two-step synthesis of 4-bromoacetanilide has been implemented that showcases a 'green' electrophilic aromatic bromination reaction using both a protecting group and redox chemistry.[32] An improved undergraduate synthesis of imidazolium room-temperature ionic liquids has been reported by faculty at Wilfrid Laurier University in Ontario.[33] Most recently a third-year undergraduate laboratory course developed at the

University of Toronto included an advanced green chemistry decision-making exercise in which students were required to devise and execute a synthetic plan to a target azlactone structure that incorporated green chemistry principles. Students were allowed to consult the primary literature; however, detailed guidance from course instructors was not provided.[34]

11.7 Future Directions and Challenges in Green Chemistry Education

The green chemistry experience in Canada is essentially a decade old. Despite rapid advances in both research and teaching efforts around the country, it is essentially flourishing in isolated pockets and is currently at a crossroads. Speaking from our own experiences, no one chemistry department has strength in both teaching the subject and carrying out research in it at the highest level. If it is offered at all as formal instruction, departments continue to keep the subject on the fringe of the curriculum by offering it as a single optional elective course. If it is mentioned in traditional courses it is usually in the form of 'add-on' examples to the curriculum. An example of this is in an Intermediate Organic Chemistry course (CHEM 1252) offered at Northern Alberta Institute of Technology (NAIT), where the final unit of the course covers the 'application of the principles of green chemistry to organic chemistry.' The second-year organic laboratory experience described here shows that this approach appears to have limited pedagogical impact. Currently, research strength in green chemistry exists in three major centres in Canada where faculty members declare themselves as green chemists: McGill University, Queen's University, and St. Mary's University. Moreover, the mistaken idea that environmental chemistry and green chemistry are synonymous still persists. Hence, there is currently an impasse in simultaneously breaking through the glass floor, *i.e.*, translating green chemistry from research to teaching; and breaking through the glass ceiling, *i.e.*, translating green chemistry from teaching to research. Surprisingly, the subject still remains an elective even at McGill University, the birthplace of green chemistry in Canada, even though it started as a faculty-led initiative. What research faculty may not have yet been convinced of is that knowledge in green chemistry strengthens and improves understanding of the fundamentals of chemistry. So far, green chemistry has been sold as addressing environmental concerns and toxicity/hazard issues associated with chemicals and its benefits are often couched in altruist and activist language. Over the last decade, our collective experience has shown us that students exposed to this subject are better able to master and integrate fundamental ideas regarding reaction mechanisms, organic synthesis, and physical chemistry into a coherent picture for each reaction they learn about, whether it is an organic, inorganic, or polymerization reaction, or some hybrid of these. In addition, they dramatically improve their knowledge and research skills in finding and validating information from the vast and ever-expanding chemical literature. By the very nature of its multi-disciplinary and multi-dimensional nature, students studying green chemistry acquire valuable written and oral communication skills, and gain confidence in expanding their knowledge to include diverse subjects such as patent law, regulatory affairs, and business that are highly marketable in the modern world today. In short, students become both smarter chemists and saleable chemists not just to the chemical industry, but also to other sectors of the economy. We believe it is these points that have not yet been sold strongly enough to the majority of research faculty who remain sceptical of the subject and its motives.

The grassroots initiative developed by students at the University of Toronto shows that they anticipate what is needed to address modern problems, and are prepared to get the necessary education and skills themselves. However, despite their enthusiasm and energy, they were not able to convince research advisors to commit the necessary resources, time, and guidance for participating students to carry out the proposals they came up with in their inaugural Future Leaders GCI Challenge.[27] This outcome is partly due to the faculty scepticism highlighted above, but may also be exacerbated by the fact that no full-time research faculty member currently sits as a permanent member on the GCI executive board, though it does freely consult with faculty for guidance and participation in workshops, judging and outreach activities. Another key problem is that because it is exclusively a student group composed of transient members there will be a constant need to groom successors to keep the group alive over the long term well beyond the founding students' graduations. Having faculty members on board would help to ensure the group's continuity and for green chemistry to gain a strong foothold in the departmental philosophy in both the research and teaching spheres. A formal undergraduate or graduate level green chemistry course with that name has not yet been considered to be part of the course offerings in that department, although, as we have highlighted in this chapter, green chemistry concepts are taught in CHM 343H: Organic Synthesis Techniques. This is in sharp contrast to the UTSC situation which is another campus of the same university. At that campus, chemistry is taught as part of the Department of Physical and Environmental Sciences rather than in a standalone department, and so is already tuned to offer a course with the name 'green chemistry' as part of its curriculum.

The pros and cons of launching stand-alone green chemistry courses *versus* 'greening up' traditional ones have already been discussed in a book dedicated to green chemistry education in lecture and laboratory.[11] A key observation made in that comparison was that teaching students to balance chemical equations the first time they learn about new transformations in traditional organic chemistry courses can go a long way in priming their interest and knowledge for future studies in green chemistry. Such a change would have a minimal impact on course costs and instruction time, and is straightforward to implement. The identification of by-products is not only essential for determination of basic metrics such as atom economy and waste production, but also greatly facilitates the understanding of reaction mechanisms for various classes of reactions. Introduction of a token green chemistry course into the curriculum in isolation from all other departmental offerings (while maintaining traditional courses) defeats the aims and philosophy of green chemistry, which ultimately emphasizes multi-disciplinary problem solving. Although this approach may require less work (since one instructor would be responsible for mounting that course compared to integrating green chemistry principles throughout all courses offered by a department), it can propagate the negative impression that all traditional chemistry courses are 'non-green', and therefore involve potentially 'harmful' chemistry. This type of labelling may set up a conflicting tension that impacts the undergraduate perception of chemistry as a subject of study, particularly since modern students are well-versed in environmental and health impact issues and social responsibility. If a department is contemplating any kind of green chemistry instruction, our experience suggests that the best approach is to align all courses with the basic principles and themes of green chemistry in a general sense.

We have pointed out future directions for curriculum development in green chemistry in a previous review.[22] Key areas that need addressing in a broader sense in green chemistry education include the following:

- Providing convincing evidence that it is cost-effective for a department to practice green chemistry experiments over traditional ones in the undergraduate laboratory by carrying out a full cost analysis that includes disposal costs using both approaches
- Designing undergraduate experiments where the product of one experiment is used in another, rather than immediately discarding that product as waste once the period is over (essentially getting more mileage out of their efforts); for example, synthesize molecule A in one lab period and then determine several physical properties of it in the following lab, or use it as a starting material in a subsequent reaction with a view to carry out a sequential series of reactions in a mini-total synthesis exercise where each reaction is complementary to a named organic reaction students learn about in lecture[35]
- Ensuring that lab exercises have an accompanying metrics analysis to go along with the experiments conducted as this is the only means of providing definitive proof that a given protocol is indeed green compared to past literature procedures; publications of green experiments in the *Journal of Chemical Education*, for example, do not demonstrate this routinely
- Addressing energy consumption issues in carrying out chemical reactions by various heating or cooling operations using various kinds of apparatus.

One final suggestion that can help get green chemistry more traction in Canada is for the Canadian Society of Chemistry (CSC) to have a permanent green chemistry division covering both teaching and research innovations that is represented in all of its national conferences. Up to now the CSC has supported green chemistry by co-sponsoring the 3rd International IUPAC Conference in Green Chemistry held in Ottawa in 2010 and the CGCE awards since 2010 mentioned earlier in this chapter. At national meetings symposia dedicated to green chemistry are left to the discretion of conference organizers and are included within the chemistry education, organic chemistry, or industrial chemistry divisions. The most notable have been the following symposia: Teaching Green Chemistry in Lecture & Laboratory (93rd CSC 2010, Toronto); Green Chemistry and Catalysis in Honour of Tak-Hang (Bill) Chan (94th CSC 2011, Montreal); and Green Chemistry I: Materials, Green Chemistry II: Catalysis, Green Chemistry III: Synthesis & Processing Methods (96th CSC 2013, Quebec City). Though these have been well received there still remains the perception that the subject is on the fringe of established chemistry circles. Following the lead of the American Chemical Society, which has held a dedicated Annual Green Chemistry and Engineering Conference over the last 18 years separate from its two annual national meetings, a decision by the CSC to emulate this model would go a long way to legitimize the subject and entrench it in both teaching and research endeavours. Having a dedicated green chemistry division that is represented annually would send a strong message to the Canadian chemistry community at the national level that the subject has arrived and would help to break through the problems and misconceptions already discussed in this chapter.

11.8   Appendix: Green Chemistry Student Survey

Two questions were put to students attending courses in green chemistry at all of the eight departments listed in Section 11.2. The responses are given here.

Question 1: Would you recommend a green chemistry lecture and/or lab to your fellow classmates?

It is interesting (to) have real life applications in industry and chemistry in general. It gives a good overview of chemistry and its environmental impact. However topics are a little scattered and some harder topics are explained a little fast. Other than that most topics are understood during class.

Important for chemists/researchers to understand the global impact of their work and think about often simpler and more cost effective methodologies.

I believe every potential chemist should take a green chemistry class. It makes you aware of how profound our role is in the scope of environmental impact. It's also very enlightening in general.

This course allows students to see how harmful using certain substances can be and provided methods in order to combat these problems. Due to the large amount of waste and problems caused by the chemistry industry, it is necessary that chemists are aware of the situation.

As someone with an interest in the environment anyway, this course was perfect. I think even if not environmentally minded it is important for us as chemists to take responsibility for what we do and the first step is learning. I had no idea the scale of the waste, but also didn't know any alternatives, so the course has been really good for both of these reasons. It is an interesting topic and a lot of aspects of green chemistry and its applications are not indicated in other courses.

It is eye opening to the wastefulness of most chemical procedures. It also helps to better understand why certain syntheses of drugs, for example, don't make it to large-scale production as a result of its effect on the triple bottom line.

I think this course is important for anyone that is taking any degree related to industrial synthesis as it gives a view of the aspects that need to be considered for taking care of the environment, reducing chemicals and waste, and developing new greener pathways, helping the industry to develop in a 'better way'.

After this course you gain a different point of view of chemical processes; it is not only important for products but also the path to make the products. In the world we live in and in all future research, there will need to be an emphasis on creating green synthesis plans. Taking a green chemistry class as an undergrad brings awareness of where processes are lacking in greenness, as well as an awareness of how much is wasted in the lab. If you start thinking about the problems early on, you have more time to develop solutions.

It is interesting and different from other chemistry courses. Not hard to understand and makes you conscious of the environmental impact of reactions.

I really enjoyed the green chemistry course. It was very interesting. I think a lab would complement this course nicely, giving practical experience to the knowledge obtained in the course. This course works well as fourth year course allowing for the necessary background to be obtained as well as giving real life applications to chemistry.

I feel it is important to educate fellow chemistry students about the importance of optimizing experiments in terms of environmental impact, safety, and waste generation. It will also benefit them in the future by looking for alternative reactants that may either be safer to use, or be cheaper to use, or will yield less waste.

The green chemistry course helped me realize the amount of wasteful and hazardous products/by-products being formed in many synthetic labs and reactions. Being aware of the Twelve Principles of green chemistry and working to apply them requires researchers to really understand the mechanisms underlying any synthesis and reaction, so not only has green chemistry taught me to be more efficient with reactions, but also pushed me to thoroughly understand them in order to apply any changes and improvements.

I believe it is important to understand both the laboratory experiments and the significance of large-scale productions as they occur outside of our usual learning boundaries. The course broadened my chemistry knowledge by exploring many alternatives and methods. Most importantly, it addressed many environmental impacts that I never thought of.

Question 2: How has your overall perception of the subject of chemistry changed as a consequence of taking a green chemistry course?

(I) feel as though we as chemists are incredibly wasteful. Though we have tools at out disposal we are resistant to change and need desperately to improve our methods. I now realize that our future and the future of the next generations is going to be very dark indeed…and full of waste; unless we find very efficient methods to deal with waste, pollution, dissipation of energy, *etc.*

It has changed greatly. Mostly for the reason stated above. I also believe chemists don't take enough responsibility for the 'green-ness' of their chemistry practice.

It made me realize how much of an impact green chemistry research can have on our society and environment. It made me appreciate the social/financial sides of chemistry that will play a more and more important role in the advances of technology. I feel like I got a better scope of the impact that humans and especially chemists have on the environment. It was interesting to see how we can play such a HUGE role in rethinking reactions to change the environmental impact they have.

I realized that simple things can be changed in my day-to-day lab work if I only CHOOSE to make that decision to try to be greener. I think using different solvents could be the easiest and have the most effect in terms of my "footprint".

It helped me to understand how important it is to choose the right chemicals and pathways for a chemical synthesis, to reduce the cost, waste (hazardous substances), social negative impact and take care of the environment.

I have a better insight into the more negative aspects of chemistry as a result of this course.

I started to be aware of the important role of catalysts and new reaction techniques.

I recognize the amount of waste and toxic chemicals that are generated during any kind of laboratory procedure. This is something that I really never considered until I took this course. As a chemist, I want to use my knowledge to benefit the human race. I believe that part of that is minimizing my impact on the environment.

I feel that many chemists are still unaware of the possibilities of optimizing their work by applying the principles of green chemistry. Chemistry labs structured towards students also do not apply the principles of green chemistry which produce a lot of waste.

It has helped motivate me to understand and question the WHY's and HOW's of any chemical reactions.

I realize the importance of consideration for every substance used in each stage of any reaction. In a commercial operation, there is more than just the final product – by-products, wastes, cost effectiveness, efficiency, environmental impact, health and safety, *etc.* In a lab, the technician often deals with our waste, while most of our focus is on the progress and results.

Being able to analyse and evaluate experimental protocols using metrics. Knowing the difference between green chemistry and environmental chemistry.

The same two questions were also put to 413 undergraduate students (see Section 11.5.1) and the responses are given below.

Question 1: Would you recommend a green chemistry lecture and/or lab to your fellow classmates?

(It is) just as useful for learning general concepts/laboratory techniques. It's the future of chemistry. It is more ethical.

Chemistry, green or not, is damaging. I hope to learn about chemistry without being considered to be green or not. Green chemistry will inhibit the variety of experiments available.

I don't feel I had enough experience with green chem. labs to confidently answer this.

Based on the one lab we did, I would recommend it as it wasn't any more difficult and it is environmentally safe.

It allows us to keep the world clean and to apply it to industry standards.

I think ecologically conscious alternatives are essential for science. Therefore everyone should learn about it.

The industry is most likely to move towards green chemistry as a result of environmental and government pressure.

It is very important to understand the effects of the harmful substances used in the lab. I think it's very useful and conscientious to students to use and understand green alternatives.

I would recommend anything 'green' to fellow classmates. I would like to see it more widely used and normalized. I feel that green chemistry is becoming increasingly important in today's world and having a good understanding of green chemistry can only help individuals in the workplace.

I think it should just be added to an existing course instead of a course on its own.

I think it is more cost effective (probably not in all cases) if you reduce your materials use; decreasing environmental impacts is becoming increasingly popular today, so it is good to know we can contribute as well.

As sustainability becomes more an issue, scientific and technological advances should have a degree of sustainable efforts.

It seems like an interesting field, but one lab didn't make me feel confident that I can perform all of the labs in green chemistry.

It was interesting to realize how much waste was actually produced in the more traditional experiments.

(It is) more simple for understanding mechanisms and reactions (and) less harmful to the environment.

Very random; wished we would have learned something in class about it first.

Everybody in chemistry should know how to perform reactions that make little waste.

I don't really know since we only did one lab in green chemistry and the green aspect was really underplayed.

We only had one experiment; it did not feel drastically different from other labs. It is a new approach for those who have only started on the organic chemistry laboratory course. It also gives students the idea that being aware of the environment is important, even in a laboratory setting.

It will help sustain (the) environment; we need to do more of this.

From the short period in the lab, little was learned about green chemistry. It is beneficial to the environment but not something of interest to me.

It improves understanding of aldol chemistry and mechanisms of reactions. Everyone must be more aware of ways to reduce environmental waste in today's world. It is especially relevant in chemical laboratories where large amounts of highly toxic waste may be made.

(It) applies to industry after university.

I don't really feel that I learned a ton about green chemistry specifically, so I don't see how different a green course would be. However, I feel that learning about environmentally friendly techniques would be valuable.

It's fun and it's important to do our part in helping preserve the environment starting with education.

Green chemistry reduces wastes and impacts that chemistry has on the environment. If the reaction yields the same product, then reducing harmful impacts will always be welcomed and helpful.

It is good for at least one lab to show students proper disposal, use and implementation of green chemicals. The twelve rules of green chemistry helped me understand the positive attributes of green chemistry in the lab.

I do not feel like green chemistry is major part of learning. A separate course is not required but it would be beneficial to implement some parts in current courses.

Because we need it. Many/most people are wasteful by default and this needs to be combated. A green chemistry course is an excellent idea.

Not exactly a 'must have' as waste produced in chemistry labs pales in comparison to other sources.

Education about sustainable practices is important to me because I care about the environment.

It is always good to practice green chemistry to harm the environment as little as possible, but the green chemistry in this course was very limited and not clearly defined.

It is an interesting perspective to organic chemistry and the fact the reactions are relatively environmentally friendly is pretty cool.

Lots of waste is generated by chemistry experiments, so we need to learn safer ways to dispose of the waste and learn safer, more environmentally friendly ways to do experiments.

Yes, because green chemistry is more environmentally friendly, so it would be better than a chem. course that isn't environmentally friendly.

It would be a good way to see a more practical application of the work you do in chem. labs and how labs work towards affecting the environment less.

I think it's important to understand our effect on the environment, especially in the lab. Many people don't know or don't care about the environment, but they should.

We did only one lab pertaining to green chemistry. There was nothing else to convince me to take a green chemistry course.

It's important for the environment and by being less wasteful one can maximize on one's resources. Everyone being aware of the impact of our production of everyday products, our consumption and waste of them, and what realistically needs to be done to accomplish 'greener' practices is becoming increasingly important in today's society.

If green chemistry was more of a priority in organic chemistry labs and integrated more efficiently, then I might recommend a green chemistry course to a friend. But, it seems like we do not have the necessary understanding to make it seem worthwhile.

Question 2: How has your overall perception of the subject of chemistry changed as a consequence of performing a green chemistry experiment?

Using less wasteful methods is always a bonus in any field. However, there was not enough exposure for us to determine an effect.

One lab has not significantly affected my experiences, but I can appreciate and understand how much of an impact a green chemistry course is making.

I now know that green chemistry exists and the key points behind green chemistry.

I had no conception of how much waste must be incurred as a result of chemical procedures. Now I do.

If simple org. chem. synthesis procedures can be made 'green', then the chemical industry as a whole should invest in greener initiatives.

That regular chemistry is very wasteful and can have negative consequences on the environment. More green chemistry should be implemented.

I have a more negative view of how chemistry is being done (traditional). I think more green practices should be implemented.

Continue it, but elaborate more in lecture.

I did not realize how much a single class wastes. Not much because we did such minimal green chemistry, but I now understand I am aware that organic reactions can occur in a 'greener' manner.

I didn't know a 'clean' way of doing some reactions existed, since I didn't know a 'harmful' way existed.

Not much, as green chemistry during the course of this lab was very limited.

It hasn't changed significantly, but I now understand that green chemistry should be an integral part of the chemistry field.

Chemistry is now environmentally concerned.

I am more excited about green chemistry and exploring the changes it will bring about in the world of chemistry.

That chemistry labs can indeed be done to better the environment and it is a good method that should be implemented in all lab experiments.

It made me realize how much waste is produced in labs.

It has changed from thinking there was not much thought put into it to seeing it as easily and potentially very green.

I liked learning about it, but I had a very hard time comprehending it all. But, would like to learn how to understand it better.

(It) made me aware that a huge amount of waste is generated as a consequence of org. chem. labs.

It made me think about how much waste chem. labs create, and just how toxic that waste can be for people's health and the environment.

I never knew it existed before.

It's not all 'throw solutions together to get a product'. It's more 'get the product efficiently'.

I'm glad that not all reactions are harmful to the environment. (It gave me) a more developed knowledge of organic chemistry.

Before, I thought chemistry lab work wasted a ton of chemicals that could harm the environment. Now, I understand that work can be done without compromising the environment.

I started feeling like a global citizen because I was no longer worried about just my grades, but about helping out the environment as well. It helps the world picture.

It hasn't changed my perception due to the lack of acknowledgement in class lectures.

I have more respect for the subject due to measures taken to be more environmentally friendly.

It's great. It should be introduced to all chemistry courses, not just in university.

I see there is a lot of waste made in the chemistry labs that we do and if we could perform more green chemistry we could minimize this.

I have a greater appreciation for the chemistry community for trying to reduce waste. I also see that there are always alternatives in chemistry to create more environmentally friendly reactions.

It changed my thoughts a little as I know now that during the experiment nothing will be harming the environment.

It made me think more about the amount of waste that is created in a lab which is somewhat concerning; but, I appreciate what green chemistry is trying to accomplish. It has not changed that much. I agree that experiments should be more green but there is already so much pollution in the world that green chemistry will have no effect.

It has changed a little, but not enough green lab work was done to influence my perception 100%.

I am now more aware of how much waste product is made in a single experiment, and that there have been ways introduced or proposed to reduce such environmental hazards.

It hasn't. The mention of green chemistry is surprisingly small; no appreciable amount of time was taken to discuss it either.

I feel chemistry is making a step in the right direction. Green chemistry is a good way to learn where minimizing environmental damage. I would be more likely to continue taking chemistry courses if they were more environmentally friendly.

Green chemistry at a 200 level offers little to no effective application to regular processes unless it is clearly stated or researched.

I understand that chemists are able to be environmentally conscious in their work. Chemistry can be greener than I thought.

That there are many ways to do an experiment, for example, different reactants could be used for an addition reaction to occur, wastes can be reduced, using alternative methods and chances of an accident occurring in a lab can be minimized.

I think it is more important and more attention should be given to it, more legitimately than we did.

No, chemistry is chemistry...green or not.

(It gave me) better impressions of how green chemistry affects companies. It is good to know hazardous properties of your compounds and the extent of their impact in the environment.

I understand that there might be some environmentally friendly methods to carry out some reactions in industry.

I understand it is possible to be more environmentally friendly when performing chemistry experiments. It is also great to know that some initiative is being taken to limit our environmental impacts.

There was only one experiment involving green chemistry, so my knowledge of it could be expanded.

It hasn't. Everything I knew about going 'green' and unhealthy chemicals I learned on my own. One lab did not enhance anything.

Not that much. I'm just more cautious of what I put down the sink now.

(It) made me realize there are different ways that may be better to do chemistry experiments.

My perception has become broader as a result of taking some green chemistry. (I) didn't realize how much waste there is.

I had no idea quite how wasteful chemistry could be; increased awareness in terms of my use of chemicals in the lab.

Many of the labs don't appear to be very green but some modifications can be made to make the labs more sustainable...should move in this direction. (I) didn't think about it before this.

I feel that I appreciate chemistry more after participating in green lab work. Until cleaner energy methods are developed and commercialized these ideas about green chemistry seem unimportant to the general overall picture.

(It) showed me how chemistry can be environmentally friendly as opposed to my previous perception.

I appreciate that chemists are trying to lessen their impact on the environment.

It's a nice change that not only prevents the pollution but also protects people doing experiments.

(I have) a greater appreciation for the elegant integration of green chemistry into syntheses, *etc.*

References

1. Green Chemistry Canada Network: http://www.greenchemistrycanada.ca (accessed November 2013).
2. *Presidential Green Chemistry Challenge Award Recipients 1996 – 2012*, US Environmental Protection Agency, EPA-744-F-12-001, p. 59.
3. (a) C. J. Li, *Acc. Chem. Res.*, 2002, **35**, 533; (b) C. I. Herrerias, X. Yao, Z. Li and C. J. Li, *Chem. Rev.*, 2007, **107**, 2546; (c) C. J. Li, *Chem. Rev.*, 2005, **105**, 3095; (d) C. C. K. Keh, C. M. Wei and C. J. Li, *J. Am. Chem. Soc.*, 2003, **125**, 4062; (e) C. J. Li and W. C. Zhang, *J. Am. Chem. Soc.*, 1998, **120**, 9102; (f) C. J. Li, *Chem. Rev.*, 1993, **93**, 2023; (g) C. J. Li, *Tetrahedron*, 1996, **52**, 5643; (h) C. J. Li. *Organic Reactions in Water: Principles, Strategies, and Applications*, ed. U.M. Lindstrom, Blackwell Publishing, Oxford, ch. 4, pp. 92–145; (i) C. J. Li and T. H. Chan, *Organic Reactions in Aqueous Media*, Wiley, New York, 1997.
4. (a) C. J. Li, *Acc. Chem. Res.*, 2010, **43**, 581; (b) N. Uhlig and C. J. Li, *Org. Lett.*, 2012, **14**, 3000.
5. (a) C. M. Wei and C. J. Li, *J. Am. Chem. Soc.*, 2003, **125**, 9584; (b) C. M. Wei, Z. Li and C. J. Li, *Synlett*, 2004, 1472.
6. (a) C. J. Li, *Acc. Chem. Res.*, 2009, **42**, 335; (b) Z. Li and C. J. Li, *Pure Appl. Chem.*, 2006, **78**, 935; (c) C. J. Li, *Sci. China: Chem.*, 2011, **54**, 1815; (d) C. J. Li, *Top. Curr. Chem.*, 2009, **292**, 281.
7. Green Centre Canada: http://www.greencentrecanada.com (accessed November 2013).
8. (a) L. Phan, J. R. Andreatta, L. K. Horvey, C. F. Edie, A. L. Luco, A. Mirchandi, D. J. Darensbourg and P. G. Jessop, *J. Org. Chem.*, 2008, **73**, 127; (b) Y. Liu, P. G. Jessop, M. Cunningham, C. A. Eckert and C. L. Liotta, *Science*, 2006, **313**, 958; (c) C. D. Ablan, J. P. Hallett, K. N. West, R. S. Jones, C. A. Eckert, C. L. Liotta and P. G. Jessop, *Chem. Commun.*, 2003, 2972; (d) P. G. Jessop, M. M. Olmstead, C. Ablan, M. Grabenauer, D. Sheppard, C. A. Eckert and C. L. Liotta, *Inorg. Chem.*, 2002, **41**, 3463; (e) P. G. Jessop, D. J. Heldebrant, X. Li, C. A. Eckert and C. A. Liotta, *Nature*, 2005, **436**, 1102; (f) L. Phan and P. G. Jessop, *Green Chem.*, 2009, **11**, 307; (g) P. G. Jessop, S. M. Mercer and D. J. Heldebrant, *Energy Environ. Sci.*, 2012, **5**, 7240; (h) S. M. Mercer and P. G. Jessop, *ChemSusChem*, 2010, **3**, 467; (i) P. G. Jessop, L. Pham, A. Carrier, S. Robinson, C. J. Durr and J. R. Harjani, *Green Chem.*, 2010, **12**, 809; (j) P. G. Jessop, L. Kozycz, Z. G. Rahami, D. Schoenmakers, A. R. Boyd, D. Wechsler and A. M. Holland, *Green Chem.*, 2011, **13**, 619; (k) S. M. Mercer, T. Robert, D. V. Dixon, C. S. Chen, Z. Ghoshouni, J. R. Harjani, S. Jahanjiri, G. H. Peslherbe and P. G. Jessop, *Green Chem.*, 2012, **14**, 832.
9. (a) L. Phan, H. Brown, J. White, A. Hodgson and P. G. Jessop, *Green Chem.*, 2009, **11**, 53; (b) A. Holland, D. Wechsler, A. Patel, B. M. Molloy, A. R. Boyd and P. G. Jessop, *Can. J. Chem.*, 2012, **90**, 805.
10. Memorial University Chemistry Department: http://www.chem.mun.ca/zfac/fmk.php (accessed November 2013).
11. J. Andraos, *Green Organic Chemistry in Lecture and Laboratory*, ed. A. P. Dicks, CRC Press Taylor & Francis, Boca Raton, 2012, ch. 2, pp. 29–68.

12. P. Anastas, F. Wood-Black, T. Masciangioli, E. McGowan, and L. Ruth, *Exploring Opportunities in Green Chemistry and Engineering Education: A Workshop Summary to the Chemical Sciences Roundtable*, National Research Council, Washington, D.C., 2007.

13. (a) A. P. Dicks, *Green Chem. Lett. Rev.*, 2009, **2**, 87; (b) A. P. Dicks, *Green Chem. Lett. Rev.*, 2009, **2**, 9; (c) A. P. Dicks. in *Green Organic Chemistry in Lecture and Laboratory*, ed. A. P. Dicks, CRC Press Taylor & Francis, Boca Raton, 2012, ch. 3, pp. 69–102.

14. J. M. Stacey, A. P. Dicks, A. A. Goodwin, B. M. Rush and M. Nigam, *J. Chem. Educ.*, 2013, **90**, 1067.

15. (a) K. J. Koroluk, D. A. Jackson and A. P. Dicks, *J. Chem. Educ.*, 2012, **89**, 796; (b) K. J. Koroluk, S. Skonieczny and A. P. Dicks, *Chem. Educ.*, 2011, **16**, 307; (c) E. Aktoudianakis, R. J. Lin and A. P. Dicks, *J. Chem. Educ.*, 2006, **83**, 1832; (d) E. Aktoudianakis and A. P. Dicks, *J. Chem. Educ.*, 2006, **83**, 287; (e) R. G. Stabile and A. P. Dicks, *J. Chem. Educ.*, 2003, **80**, 313; (f) L. L. W. Cheung, S. A. Styler and A. P. Dicks, *J. Chem. Educ.*, 2010, **87**, 628; (g) L. L. W. Cheung, R. J. Lin, J. W. McIntee and A. P. Dicks, *Chem. Educ.*, 2005, **10**, 300; (h) R. G. Stabile and A. P. Dicks, *J. Chem. Educ.*, 2004, **81**, 1488; (i) R. G. Stabile and A. P. Dicks, *J. Chem. Educ.*, 2003, **80**, 1439.

16. L. L. W. Cheung, E. Aktoudianakis, E. Chan, A. R. Edward, I. Jarosz, V. Lee, L. Mui, S. S. Thatipamala and A. P. Dicks, *Chem. Educ.*, 2007, **12**, 77.

17. E. Aktoudianakis, E. Chan, A. R. Edward, I. Jarosz, V. Lee, L. Mui, S. S. Thatipamala and A. P. Dicks, *J. Chem. Educ.*, 2008, **85**, 555.

18. E. Aktoudianakis, E. Chan, A. R. Edward, I. Jarosz, V. Lee, L. Mui, S. S. Thatipamala and A. P. Dicks, *J. Chem. Educ.*, 2009, **86**, 730.

19. A. P. Dicks and R. A. Batey, *J. Chem. Educ.*, 2013, **90**, 519.

20. A. P. Dicks. *Green Organic Chemistry in Lecture and Laboratory*, ed. A. P. Dicks, CRC Press Taylor & Francis, Boca Raton, 2012, pp. 257–271.

21. A. P. Dicks, *Chem 13 [Thirteen] News*, 2010, **377**, 17.

22. J. Andraos and A. P. Dicks, *Chem. Educ. Res. Pract.*, 2012, **13**, 69.

23. For Analytical Chemistry, see: (a) M. Koel and M. Kaljurand, *Pure Appl. Chem.*, 2006, **78**, 1993; (b) A. Gauszka, P. Konieczka, Z. M. Migaszewski and J. Namieœnik, *Trends Anal. Chem.*, 2012, 37, 61; (c) S. Armenta and M. de la Guardia, *J. Chem. Educ.*, 2011, **88**, 488; (d) P. R. M. Correia, R. C. Siloto, A. Cavicchioli, P. V. Oliveira and F. R. P. Rocha, *Chem. Educ.*, 2004, **9**, 242; (e) L. U. Gron, *Green Chemistry Education: Changing the Course of Chemistry*, Am. Chem. Soc. Symp. Ser., ed. P. T. Anastas, I. J. Levy and K. E. Parent, American Chemical Society, Washington, D.C., 2011, vol. 1011, pp. 103–116; (f) E. J. Olson and P. Buhlmann, *J. Chem. Educ.*, 2010, **87**, 1260.

24. For Inorganic Chemistry, see: (a) J. P. Canal, *Chem. Educ.*, 2009, **14**, 26; (b) R. A. Clark, A. E. Stock and E. P. Zovinka, *J. Chem. Educ.*, 2012, **89**, 271; (c) J. E. Huheey, *Green Chemistry: Designing Chemistry for the Environment*, Am. Chem. Soc. Symp. Ser., ed. P. T. Anastas, T. C. Williamson, American Chemical Society, Washington, D.C., 1996, 626, pp. 232–238.

25. For general chemistry, see: S. Prescott, *J. Chem. Educ.*, 2013, **90**, 423.

26. For science and non-science majors, see: (a) R. Manchanayakage, *J. Chem. Educ.*, 2013, **90**, 1167; (b) E. M. Gross, *J. Chem. Educ.*, 2013, **90**, 429.

27. University of Toronto Green Chemistry Initiative: http://www.chem. utoronto.ca/green (accessed November 2013).

28. J. Andraos and M. Sayed, *J. Chem. Educ.*, 2007, **84**, 1004.

29. J. Andraos and J. Izhakova, *Chim. Oggi/The Int. J. Ind. Chem. Biotech.*, 2006, **24**, 31.

30. S. M. Mercer, J. Andraos and P. G. Jessop, *J. Chem. Educ.*, 2012, **89**, 215.

31. M. L. Cheney, M. J. Zaworotko, S. Beaton and R. Singer, *J. Chem. Educ.*, 2008, **85**, 1649.

32. P. Cardinal, B. Greer, H. Luong and Y. Tyagunova, *J. Chem. Educ.*, 2012, **89**, 1061.

33. C. L. Williamson, K. E. Maly and S. L. MacNeil, *J. Chem. Educ.*, 2013, **90**, 799.

34. L. J. G. Edgar, K. J. Koroluk, M. Golmakani and A. P. Dicks, *J. Chem. Educ.*, 2014, **91**, 1040.

35. M. R. Dintzner, C. R. Kinzie, K. Pulkrabek and A. F. Arena, *J. Chem. Educ.*, 2012, **89**, 262.

CHAPTER 12

Green Chemistry Education in Russia

NATALIA TARASOVA[a], EKATERINA LOKTEVA*[b], AND
VALERY LUNIN[b]

[a]D. Mendeleev University of Chemical Technology of Russia, 47 Miusskaya
Square, Moscow 125047, Russia; [b]Lomonosov Moscow State University,
Faculty of Chemistry, Leninskie gory 1, building 3, Moscow 119991, Russia
*E-mail: les@kge.msu.ru

12.1   The Perception of Green Chemistry Concept in Russia as the Base for the Construction of Educational Schemes

The Russian Federation (Russia) is a big country; its territory is the largest in the world, and it contains sparsely populated areas along with megalopolises and industrially developed areas. Even more important, Russia is a multi-national and multi-cultural country. The results of the 2010 census show more than 200 nationalities in Russia of very different cultural and religious traditions, although ethnic Russians comprise 81% of the country's population. Thus the mentality and educational level of population are not uniform. As a result of the huge size of the country, environmental problems in separate locations may not individually touch the inhabitants of other areas.

Worldwide Trends in Green Chemistry Education
Edited by Vânia Gomes Zuin and Liliana Mammino
© The Royal Society of Chemistry 2015
Published by the Royal Society of Chemistry, www.rsc.org

Additionally, and because of profound political changes during last two or three decades, which have forced people to concentrate on their own economic problems, global environmental problems are not the first priority for many people. Therefore great effort is need to engage the Russian population in responsible behaviour with respect to the environment. The major part of the Russian population is not familiar with the concept of green chemistry; rather they are well informed about the ideas of resource and energy conservation in industry, especially in chemical industry.

However, there is a growing understanding of the need for change. In 2013 two important opinion polls concerning environmental issues were held in Russia. One was conducted by the Russian Public Opinion Research Center on 20–30 November 2013 and published on the website of the Ministry of Natural Resources (selection scope: 1600 respondents, method used: telephone survey).[1] During this survey, 85% of respondents noted that, currently, Russia has environmental problems, and between them 26% of respondents characterized the overall environmental situation in Russia as close to catastrophic. Forty-three per cent of respondents indicated industrial facilities as a source of an environmental hazard. Eighty-eight per cent of the population surveyed believed that sustainable (green) production technologies can help to improve the environmental situation in the country, while 56% of respondents believe that the implementation of green technologies and reduction of industrial emissions should be a priority for Russian environmental policy. It appears that the population of Russia is interested in green industries, including the chemical sector. About half of the respondents expressed a willingness to use environmentally friendly types of transport and environmentally friendly fuels, so it is possible to conclude that, potentially, there is a wish within the population of Russia to give preference to green chemical products.

The other survey[2] was held to clarify the issues related to innovation policy, competitiveness and prospects for enterprise development after Russia joined the World Trade Organization (WTO) in 2012. The questionnaire contained the question section on green chemistry and its role as potential growth area for the Russian chemical industry. One hundred and four companies which participated in the survey are engaged in the production of the wide range of chemicals. Sixty-one per cent of surveyed enterprises had identified themselves as large, while others identified themselves as medium (32%) and small (7%). Eighty-two per cent of surveyed companies manufacture products for industrial use and 18% manufacture consumer products; some of them are focused towards the Russian market, the others are export-oriented.

The commercial viability of green chemistry is estimated as relatively high by Russian manufactures of chemical products for further industrial use (53%); the companies producing household chemicals and other chemical products for public use are more sceptical (40%), only 4% of the latter have expressed strong belief in the possibility of green chemistry being profitable. This situation suggests that Russian companies have little faith in the Russian public as possible consumers of green chemistry products. The chemical enterprises with the main focus on the Russian market are generally more optimistic: 19% are confident that joining the WTO will allow implementation of the principles of green chemistry effectively from the profit-making point of view (answer, 'yes'), and another 29% (the answer is, 'probably yes') believe that it is possible. About 5% of export-oriented enterprises answered 'yes', 37% answered 'probably yes', showing a greater restraint in the estimates which could be explained by two reasons: (1) some companies have not yet faced the realities of the WTO and competition of foreign producers; and (2) the specific structure of exports. Modern Russia mainly exports fertilizers, oil and oil products, metals; these product types are not fully applicable to the green chemistry concept, making it rather vague, confusing and, as a result, not sufficiently important to be taken into account during the export activities.

Careful analysis of the data of these two surveys clearly demonstrates the great need of education and enlightenment of both public and experts communities not only to give the basic knowledge in the field of green chemistry but also to demonstrate perspectives connected with this approach. Often, even the term 'green chemistry' is misapprehended in Russia as meaning only 'chemicals produced from plants' or interchangeable with 'ecology'. The term 'sustainable chemistry' is even vaguer in Russian translation.

Modern Russian officials understand the need to move to new, greener technologies and regard the development of green chemistry as a major component of this transition, as a top priority recognized at the highest governmental level. The need to move towards a green economy was a major point in the speech of the Prime Minister of the Russian Federation, Dmitry Medvedev, at the UN Conference on Sustainable Development 'Rio+20' (June 2012). He noted that this approach is a key tool for sustainable development and environmental management.

During the meeting of APEC Ministers responsible for environmental protection in Khabarovsk (July 2012) the Russian Minister of Natural Resources and Environment, Sergei Donskoi, stated that modern Russia is taking steps towards greening the economy with a main focus on conservation of the environment and rational use of resources.[3]

Several steps have been made in that direction:

- Approval of the State Programme 'Environmental Protection' for 2012–2020 by Governmental decree No. 2552-r of 27 December 2012
- Approval of the Comprehensive Programme of Biotechnology Development for the period up to 2020 by the decree of the Chairman of the Government of the Russian Federation No. 1853p-P8 of 24 April 2012
- Following the meeting on innovative development in the field of environmental safety and the rational use of natural resources held on 17 May 2013, Dmitry Medvedev issued instructions for the Ministry of Economic Development (Andrei Belousov) and the Ministry of Natural Resources and Environment (Sergei Donskoi) to submit to the

Government proposals on drafting directives to establish environmental specifications for the purchase of items to be used as criteria for assessing the requests of those involved in the purchase of goods and services acquired by companies with state capital under Federal Law No. 223-FZ of 18 July 2011 On the Purchase of Goods, Works and Services by Specific Categories of Legal Entities[4]

- Signing of the Edict No 752 of 30 September 2013 On Reducing Greenhouse Gas Emissions by the President of RF Vladimir Putin. The goal of this edict is 'To ensure reduction of greenhouse gas emissions to 75% against the 1990 emission level by 2020'.
- The decision to organize the State Centre for Green Chemistry made during the visit of UNIDO (United Nations Industrial Development Organization) President Yong Li to Russia in October 2013.

These governmental initiatives give the hope that green chemistry education (GCE), now driven mainly by the efforts of small groups of individuals as well as several universities and organizations in Russia, will become one of the main components of education in Russia.

In this chapter we attempt to elucidate the activity of the universities situated in different parts of Russia. The diversity of the economics of Russian regions defines the characteristic features of the GCE in different universities. But there are common features as well, first of all close connection of research and educational activity in green chemistry and significant contribution of researchers working in institutions of Russian Academy of Sciences (RAS) into the development and implementation of GCE in the universities.

GCE is aimed primarily at the students and graduates studying chemistry, chemical engineering, and environmental sciences, as well as at researchers and teachers dealing with the same topics. Therefore the professional training in the field of green chemistry in various forms is necessary. As for other groups, for example, secondary school students, mass media or general public, having no professional connection with chemistry, the GCE has to be an important part of education for sustainable development (ESD). ESD introduces knowledge and skills to encourage harmonious relations between humankind and nature. It provides necessary conditions for the creation, functioning and further development of the whole system of general public education at all levels: pre-school and kindergarten, primary and secondary school, higher education, life-long professional training, informal education for local communities and, finally, informal education for the mass media. The methodology adopted by the Russian educators who are engaged in promoting ESD in the country is based on the following principles:

- Variety of forms and methods of education and upbringing
- Inter-relation with other education programmes
- Programme continuity at different levels
- Adaptability
- Consideration of local conditions

- Inseparability of general, professional and environmental education
- Practical activities.

Green chemistry constitutes an important part of the platform for sustainable development because modern people in everyday life widely use chemicals as constituents of food, household articles, medicines, cosmetics, and face chemical pollution of air, water and soil. For Russian inhabitants, the negative attitude to chemistry is generally not typical, but it increases significantly in locations close to chemical enterprises, both operating and closed, both modern and using obsolete equipment. A positive perception of chemistry as an integral part of humankind development can be provided by the use of the green chemistry concept. That is why the enlightenment of general public is a challenging and necessary component of GCE in Russia.

Classical chemistry education in Russia from the times of the USSR has been characterized as being of high quality because of its fundamental character and close contact with research institutions. In the 1990s, during the period of political transition, it was saved by the efforts of the 'old guard' of secondary and high school teachers. The graduates of famous Russian universities, such as Lomonosov Moscow State University (MSU), Novosibirsk State University (NSU) and many others easily found good jobs not only in Russia, but also in high rank foreign universities, research institutes and firms.

As for education in the field of chemistry engineering, a certain brain-drain of gifted young people from this field was observed in 1990–2000. It was connected to a total decline of industrial production in Russia and occupational prestige of this profession. To some extent such a reduction in prestige has been observed for the researchers also. Some political initiatives of the last years directed to the reform of the Russian Academies (of Science, Medicine and Agricultural Sciences) are making the modern situation uncertain and the perspectives vague.

Nevertheless, the higher education system for engineering in Russia is undergoing transition to a new educational paradigm: the professional training of future specialists in a holistic manner. Addressing the need for sustainable use of natural resources, energy conservation, environmental protection, prevention of technological accidents, and risk management requires the active participation of engineers, with their particular knowledge and skills. Hence, the goal of education must be training specialists within a holistic life paradigm, so that they can be responsible citizens. The special role of engineer–chemists should be mentioned, because environmentally friendly technologies and green chemistry could greatly help in the transition to sustainability.

During the former Soviet Union's period of industrialization and construction of a centrally planned economy, higher education was subjected to quite strict state control over the structure, content of curricula, and the style of teaching. In the USSR and then during the first decade of independent Russia the only way of higher education organization was specialists programme which included from 5 to 6 years (depending on the specialty) of continuous education. Then gradual transformation to two-stage education was initiated by the educational reform. Now some universities have the traditional for

European and U.S. universities two-stage system, including preparation for Bachelor's degrees and then Master's degrees. Until now, the former specialists programme existed on equal terms with the two-stage programme. For some specialties (*e.g.*, medicine) specialist programmes will continue in the future as the only option. In the Soviet period the in-depth specialization of graduates and their placement was the responsibility of state ministries and higher institutions. Such an approach resulted from the fast growth of industry and the beginning of the Cold War. Today universities provide possibilities for job hunting but the choice is the responsibility of graduates. New techniques and technologies in all branches of material goods production make it evident that the role of chemists and engineers is becoming more important. Graduate specialists trained in a systemic and holistic manner can become agents of change in the practical implementation of the concept of sustainable development. The leading classical and technical universities in Russia are steadily moving in this direction.

In general, the scheme of GCE in Russia can be visualized as an 'educational tree' presented in Figure 12.1.

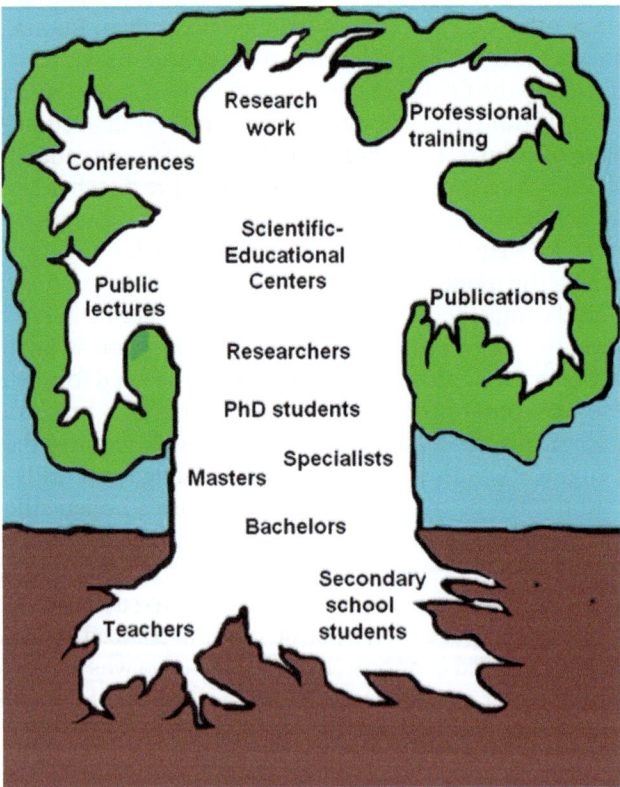

Figure 12.1   The 'green educational tree', the scheme of green chemistry education in Russia.

12.2   Green Chemistry Education in Universities

12.2.1   Methodology

Education for green chemistry as a part of sustainable development education is a very complex and innovative process. Generally Russian universities use the following approaches in the educational process that were first implemented at D. Mendeleev University of Chemical Technology of Russia (MUCTR):

- An interdisciplinary approach, which implies the combination of traditional forms of education with innovative ideas and methodologies. Young chemists learn chemistry along with special courses on green chemistry and its parts, sustainable development and environmental risk assessment and management, as well as sustainable patterns of production and consumption.
- An over-arching approach for dealing with global problems at the local level.
- Active personal involvement and interaction between educators and students.
- Use of role-playing and other active methods of student's involvement in creative participation within the teacher–student exchange.
- Simulation games with small groups of students, which create real-life problematic situations in need of green chemistry or sustainable solutions.

The authors of this chapter are working in the oldest and the most distinguished institutions in the field of GCE in Russia. The first is the Institute of Chemistry and Problems of Sustainable Development at MUCTR, created in 1995 as Department for the Problems of Sustainable Development and transformed to Institute in 2000, headed by Professor, member of RAS Natalia Tarasova. The second is the Scientific–Educational Centre (SEC) Sustainable-Green Chemistry (SGC) of MSU created in 2006 and headed by Professor, academician of RAS Valery Lunin. Thus the experience of these two institutions will be described first. Later, the activity of other university will be reflected. The map of Russia shown in Figure 12.2 represents the location of the universities mentioned in this chapter.

12.2.2   Green Chemistry Education at MUCTR

The advancement of GCE in Russia builds on the efforts of environmental education and education for sustainable development in its best methodologies and forms. The 'greening' of higher education in the Soviet Union began in 1983 at the then Moscow Mendeleyev Institute of Chemical Technology, with the initiative of its rector, the future Minister of Education and academician G.A. Yagodin, who founded the Department of Industrial Ecology.

Figure 12.2   The locations of the Russian universities mentioned in this chapter. This map of Russia was taken from http://ru.wikipedia.org/wiki/ File:BlankMap-RussiaDistricts.png#filelinks and adapted by addition of stars and city names. This work is licensed under the Creative Commons Attribution-ShareAlike 3.0 Unported License. To view a copy of this license, visit http://creativecommons.org/licenses/by-sa/3.0/ or send a letter to Creative Commons, 444 Castro Street, Suite 900, Mountain View, California, 94041, USA.

In order to bring the knowledge of sustainability into the world of professional engineers, the university became the first school in Russia to organize the Department for the Problems of Sustainable Development in 1995 and, in 2000, the Institute of Chemistry and the Problems of Sustainable Development at the MUCTR was established. This unique educational institution now includes:

- The Higher College for Rational Use of Natural Resources
- The UNESCO Chair in Green Chemistry for Sustainable Development
- The Department of Sociology
- The Department of Risk Assessment and Risk Management
- The Higher Chemistry College of the RAS
- The Higher School of Environmental Sciences.

All technical majors in environmental protection at MUCTR are grouped under the umbrella of the Chair of Industrial Ecology. In compliance with the recommendations of the international conference, *Environmental Chemistry. Environmental Engineering* is now a compulsory course for all future engineers. The Educational Department has developed programmes and specialized courses, such as *Development and Natural Resources* for its students who are studying to become chemistry teachers.

Since 1995, two compulsory courses on sustainable development, *The Problems of Sustainable Development* and *Industrial Security and Risks* have been included into the curricula of all departments and institutions of MUCTR.

It is important to note that the Earth Charter is integrated into the text of lectures given to the students of MUCTR during a special course *Problems of Sustainable Development*. It is being taught to all students during their second year for 108 hours, of which 36 hours are dedicated to the lectures, and 72 hours to practical and individual work (one semester).

All students have to master the computer game 'Stratagem' as an after-class exercise. The game is based on the theory and practical methods of applied systems approach and management introduced by Dennis Meadows. The game requires close interaction of several participants united in a working group that helps participants master the fundamentals of system thinking and apply them concretely to the solution of each given problem. The outcome of the game depends on coordinated decision-making that integrates the needs and perspectives of every participant.

Students also watch four documentaries on the most complex problems of sustainable development (also completed after class). During their group exercises, and also as an individual assignment, students work on finding solutions to the problems raised in these films and present them to their professor. All students have to prepare and present three papers and one thesis on sustainable development issues. Out of the three mandatory papers, one must be written on the Earth Charter principles and their importance for sustainable development processes.

By applying these methods, MUCTR plays an important role in the education of new generations of professionals ready to face interdisciplinary challenges and find innovative and sustainable solutions.

Since the year 2000, several summer schools have been organized at MUCTR to update young university faculty on innovative pedagogical methodologies. The attendees were presented with programmes in such subjects as sustainability, democracy and justice, the goals of sustainability, the reorientation of existing education towards sustainable development, and the best pedagogical practices and experiences at the international and national level. The Department of Industrial Ecology at MUCTR organizes scientific–practical student expeditions that seek to provide students with practical knowledge of education for green chemistry and sustainable development. These expeditions have been very successful, and the results of several ones were included in the National Report *Lead Pollution of the Environment and its Influence on Public Health*. In April 2007, the Academic Council of the Institute of Chemistry and the Problems of Sustainable Development at MUCTR approved the professional oath that each graduate of the Institute takes during the graduation ceremony, starting with the 2007–2008 academic year. The text of the oath is built on the main principles and values of sustainability.

12.2.3   Green Chemistry Education at MSU

SEC SGC of MSU was created in 2006 to implement the concept of green chemistry in the body of chemistry education in MSU. Previously, the elements of green chemistry approaches were included into ecological courses, such as *Risk Assessment* provided at the Department of Chemical Engineering, or *Environmental Chemistry and Toxicology*, provided at the Department of Organic Chemistry.

The first task of the staff of SEC SGC was to estimate and perform the inventory of existing educational programmes in the field of green chemistry at MSU as a leading classical university in Russia and in other universities and educational institutions in Russia. This difficult task was simplified by the fact that classical universities in Russia are united into the Association that includes the Educational–Methodical Council on Chemistry. It was found that the green chemistry principles have been already included in the educational programmes of many Russian universities, as it will be described below. The inventory also shows that some basic compulsory courses of Faculty of Chemistry of MSU contain green chemistry elements. Among these the courses of *General Chemistry* (Prof. L. Kustov), *Organic Chemistry* (Prof. S. Vatsadze), and *Physical Chemistry* (Prof. V. Eremin, Prof. S. Tkachenko) for chemistry and other faculties have to be mentioned.

MSU consists of 40 faculties and several scientific institutions. Apart from the Faculty of Chemistry it includes other branches providing chemistry courses as a major part of educational programme: Faculty of Materials Science and Faculty of Fundamental Physical and Chemical Engineering. At the associated faculties, such as Physical, Biology, Bioengineering and Bioinformatics, Fundamental Medicine, Geology, and Soil Science faculties, chemistry is the essential part of educational programmes. Now while teaching basic courses of general, organic and physical chemistry the lecturers pay special attention to the topics of green chemistry, such as the Twelve Principles, biomass as the future resource of energy and chemicals, widening of the application of catalytic processes, possibilities provided by the use of 'green' solvents, *etc*. To encourage the teaching staff E. Lokteva presented the report about green chemistry on the fortnightly staff meeting of the Department of Physical Chemistry, the biggest department of the Chemistry Faculty.

The next educational step of SEC SGC was the development of the programmes of Master's and Bachelor's degrees under the common name 'Chemistry for sustainable development'. To elaborate these programmes specialized lecture courses for Master's students were created, namely the courses devoted to catalysis, the use of supercritical fluids in different branches of chemistry, and innovative for Russia, the specialized course *Methods of Green Processes Realization*. The latter course is presented by E. Lokteva both to specialist students at MSU and to Master's students at Gubkin Russian State University of Oil and Gas (GUOG).

An important part of the green chemistry approach is the implementation of practical skills and possibility to estimate short-term and long-term consequences of day-to-day activity of professional chemists. Very helpful for this purpose is INTERNET-resource Novel Organic Practicum (NOP) developed in Germany (Regensburg University, Germany) by Prof. Burkhard König and his team[5] that includes, as a significant part, the estimation of the environmental impact of different training assignments. The staff of SEC SGC have become a part of the international group and performed the translation of this practical course into Russian. The Internet-site of SEC SGC[6] allocated the advertising of this resource for Russian universities. It is important that NOP is free for the users.

The SEC SGC performs scientific work in the field of green chemistry to provide possibility for students and postgraduates to participate and master the practice in this field, especially in catalysis (Prof. V. Lunin) and the chemistry of humic substances (Prof. I. Perminova). It has to be stressed that during last 5 years the Russian Ministry of Education and Science (RMES) has been providing grants to maintain the work of scientific–educational centres in Russia. The announcement of this programme provoked generation of many SECs in Russian universities and institutions, but SEC SGC was created long before the start of this campaign and easily won the first competition for grants (2009). During 2009–2011 SEC successfully performed the work named *Development of the New Methods of Effective and Environmentally Safe Disposal of Toxic Human-Made Wastes and the Catalysts for these Methods*. In the frame of this project effective reductive catalytic process for the disposal of the different types of chlorinated wastes was developed. This process avoids dioxin formation and makes possible the production of useful hydrocarbons from chlorinated substances. More than 25 young researchers, including students and graduates were involved and mastered their experience in green chemistry (see Figure 12.3). In 2012–2013 the RMES supported another work of the SEC SGC, devoted to the development of micro-mesoporous catalysts for the

Figure 12.3 S. Kachevsky is performing PhD work in SEC SGC using modern equipment of MSU. The photo was taken by E. Lokteva. We acknowledge the agreement of S. Kachevsky for publication of the photo.

use in chemical and petrochemical processes, such as *n*-buthene isomerization to isobuthene, production of isopropylbenzene from cumene, selective
hydrogenation of acetylenes and reductive transformation of polychlori-
nated organics. Under the guidance of Prof. V. Lunin and Prof. I. Ivanova very
effective zeolite and supported metal-containing catalysts have been created.
Micro-mesoporous structure of the catalysts provides the careful tuning of
the reactions selectivity. Thirty-six young scientists (less than 35 years of age)
participated in the last project, between them 13 PhD students and 15 other
students. An important feature of such type of projects is the necessity to pay
more than 50% of wages fund to young participants. The scientific and practical results of these two projects were published as two patents, more than
17 scientific papers in leading international and Russian scientific journals,
and created the possibility for young team members to participate in prestigious conferences in the field of catalysis and green chemistry. A high scientific level of the students' research works was ensured by the use of modern
and high-quality scientific equipment, both industrially produced and made
individually for special tasks. The most expensive and rare devices are used
by researchers through the Common Use Centre of MSU.[7]

An essential role in the implementation of GCE in MSU is played by the
Moscow University Supercritical Innovation Center, MUSIC[8] (Head, Prof.
V. Lunin; Scientific Sectetary, Prof. S. Vatsadze). The main goal of MUSIC is
the training of high-quality specialists in the use of supercritical fluids (ScF)
for research laboratories, performing the study of fundamentals and developing new ScF technologies. The instrumentation of MUSIC includes a unique
mobile ScF set with a high-pressure cell equipped with an inspection window,
four stationary sets with a united control centre as well as a flow-type device.

Three practical tasks are in the process of implementation in the training
courses of physical chemistry for 4th grade students: (1) the test of solubility
of the substances in sc-CO_2; (2) the determination of phase composition and
equilibrium of two-component system; and (3) the determination of critical
point using a unique fibre-optic densitometer.

The Faculty of Chemistry at MSU serves as a base to perform the lectures of
the Open Ecological University (Rector, Prof. Valery S. Petrosyan), which has
functioned since 1987, as a programme of free additional education. This programme provides the possibility for interested students, graduates, researchers and teachers of MSU and other universities to improve their knowledge
in the fields of ecology, environmental protection and green chemistry. It is
provided as various educational projects, which include from 80 to 160 hours
(during one or two years with intermittent exam). During 27 years of functioning the Open Ecological University has performed educational projects not
only in Moscow, but also in other Russian cities (St. Petersburg, Novgorod,
Obninsk, Zelenograd). Within eight blocks of the typical programme one
is usually devoted to environmental and green chemistry. The specific feature of the Open Ecological University is an invitation to the most famous
Russian and foreign scientists in the capacity of lecturers. Thus, during the
long period of the activity of the Open Ecological University, lectures have

been given by Nikita Moiseev, Jores Alfyorov, Valery Legasov, Gennady Yagodin, Victor Sadovnichiy, Vladimir Skulachev, Valery Lunin, Natalia Tarasova and many others. One of the recent projects has been devoted to the Rio+20 Summit.

Starting from 2013, the MSU students of 3rd and 4th grades need to choose the compulsory interdepartmental course within the wide range of courses provided by various faculties of MSU. The Faculty of Chemistry has provided the course of Prof. V.S. Petrosyan *Chemistry, Human Beings and the Environment*. The short experience has demonstrated that besides natural sciences students many social sciences students have chosen this course to attend and learn the subject with the great interest.

12.2.4   Green Chemistry Education at GUOG

It is known that the concept of 'cleaner production' was developed during the preparation of the Rio Summit as a programme of UNEP (United Nations Environmental Programme) and UNIDO. The concept is intended to minimize waste and emissions and maximize product output on the base of careful expert analysis of the flow of materials and energy in a company. Therefore such activity is associated with green chemistry and green technology. In Russia, where oil and gas production constitutes the basis of the economy, the Fund 'National Environmental Management and Cleaner Production Center of Oil and Gas Industries' was established in 1999 in GUOG as a first specialized centre of cleaner production for oil and gas industry in UNIDO practice. The Center, as well as the Department of Industrial Ecology of this university, is guided by Prof. S.V. Meshcheryakov. The placement of the Center in the university makes it possible to involve students of different specialties in the projects and provides practical training on real tasks. Indeed the prevention of wastes formation in the approach of cleaner production contemplates the elimination of chemical, technical, technological and even organizational problems of particular processes or industry, so the participation of the persons skilled in different areas is necessary. A turn-key approach is characteristic for the activity of the Center; it consists of a search for the problem source, proposal of several variants of complex decision, expert estimation and selection of the best one, then its implementation on the particular industry and monitoring of the results. Therefore the activity of the Center is focused on the training of specialists in environmental management and audit on environmental safety and toxic wastes management according to IRCA (International Register of Certificated Auditors) recommendations; environmental audit of oil and gas industries; and consulting in the search and estimation of new technologies of oil-polluted soils recuperation, waste water cleaning, disposal of oil sludge and cuttings, de-NOx, and the selection of the technologies for particular industries of oil and gas complex.

The results of the Center's activity obtained with the participation of students are very impressive. The economic effect due to implementation of the projects on Russian enterprise Astrakhangasprom was about \$850 000, on Orenburggasprom, about \$300 000. New technology of combustion gas cleaning for waste incinerators provides the content of NO_x in purified gas of 50 mg m^{-3}, which is lower than European standard (200 mg m^{-3}).

12.2.5　Green Chemistry Education in Northern and Siberian Universities

Starting from 2009, in the frame of the national project *Education* in each of the federal districts of Russia large innovative educational–research complexes, so-called 'federal universities', were created. Now nine federal universities have united the most distinguished universities and institutions that existed in corresponding regions; but the state universities still plays significant role in Russian education. The extent and content of activity in the field of GCE differ much depending on the local scientific directions and, to a significant extent, on personal interests of leading scientists.

The involvement of renewable resources in production of chemicals is one of the pillars of green chemistry. Russia has the world's largest forest reserves, situated mainly in the north of Russia, in Siberia and Far East. The timber industry is very important for Russia and could be used as a platform to start green technologies, and therefore the GCE in these regions makes emphasis on biomass transformation.

The powerful industries of fuel and energy, mining and smelting, as well as timber industry complexes are situated in the Russian North and Siberia Regions. Many of them use obsolete technologies, which are not consistent with modern environmental demands. For these reasons the pollution of soil, water and air is characteristic for huge territories, not to mention ample solid wastes that are usually stored in piles.

Thus there is a great need to have wide use of waste-free technologies for natural feedstock processing that will meet modern criteria of green chemistry. Also, environmentally friendly technologies of existing and future wastes disposal need to be implemented.

The Northern (Arctic) Federal University (NAFU) situated in Arkhangelsk is the leader in GCE in the Russian North. The elements of green chemistry have been included in the educational programmes *Fundamental and Applied Chemistry* in the field of chemistry, and *Energy and Materials Saving Processes in Chemical Technology, Petrochemistry and Biotechnology* in environmental science. The Master's educational programme *Industrial Ecology and Nature Resources Conservation* was established in the frames of the latter.

The principles of green chemistry are reflected in the educational courses Environment Monitoring and Protection, Chemical Engineering, Modern Chemistry and Safety, Analytical Chemistry of Environmental Objects, Physical Chemistry of Plants Polymers, Methods of Natural Chemicals Analysis, Chemistry of Wood and Synthetic Polymers, Physical–Chemical Bases of Environment Protection Processes in Pulp and Paper Industry and some others.

The Russian North is a wood-producing area, so the most important section of green chemistry is the widening of biomass use. So, many of these courses are based on the new environmentally benign methods of the synthesis of organic and inorganic substances and materials by ScF extraction and chromatography; using ionic liquids, water or ScF as solvents, involvement of biomass as renewable resource into the synthesis of chemicals and energy production.

Since 2009 important part of scientific and educational activity of the NAFU is the common use centre 'Arctic' which is well equipped with the devices for ScF extraction, chromatography, and chemical reactions performing. It helps in the analysis of biomass of different plants composition and properties, and synthesis of new materials in ScF carbon dioxide. Not only postgraduate or doctoral students but also students working on Bachelor's and Master's programmes have free access to such modern equipment during seminars and training as well as for preparation of their own projects.

It is interesting that both MSU and NAFU were named after the same person, the first world-known Russian natural scientist, poet and philosopher Mikhail Lomonosov, who was born in Arkhangelsk Region in the 18th century to a fisherman's family. He was about 19 when he decided to continue his elementary education in Moscow and performed a three-week hike with fish-loaded string and a sleigh, having little luggage except for two study-books. As the heirs of Lomonosov the staff of NAFU started a unique project, *Floating University*. On the board the scientific ship *Vladimir Molchanov* Bachelor's, Master's and PhD students perform research work guided by the teachers from NAFU and MSU, Institute of Environmental Problems of the North (Urals Branch of RAS), Arctic and Antarctic Research Institute, and Northern Department on Hydrometeorology and Environment Monitoring. Mainly such research work is directed to environmental and analytic issues and includes complex hydrochemical, physical–chemical, radiological, biological and other investigations of the water area of Arctic seas, continental and island territories to estimate the influence of anthropogenic factors on the state of ecosystems. During the field period of 2012–2013 the participants of the project Floating University navigated more than 30 314 km in White, Barents, Greenlandic, and Kara seas and investigated the state of the environment of 11 Arctic islands. The 26 406 specimens collected during the navigation were analysed by the participants. The picture (Figure 12.4) presents the moment of such expedition.

Many research and educational institutions of Siberia participate in the development and realization of sustainable development of this region. Renewable forest resources provide a great potential for the development of Siberian economics. Recently, large-scale timber industry complex of Siberia was oriented to the export of industrial wood and timber. Now due to high transportation tariffs the export of industrial timber significantly fell down.

To widen the market for woodworking products there is a great need for the implementation of a deep wood processing to form cellulose, activated carbon, valuable chemicals, and biofuels, because the transportation cost for

Figure 12.4   Research work of students and PhD students of NAFU on the board the scientific ship *Vladimir Molchanov*. The photo is reproduced by kind permission of the author, A. Sazonova, and model, P. Kaplicin.

this type of products is economically sound. Note that additional amounts of products can be formed at processing of wood residue, forming in ample amounts during timber harvesting, mechanical treatment of wood and its chemical treatment on pulp-and-paper mills and biochemical plants. After the ratification of the Kyoto Protocol the urgency of such work significantly increased.

In the Siberia Region some large industries for deep wood processing are situated, such as Yenisei, Bratsk, and Selenga pulp and paper mills; Kansk and Angara biochemical plants, rosin extraction and other wood-processing industries. A new large-scale lumber complex is under construction in the Lower Angara region. It is evident that there is a growing need in high quality human resource development to fulfil demands of industry and research companies.

The Siberian State Technical University (SSTU) and Siberian Federal University (SFU), situated in Krasnoyarsk, Altai State University (Barnaul), as well as research institutes of the Siberian Branch of RAS (SB RAS), situated in Novosibirsk, namely Boreskov Institute of Catalysis (BIC), N.N. Vorozhtsov Novosibirsk Institute of Organic Chemistry, Institute of Solid State Chemistry and Mechanochemistry; Irkutsk (A.E. Favorsky Institute of Chemistry), Krasnoyarsk (Institute of Chemistry and Chemical Technology), Biysk (Institute for Problems of Chemical & Energetic Technologies) and others form the principal cluster for the education and training of qualified specialists in this field, and for research work directed at the development of environmentally benign methods of wood processing. With the deep belief that people are a determinant factor for the development of the world as a whole and each country, and that the best way for researchers to improve their own education is to teach others, in Russia it is common for high-grade research specialists working in research institutes to perform educational activities in local universities by giving lecture courses, but even more widespread is the supervision of the scientific work of university students or graduates.

Modern global tendencies in deep processing of wood biomass are based on green chemistry approach, including simultaneous use of all main wood components (cellulose, lignin, hemicellulose, extractive substances), and also wood wastes (bark, pulp lignin), non-standard and low-value wood.

The students of SSTU and Altai State University (ASU) are deeply involved in research work in green chemistry. Thus, the branch of the Department of Chemical Engineering of Wood and Biotechnology is working at the Institute of Chemistry and Chemical Technology (ICCT) SB RAS. Several lecture courses such as *Chemistry and Technology of Secondary Products of Chemical Treatment of Wood*, *The Technology of the Sorbents from Plant Raw Materials* are provided by the researchers of this department. The leading scientist of ICCT, Prof. Boris Kuznetsov, is also the head of the base department of analytical and organic chemistry of SFU. This department, together with SEC 'Biomass Chemistry' of SFU and ICCT, performs education of Bachelor's degrees, specialists and Master's in organic and analytical chemistry. The labs of ICCT are used for the experiments during preparation of diploma projects, where green chemistry forms a methodological basement to develop the new generation of the technologies for deep processing of plants biomass to produce marketable substances and biofuels.

The research work performed with the help of students and PhD students of SFU is connected with (1) delignification of wood using such non-toxic 'green' agents as oxygen or hydrogen peroxide instead of toxic sulfur and chlorine-containing substances; and (2) the development of non-corrosive delignification catalysts. Thus the new method of pure cellulose (less than 1 mas.% of residual lignin) production in mild conditions with high yield was proposed which is based on oxidative delignification of wood with hydrogen peroxide in the reaction media formed by water and acetic acid, using titanium suspension as a catalyst.

The participation of students in the development of methods for utilization of wood wastes, low-value and non-conditional wood (for example, injured by fire or plant pests) to produce charcoal or carbon sorbents that find applications in different branches of industry as well as in environment protection contributes to the formation of an environmental mentality of new generations of researchers.

Interdisciplinary training of students is provided during the development of complex methods of wood processing involving integrated extractive, catalytic, thermo-chemical and biotechnological treatments. Such training gives rise to the knowledge and practical skills in chemistry, engineering, environmental and computational science and mathematics. An efficient platform for education of SFU students was provided during the fulfilment of the integrated process of biomass transformation to the wide range of valuable products: new polymer materials based on polysaccharides and polyphenols, bio-ethanol and important glucose derivatives. This method includes catalytic separation of wood biomass on cellulose and polyphenols (soluble lignin); the use of cellulose to produce new sulfonated functional polymers and composites or glucose; enzyme assisted transformation of glucose to bio-ethanol or catalytic conversion to carbonic acids and sorbitol. Simultaneous extraction of non-cellulose polysaccharides provides raw materials for the production of biologically active substances and unique nanobiocomposites. Thus a wide range of valuable products is produced from the waste wood, and this is the impressive example of the possibilities of green chemistry in tackling the environmental problems, providing sustainable development of wood-producing regions of Russia such as Siberia.

Catalysis science is one of the main constituents of green chemistry. Simultaneously catalytic processes form the major part of all industrial chemistry. One of the most distinguished Russian research institute is this field is BIC, situated in the Research and Educational Cluster near Novosibirsk. More than 35 researchers from BIC participate in the training of specialists in chemistry, including green chemistry and catalysis. Most of them are ensuring the close and time proved connection of the BIC with the main local educational centres: NSU and Novosibirsk State Technical University (NSTU). They give lectures, seminars and practical classes; participate in the organization and renovation of the educational process at the Faculty of Natural Sciences (FNS) of NSU.

NSU was established in 1958 as an essential part of the SB RAS; it embodies the idea of the deep integration of science and education. Since its beginning, NSU has pursued three fundamental principles. The first consists in the idea that teachers need to be experts engaged in real science. The intellectual basis of NSU includes more than 5000 scientists from 30 research institutes of the SB RAS. The second principle is a thorough mathematical base provided to the students of all departments. According to the third principle the students master theoretical disciplines at university during their first three grades and do their practical research at academic institutes of SB RAS during the final years.

The Department of Catalysis and Adsorption (Head, Prof. V. Bukhtiyarov) is of special importance for the NSU and BIC. It is only 5 years younger than NSU itself. The foundation of this department was initiated in 1964 by Academician G. Boreskov, a world famous researcher in catalysis, the founder and first director of the BIC. The department is located in the BIC, which provides all everyday needs for efficient training of 25–30 fourth-, fifth- and sixth-year students annually, specializing in catalysis and technology of catalytic processes.

The NSTU—one of the largest research and educational centres of Russia—trains and retrains qualified specialists for research and industrial complex of Siberia and the Far East. The personnel training in the field of technology of catalytic processes is of special importance for practical application of fundamental knowledge in catalysis as one of the pillars of green chemistry. Such an approach is provided for the students specializing in the field of environmental engineering in the fuel and energy complex. This educational direction started at NSTU in 1998 as essential step in the realization of the joint specialists training programme at NSTU and institutes of the SB RAS. The basic training is conducted by the Department of Environmental Engineering (Aircraft Faculty) founded at the BIC, and the Department of Chemical Technology Processes and Apparatuses (the Faculty of Mechanics and Technology).

The BIC researchers present courses in industrial ecology and technology of the environment protection, manufacturing science and catalytic methods. For the training of young scientists of the highest qualification BIC collaborate with graduate schools of SB RAS and NSU. Annually 20–30 graduates are trained in BIC.

Graduation studies of the students and PhD students in chemical kinetics and catalysis, physical chemistry, chemical engineering are conducted at the laboratories of the BIC and supervised by experienced researchers. Their results are usually published in scientific journals. All these approaches allow the students to prepare their graduation works at the highest scientific level and become widely informed and called-for specialists in chemistry and catalysis.

An essential part of educational activity of NSU, NSTU and BIC is connected with SECs. Educational, scientific and innovational complexes give the possibility of improving professional education by providing the unity of educational, scientific and technical innovational activity at every stage of specialists training.

Starting from 2008, seven SECs were formed by NSU and BIC, with NTSU participating in some of them. Mostly they work in the fields of catalysis and safe methods of energy production. The first SEC, *Catalysis*, was organized jointly with BIC, NSU and NSTU in 2008 and was carried out by Prof. V. I. Bukhtiyarov (BIC), Prof. V. V. Sobyanin (NSU), Prof. V. V. Larichkin (NSTU). Between the others mention can be made about SECs (1) *Physical Chemistry, Electrochemistry and High Energy Chemistry for Ecology, Chemical Industry and Energy of Future*; (2) *Energy Efficient Catalysis*; (3) *Catalysis for Sustainable Energy*; (4) *Catalysis for Atmosphere Protection*, and the Competence Center *Catalysts for Energy Efficient Technologies* (SEC *Chemical Information On-Line*).

The collaboration in the frames of SECs widen possibilities for students and graduates to perform practically important projects in different regions of Russia and provides interdisciplinary researches that are significant for the innovative development of the country.

The National Research Tomsk Polytechnic University (NRTPU) is a significant Siberian university which was included in UI GreenMetric World University Ranking, a 'green' rating of higher education institutions of the world, published by the University of Indonesia. In this university GCE is put into practice mainly by involving students in research activity in the fields of resource efficiency and medicinal chemistry. The first direction includes the development of innovative approaches for effective suppression of fires, reducing fuel consumption and toxicity of exhaust gases, developing water and sewage treatment complexes; engineering of wind-solar power stations. The Institute for Power Engineering of NRTPU realizes the projects *Resource-effective Generation*, *Intellectual Power Supply Systems*, and *Active Consumer* directed towards power saving.

Green chemistry approaches for medicinal chemistry were provided by the Department of Organic Chemistry and Biotechnology (Head, Prof. V.D. Filimonov; Prof. E.A. Krasnokutskaya) at the Institute of Natural Resources, which is a part of NRTPU. The students take part in the series of projects that are united under the name *Green Chemistry and Future Medicine*. The synthesis of 1-phenyl-2,3-dimethyl-4-iodinepyrosolone-5 (an anti-inflammatory agent which is used in Russia as a medicine against tick-borne encephalitis) was developed in this department in 1980s, and later it was modified to substitute methanol—used previously as a solvent—by water, or to perform synthesis in the absence of any solvents, by mechano–chemical reaction. Students of several grades studied new methods of synthesizing a drug to treat infection caused by the liver fluke *Opisthorchis viverrini*. As a result of common efforts a green method for production of this medicine, on the basis of natural raw materials, was found. Another direction of the project is the grafting of organic substituents to nanoparticles of various natures to provide targeted drug delivery. Students specializing in chemistry are very much interested in the development of medicinal topics, and this interest is maintained by the special grant programmes of RMES and several other foundations. The results obtained form the basis for the new educational module *Innovative Development of Chemical Engineering of Biologically Active Substances* in the Master's educational programme *Innovative Development of Chemical Engineering*.

12.2.6   Green Chemistry Education in Central and South Russia

Supercritical fluid technologies are the basis of the education in the field of green chemistry in Kazan National Research Technical University. For the first time the training course *The Bases of Supercritical Fluid Technologies* was implemented for students of specialists programme studying the specialty *The Energetics of Heat Engineering*. It is provided during the 5th grade and involves 36 lecture hours, 14 training hours, 22 hours of lab work, and 64 hours of self-study. For Bachelor's education the course *Supercritical Fluid Technologies* is provided as the elective course in the professional cycle, whereas for the Master's degree three courses on ScF technologies are provided: *Heat Transformation in Thermodynamic Systems Including Sub- and Supercritical Media*, *Supercritical Fluid Technologies*, and *Energy and Materials Saving in the Frames of Supercritical Fluid Technologies*. Also this topic serves as a base for the course *Physical–Chemical Bases of Innovative Technologies* for the students studying in five directions, namely material science and technology of technical chemistry, petrochemistry, nanomaterials, coatings and surfaces, and composite materials.

In 2013 International Educational Master's programme *Supercritical Fluid Technologies of the Processes of Raw Hydrocarbon Deep Conversion* was developed. It was launched in 2014. This programme is very appropriate and innovative for Tatarstan, as in this part of Russia, petrochemistry is an economic priority, and based on the competence approach; the results will be estimated in credits for the whole course and each discipline.

In addition to lecture courses, single lectures explaining the importance and the essence of green chemistry are provided during summer schools for secondary school students and in the regular lecture courses on chemical and technological disciplines for Bachelor's, Master's and even PhD students on demand of principal lecturers.

Another Russian university where GCE serves as a basis for innovation policy is Astrakhan State University (ASU) situated in the south of Russia in a mostly agricultural region. Since 2004 the UNESCO Chair 'Learning Society & Social Sustainable Development' headed by the University Rector, Prof. A. Lunyov, has worked at this university. Between other top-priority directions, the department performs activities towards setting up an integrated system for research, teaching, and documentation in different fields of scientific knowledge, and increasing the role of Universities of the Caspian Region in tackling the problems of the regional sustainable development.

In 2010 SEC *Green Chemistry* was created at the Faculty of Chemistry to develop and implement advanced technological approaches to the creation of innovative new products, including improving the quality of life, and to promote 'green' approaches in the field of education, research and innovation. The first educational task of SEC was the implementation of the Master's programme *Green Chemistry*.

The centre consists of three small innovative enterprises: scientific-production enterprise, Astra-Phytos, LLC; scientific-production enterprise, Grin-Extra, LLC; scientific-production enterprise, Aro-Miss; and one research laboratory, Synthetic Azaheterocyles and Intermediates for its Synthesis (Head, Prof. A.V. Velikorodov). Business administration of the SEC is performed by Prof. A.G. Tyrkov.

Among 44 people involved in the activity of SEC *Green Chemistry* there are four PhDs and 30 undergraduate students. The centre fulfils educational, scientific and innovative aspects of educational process at the Faculty of Chemistry of the ASU. Thus, the centre provides an instrumental foundation for the realization of intra- and inter-disciplinary projects, research training of secondary school students in chemistry, professional and advanced training for the staff of local industries (in chemical engineering and chemistry). But the main task is to provide research facilities for final qualification works of Bachelor's and Master's students graduating specialties chemistry, green chemistry, organic chemistry, and specialist students studying fundamental and applied chemistry, as well as PhD students specializing in organic chemistry.

The objectives of the centre include implementation of researches in such areas of green chemistry as the use of supercritical fluid technologies in the

synthesis of new chemical products; preparation of new medicines components based on azaheterocyclic compounds; and the integration of the results
obtained in the educational, scientific and innovation activity of faculties of
ASU, to promote the ideas of green chemistry among schoolchildren students,
teachers, young scientists. According to local needs, innovative products
development in the centre is focused primarily on the extraction of chemicals from introduced or endemic plant raw materials. Recently, SEC has created technologies to produce essential and fatty oils from grape, watermelon
or pumpkin seeds, rose hips, and especially from mountain mint (*Lophantus
anisatum*), which is known as 'northern ginseng', on the base of supercritical
fluid extraction. Together with LLC scientific-production enterprise Vulkan
students and specialists of the centre built up the technology for production
of sugar-free (based on lactitol) candies based on mountain mint. Work is
under way to establish a new pharmaceutical substance, Immunoflan, which
has a pronounced immunotrophic activity, contains a balanced composition
of various phenolic compounds and flavonoids.

SEC *Green chemistry* can be regarded as a launch pad for a unified educational environment, school–university–postgraduate activity, as well as a
base for the fulfilment of innovative projects within the CDIO standards by
students. The results of the projects are demonstrated at specialized exhibitions, and competing for external financial support in the frames of different
granting programmes, such as UMNIK and Start, provided by the Foundation for Promotion of Small Enterprises in Science and Technology. The
most interesting and innovative students' projects were directed towards the
development of the inhibitor of acid corrosion IC-1, the tomato growth regu-
lator Nitrofungin-A, and the herbicidal composition Nitrozol-A.

As in the other universities, the holistic principle constitutes the base of
chemistry education in Southwest State University (SSU) situated in Kursk.
The environmental issues as well as green chemistry are involved in teaching
courses of nearly all chemistry disciplines provided by the Department of
Chemistry of SSU (Head, Associate Prof. O.V. Burykina). During the course
of inorganic chemistry attention is paid to the impact of chemicals (sulfur
dioxide, carbon monoxide, ions of heavy metals, pesticides, surfactants) on
vital functions of living organisms.

In the course of organic chemistry environmental impact of organic substances (especially those that are difficult to dispose and can accumulate in
nature) on the environment is discussed in details to aim students on the
search of greener and environmentally safe synthesis methods of valuable
organic products rather than utilization of the wastes.

The basis of the course of analytical chemistry and the course *Analysis
of Natural and Waste Waters* is analytical control of environmental objects
using express methods of soil and water analysis as well as physico-chemical
methods of wastewater purification. To involve students into practice of environment protection and implementation of waste-free and resource-saving
technologies project approach is used. The students are united in several
small groups, aimed at the decision of specific scientific tasks. The choice

of the task is stipulated by the training level and specialty of students. Thus, first-grade students studying environmental science and nature management perform the analytical monitoring of local natural and waste waters, soil and air. In some projects the quality of water and soil nearby unofficial landfills (Starkovo and Chaplygino) and in V. Alekhin Central-Chernozem State Reserve was compared. Third-grade students participate in the development of non-traditional sorbents for the treatment of waste water to extract heavy metals ions and industrial dyes. Scientific and practical results of the performed projects often are presented at the conferences of different levels, from university to All-Russia and even international level. For this purpose the Department of Chemistry organizes the annual conference *Current Trends in Chemistry and Ecology*. The work on non-traditional sorbents mentioned above was selected for the federal stage of the All-Russia Students Forum (St Petersburg, 2013). In 2011 the project was directed at the removal of hardness salts, and $Fe(II)$ and $Fe(III)$ salts from tap water in the Kursk Region using local minerals and was presented at the exhibition *Healthy Lifestyle–2011*.

Medicinal topics are also the fruitful platform for the training of students in green chemistry. In the framework of the SSU programme *Activization and Involvement of Young People in Scientific Activity of the University* students participate in the development of new biocompatible materials.

12.3 Green Chemistry Education in Secondary Schools

At secondary school level GCE has to be a necessary element of more wide sustainable education, because chemistry education (as well as ecology) starts in Russian secondary schools only from 8th grade (from 7th grade in some gymnasiums). However, the teaching of the scientific discipline 'Outworld' starts from the 1st grade of the secondary schools. For the 5th to the 7th grades the educational programme contains the discipline 'Nature Study'. It seems that the bases of sustainable education have to be included into the educational programme on the early stages of secondary schools.

An interesting project was launched in Moscow by the efforts of Moscow Institute of Open Education (MIOE), which performs professional development of Moscow secondary school teachers, and Prof. G. Yagodin. A new discipline 'The Ecology of Moscow and Sustainable Development' has been implemented for 10th grade students of secondary schools. G. Yagodin has prepared the textbook for this course, and the members of SEC SGC translated into Russian and published the additional tutorial *Global Climate Changes* by F. Zecchini. The course provides wide dissemination of the ideas of sustainable development in all Moscow secondary schools (the total number is more than 1700).

The activity of the Open Ecological University mentioned above is also addressed to school students. Starting from 2005 the University carried out some events in the secondary school #864 in Moscow district Yasenevo, which is specializing in environmental direction. Thus in 2005 the workshop *The Ethics of Chemistry Teaching in Secondary Schools*, carried out in this school, included the topic of green chemistry in connection with social responsibility of chemists, in general, and school teachers in particular.

During recent years SEC SGC together with MIOE has been providing lectures about green chemistry for secondary school teachers of chemistry, biology and environmental sciences. Usually, the topic of green chemistry causes lively interest in the teachers' auditorium. They often contact the presenting lecturer with requests to establish tasks for scientific works of school students in the field of environmental science and green chemistry.

Starting from the first Festival of Science, organized in 2005 in MSU, which was significantly widened later and is now known as annual Russian Festival of Science (the 4th event started on 8 February 2014),[9] the specialists of the SEC SGC provide for the participants lectures and master-classes in green chemistry, usually followed by excursions to the laboratories of the Center. During these excursions the staff demonstrate to attendees amazing possibilities of modern physical methods to investigate the structure and properties of materials. The SEC GC and Faculty of Chemistry have high-quality instrumentation to perform scanning and transmission electron microscopy, infrared or electronic paramagnetic resonance spectroscopy, X-ray photoelectron spectroscopy and many other modern techniques. The researchers show schoolchildren how all these instruments can help in the investigation of catalysts and other new materials such as nanotubes. These demonstrations are very attractive to high grade school students, especially when researchers use some tricks, for example, demonstrate SEM images of mosquito, or an EPR spectrum of human hair. Quite often the participants of such excursions 'into science' later become students at the MSU and address the centre staff for directing their graduation and PhD studies.

In order to achieve success in sustainable education, there is an urgent need for education programmes based on sustainable development curriculum for use in teacher training. Such training will allow teachers to use interdisciplinary approaches to integrate sustainable development and green chemistry principles into the different subjects of the formal school curriculum.

In addition to its activities at the higher education level, MUCTR set up a long-term patronage programme to promote education for sustainable development in secondary education by creating centres of environmental monitoring in secondary schools and colleges in various parts of the country. In Moscow and Tomsk, the professors of the UNESCO Chair in Green Chemistry for Sustainable Development of the MUCTR developed and held a series of workshops and seminars on global issues related to sustainable development. As a rule, these seminars were held during the summer and winter school holidays, and autumn and spring breaks. The duration of each workshop varied depending on the individual requirements of each group and the level of their preparedness. Among other topics, such issues as environmental protection and its relation to poverty, and examples of unsustainable economies and poor environmental management were addressed. As the result of this sustained effort, ESD practices and methodologies became an indispensable part of the educational process for every teacher, faculty member, and student from Tomsk Municipal College who took part in these workshops.

The creation of the Centres for School Environmental Monitoring turned out to be one of the most effective tools for the integration of sustainable development issues into the formal and informal educational process in secondary education. These centres provide college students (15–18 years old) with the opportunity to carry out systematic scientific research by working on various projects. The programme is developed with consideration of the educational and psychological levels of students' development. The experience of the first Centres for School Environmental Monitoring and the information gathered by the MUCTR was very useful for further expanding this programme. For the last several years such centres were organized in more than 40 Moscow schools.

Very impressive for young people is the participation in the competitions of projects, especially in high-rank ones. In Russia many contests of the projects performed by the students of secondary school of local, regional and All-Russian levels are organized each year. There are, however, the most prestigious competitions, and some of them comprise the topics of green chemistry and sustainable development. For example, in 2014 MSU and Intel Company organize the contest 'Researchers for the future'. Rector of MSU Prof. V. Sadovnichy heads the organizing committee of the contest. The winners of the superfinal in the division 'Chemistry and Nanotechnology', E. Mozgunova and D. Maksimova, performed a project in the field of green chemistry, *Construction and Synthesis of Macromolecular Microbicide and Antitumoral Chemical on the Base of Urocanic Acid*, and will participate in Worldwide competition Intel ISEF in the USA.

The auditorium of the All-Russian Child's Competition of Research and Creative Works 'First Steps in Science' are students of junior and upper junior grades of the secondary school. This contest was launched in 2007 under the logo *Amat victoria curam*! (translated from Latin it means, victory likes endeavour). In 2013 three works were awarded in the section 'Chemistry, Ecology, Safe Living'. The projects were: *How to use Rubbish* presented by F. Volozhin from Angarsk; *At Peace with Radiation* by E. Glyantseva from Novouralsk; and *The Influence of Exhausts Gases of Vehicles on Soil Pollution and Lead Content in The Mushrooms* by L. Shchukin from Porotnikovo village.

Specialized Educational–Scientific Centres (SESCs) for secondary school students are established at some Russian universities. The main goal of such structures is to attract gifted children from small cities and villages where only limited possibilities for advanced education can be provided by the secondary schools. Such SESCs function as a part of MSU, NSU and some other high schools. Teaching of green chemistry theory and practice to such gifted young people provides high output, because nearly all graduates of such SESCs continue their education at universities and eventually become leading scientists and specialists in different branches of economic.

P.A. Kirpichnikov college of the boarding school type for the gifted children with in-depth training in chemistry is working in the structure of Kazan National Research Technical University. The students can choose chemical–biological or physico-chemical education profiles, but in both sections elementary knowledge about green chemistry is provided during educational courses. Green chemistry is one of the substantial topics of the lectures given by professors of the university; some elements of green chemistry are included into lab training 'Introduction to Specialty'. The tasks connected with sustainable development and green chemistry are posed during chemicals Olympiads, *e.g.*, Scientific Olympiad in Chemistry and the intellectual games 'Chemistry Brain Ring'. Students' works in this field participate in the Annual Conference of Scientific Works in Chemistry.

SSU collaborates with secondary schools of Kursk Region and involves best school students into scientific projects performed by university students. Some chemistry circles organized in SSU for school students provide deep study of chemistry and performing the scientific projects. Such activity gives rise to the interest of schoolchildren to the chemistry and environmental science and prompting them to receive higher chemistry education.

To summarize, it can be said that GCE in secondary schools primarily is targeted at teachers of the disciplines, connected with chemistry (chemistry, biology, ecology, outworld and nature study), and at the gifted school students studying deepened programmes oriented on natural science. For the other students mainly the bases of sustainable education are provided.

12.4 Professional Training and Enlightenment of the General Public in the Field of Green Chemistry

Sustainable education at all levels—life-long, formal and informal—should also be available to all citizens of Russia in order to achieve a more ambitious goal: the transformation of the mainstream lifestyle of unsustainable consumption and production to a sustainable way of living. In Russia, for most citizens, the understanding of environmental problems is paradoxically connected with irresponsible behaviour with respect to the environment. While on a higher governmental level the edicts were approved to stimulate environmental protection, local authorities on the mundane level rarely provide effective actions in this direction. Whereas in Stockholm the power system is based on the energy from rubbish, in Moscow even the possibility of waste separation is not provided for the inhabitants. Therefore enlightenment in these problems is strictly necessary. Many non-governmental organizations (NGOs) and individuals perform actions in this direction. Between them we can mention Prof. L. Fedorov, the founder and head of the Union for Chemical Safety, who regularly disseminates the electronic bulletin *Environmental*

Safety that describes the problems and achievements in the environmental protection mainly in Russia, but also provides examples from foreign countries.

Enlightenment in the field of green chemistry is the essential part of sustainable education. This can be done through public lectures, the creation of new learning centres, seminars and workshops, mass media such as TV and radio programmes, and audio–visual means, as well as local and national round tables and conferences. The research, development and industrial implementation of new 'nature-friendly' technologies is one of the most significant components of the whole system of sustainable education.

The primary target groups for green chemistry training are specialists in chemistry, managers of all levels, and teachers of high and secondary schools.

The MUCTR's experience in convening ongoing training seminars and workshops on environmental management was incorporated by the Government of Moscow by Decree N990 issued in 1990, in which both environmental and sustainable development professional training were pronounced compulsory for professional and business managers at all levels.

In accordance with the Decree, the MUCTR, as a member of the Moscow Association of Environmental and Sustainable Development Education, provides professional training by giving the course *Environmental Protection at Industrial Enterprises*. Since 2000, about 3000 representatives from various plants and factories in the Russian capital have received this training.

The Director and faculty of the Institute for Chemistry and Problems of Sustainable Development at the MUCTR have developed a programme for the promotion of sustainable development at the regional level. They created a partnership with the Inter-Regional Association for Economic Cooperation of the Subjects of the Russian Federation, called The Siberian Agreement. This association constitutes an NGO that brings together nineteen regional subjects of the Russian Federation, and focuses its activities on life-long environmental and sustainable development education, raising the awareness of the broader Russian public through formal and informal ESD. Week-long training courses were held in various Siberian cities: Tomsk, Krasnoyarsk, Novosibirsk, and others. The courses were held by professors of the institute headed by its director; they were mainly focused on secondary school teachers and university professors, although, on one occasion, Prof. Tarasova's audience consisted of kindergarten teachers. This activity is performed in close cooperation with UNESCO and serves as an experimental testing ground for new educational methods and curricula. In order to train the diverse and numerous educational communities in Russia, educators at the institute developed courses on natural protection, sustainable development, and green chemistry to be taught in hundreds of schools across nineteen regions of the Russia.

Since early 2000, the institute has been convening a series of training workshops and seminars in Moscow and other regions; this is an ongoing process. The courses are designed primarily for secondary and high school science teachers and are built around the new educational curricula developed at the

Institute. Interactive approaches and experience sharing are widely applied in these courses, as well as simulation role-playing, discussions of global problem issues, and social and economic aspects of sustainable development.

The Faculty of Chemistry of MSU closely cooperates with major chemicals producers in Russia, performs special educational programmes for them, and the basic elements of green chemistry are included in such training courses. Moreover such huge business companies invite the specialists of the SEC SGC to enlighten the people living proximate to chemical plants about the potential of green chemistry. For example, in April 2013 E. Lokteva provided such lectures in Kirov and Kirovo-Chepetsk on a request from URALCHIM company, a world known fertilizers producer. A major part of audience was representatives of local NGOs, generally having a negative attitude to chemistry. After the lectures they were impressed with the possibility of green chemistry and ready to collaborate with SEC SGC and with local chemical enterprises.

12.4.1 Conferences, Workshops and Exhibitions as a Part of Professional Training

The Russian educational and research community is very active in the organization of conferences in the fields of green chemistry and sustainable development as a whole and in different branches of green chemistry. Even the announcement of such meetings attracts the attention of the people working in various segments of chemistry to the significance of green chemistry issue giving rise to the number of scientists involved in this area.

The organization of green chemistry conferences, as well as seminars and round tables in the frames of wider conferences and congresses is an important activity of SEC SGC of MSU. In 2008 the Second IUPAC Conference on Green Chemistry attracted about 200 participants from 20 countries. It was organized together with IUPAC and sponsored by RAS, the Russian Foundation for Basic Researches and Organization for the Prohibition of Chemical Weapons (OPCW). During a boat trip from Moscow to St. Petersburg, with short stops in old Russian towns for excursions, the participants, including members of the Green Chemistry Sub-Committee of IUPAC, were involved in interesting discussions concerning biomass transformation, catalysis, the use of supercritical solvents both as reaction media and reagents, education in green chemistry and many other topics. K. Koshika from Japan ('Secondary battery with electrode on the base of hydrophilic radical polymer: the estimation of green chemistry principles consistency') and V. Skorkin from Russia ('Catalysts for hydrogenation of conjugated dienes on the base of dendrimers') were awarded with IUPAC prizes for the best posters from young researchers. This conference has given impetus to the development of GCE in many Russian universities, the affiliations of the Russian participants.

The topic of education in green chemistry was discussed at the XVIII (2007) and XIX (2011) Mendeleev Congresses on General and Applied Chemistry. The satellite symposium *Green Chemistry, Sustainable Development and*

Social Responsibility of Chemists of the XVIII Mendeleev Congress headed by N. Tarasova and V. Lunin attracted 100 Russian and more than 20 foreign specialists. The plenary reports were presented by Prof. V. Petrosian (MSU), *The Ethics of Chemistry Teaching in 21st Century*, and by V. Kuznetsov (MUCTR), *The Knowledge in Environmental Chemistry as a Factor for Social Responsibility Increase*.

E. Lokteva and V. Lunin presented the lecture *Green Chemistry Approach in Petrochemistry* at the conference, and the School for Young Researchers presented *Current Trends in Petrochemistry* (Moscow Region, 2012). Green chemistry is an essential part of this series of events. At some of them, special small-groups training for young scientists was provided. It was obligatory that the members of each small group were the representatives of different institutes or universities to simulate brain-storming with unfamiliar persons. Working in the groups of three to four persons they had to offer and develop, in one hour, a new scientific project in the field of green chemistry, prepare one or two slides for presentation, and report the idea to the jury. The most promising project and speaker were awarded with a small prize. Such a business game provides good practice for creating ideas, exchanging experience, preparation of projects in a limited time period and reveals the characteristic features of each student in the group (leader, speaker, generator of ideas, *etc.*).

Green chemistry was included as an important part into the programme of the First Russian Congress on Catalysis (2011). The important part of the congress was the Round Table on the problems of education in catalysis and green chemistry, moderated by V. Lunin. New interactive demonstration material (Power Point presentation) for the lecture course *Basic Principles of Catalysis* presented by Prof. A. Stakheev (N.D. Zelinsky Institute of Organic Chemistry, Moscow). Interactive presentation of the kinetic curves for the reactions of different order provides the possibility to change the appearance of the curve by varying the reaction parameters. Such visual presentation causes lively interest of students and helps in a deep understanding of the difficult subjects such as kinetic and catalysis.

Various branches of the green chemistry were creating a platform for such interesting events as a series of Scientific-Practical Conferences *Supercritical Fluids – Fundamental, Technology, Innovations* (2011, Irkutsk; 2012, Ivanovo; 2013, Kaliningrad). Together with MSU the main organizers were RAS, CC 'Schag', and the editorial board of the Russian journal *Supercritical Fluids – Theory and Practice*.

All events attract famous green chemists from all over Russia and some foreign partners. Famous Russian researchers in organic chemistry, V. Charushin and O. Chupakhin from Yekaterinburg; in wood chemistry, K. Bogolytsyn from Arkhangelsk and Boris Kuznetsov from Krasnoyarsk; in catalysis, Valery Bukhtiyarov from Novosybirsk and Alexey Pestryakov from Tomsk; in carbon materials, Vladimir Likholobov and Jury Krjagev from Omsk; in ScF technologies Rafinat Yarullin from Kazan more than once presented lectures during the conferences mentioned above. Young researchers under 35 years of age, graduates and students constituted more than a half of participants of these events. Close contact between distinguished scientists and young researchers during such conferences encourages deeper understanding of the subject, generates new ideas for future work, and give chances for the employment of young participants.

Two years ago the SEC GC started collaboration with Moscow Exhibition Center EXPOCENTRE. The biannual exhibition *International Chemistry Assembly* in 2012 was generally devoted to the green chemistry. Due to high importance, for the first time an exhibition, not a conference was awarded with IUPAC logo. Many events of the business programme concern green chemistry topics, such as the conference *Supercritical Fluids Technology*; the conference and exhibition *Chemical Security*, to commemorate academician of the RAS Valery Legasov; the workshop *Innovative Educational and Public Awareness Programmes in the Field of Green Chemistry: Making Green Chemistry Part of the General Courses of Lectures*; and a round-table discussion *Promoting Green Chemistry Through the Internet and Mass Media*. Great attention was given to the First All-Russia Competition of innovative works of young scientists in the field of green chemistry, organized in the frames of the exhibition. The winners were awarded IUPAC prizes as well as prizes from the Organizing Committee and EXPOCENTER.

Green chemistry became an important issue of another biannual exhibition *Chemistry – 2014*. This event also involved business-programme of sustainable direction: All-Russian symposium in Green Chemistry; International Chemical Forum *Health, Safety, Environment*; seminar 'Supercritical Fluids Technologies – the Technologies of Green Chemistry', and the Second All-Russia Competition of innovative works of young scientists in the field of green chemistry. This year the winners included young scientists from Syktyvkar (J. Bykhovtseva, 'The production of medicine materials on the base of biodisposable and renewable raw-materials from plants'), Moscow (E. Kudryavtseva, 'The development of informational–educational programme "Responsible Care" for Russian chemical industries as one of basic constituents of green production'), Omsk (E. Martynenko, 'New ways of chlorinated polymers disposal'), Barnaul (S. Autlov, 'The synthesis of microcrystalline cellulose under the influence of microwave radiation'), and Kazan (A. Fazlyev, 'Regeneration of ionic liquids from water solution using pervaporation membranes'). The affiliation of the winners reflects the geography of the most prominent organizations providing GCE and described above. The winners were awarded with the possibility to present their work as a short oral report at the All-Russia symposium on green chemistry, as well as organizers, UNIDO and EXPOCENTER diplomas. The representative of UNIDO Mrs P. Schwager personally presented UNIDO diplomas and small gifts to the winners at the closing ceremony in Moscow.

The lecture about the possibilities of supercritical technologies became traditional at jubilee and annual scientific conferences in Nignekamsk Branch of Kazan National Research Technical University, and is provided sometimes in the Volgsky Branch.

In 2010 and 2012 All-Russian School-Conference of Young Scientists *Supercritical Fluid Technologies in the Decision of Environmental Problems. Plants Biomass Extraction* was held on the base of North (Arctic) Federal University. Taking into account the major local directions of scientific work, the scope of the School-Conference included such issues as the properties of supercritical fluids, ScF extraction of plant biomass, chemical and biological processes in ScF media, ScF chromatography in analysis and separation of natural extracts.

The Department of Pure and Applied Chemistry of NAFU provides on the regular basis the training courses on environmental safety for officials, managers and specialists from the industries of northwest of Russia. During last 5 years 476 representatives of the industrial firms completed their training in accordance with this programme.

Organization of specialized conferences and school-conferences for young researchers is an important part of BIC activity in the field of green chemistry education. Only for the last few years BIC jointly with leading institutions in the field of chemistry and catalysis organized twelve scientific conferences, congresses, seminars and schools for young scientists, including the 2nd international conference *Catalysis for Renewable Sources: Fuel, Energy, Chemicals* (22–28 July 2013, Lund, Sweden), the international conference *Catalysis for Renewable Sources: Fuel, Energy, Chemicals* (28 June to 2 July 2010, Tsars Village, St. Petersburg suburb, Russia), then international conference *Nanostructured Catalysts and Catalytic Processes for the Innovative Energetics and Sustainable Development* (6–10 June 2011, Novosibirsk, Russia), the Russian conference-school for young scientists *Chemistry under the Sigma Sign: Researches, Innovations, Technologies* (14–22 May 2012, Omsk, Russia) and many others.

12.4.2 Cooperation with Foreign Partners and Publications

The international consortium of Russian and foreign universities was created to focus on developing new educational programmes for specialist training in energy management. The new programme should include progressive achievements in the field of energy conservation and more efficient use of all forms of energy combined with environmental monitoring and control.

The consortium consists of four European universities, led by the University of Genoa, Italy, and eleven Russian universities: MUCTR of Russia, Tambov State Technical University, Vladimir State University, B. Yeltsin Ural Federal University and others; some associated partners representing businesses and organizations from different regions are also involved.

The project is designed to:

- Implement effective curricular reform oriented towards the labour market requirements and the needs of the links of higher education with society

- Develop a constructive model of professional recognition of new qualifications in energy saving and environmental protection education (for
  civil engineering, industrial chemistry, chemical engineering, nature
  management and environment protection sectors)
- Deliver appropriate education and regulatory documentation of universal character to be approved and assessed by stakeholders' institutions
  chosen among associations of entrepreneurs and public authorities
  having environmental tasks.

The new qualification proposed is the Master's degree in innovative technologies for energy saving and environmental protection, 'Green Master'. It
will be partially based upon the existing similar postgraduate course at the
participating European universities.

This project sets the ambitious and challenging task, leading to a mas-
sive change in the energy potential of the economy. That is why the project
is supported by the Tempus programme. The action plan is developed over
three years providing programme graduates with a Master's degree having
knowledge not only in engineering, but also in the field of energy saving and
environmental control.

The green chemistry approach connects French and Russian chemists
working in ICCT and SFU (Krasnoyarsk) in the Research Consortium 'Catalytic Processing of Biomass to Valuable Products'. They are performing
the project *Processing of Wood and Agricultural Wastes* in the frames of Arcus
(Actionen Region de Cooperation Universitaire et Scientifique) Programme.
ICCT and SFU collaborate also with Engineering Center of Johnson Matthey
(a UK company) on the project *Thermochemical Biomass Transformation*. All
listed projects stipulate foreign training of PhD students and their scientific
advisors, as well as PhD thesis preparation under the common guidance of
Russian and foreign professors.

The person who is deeply involved in educational and promotional activity in the field of green chemistry in Russia is Prof. Martin Poliakoff from
Nottingham University, UK. As a person with Russian roots, he has been
one of the pioneers of GCE in Russia. In 2000–2001 two workshop for young
researchers, *Green Chemistry and Catalysis*, were organized by Prof. M. Poliakoff together with MSU researchers, the first one in Nottingham, and the second in Moscow. The article in the magazine *Chemistry and Life – 21st Century*
with the title 'Green chemistry – another industrial revolution' by M. Poliakoff was nearly the first article on the topic published in popular Russian
editions.

In the process of SEC SGC formal establishment the primary task was
to perform inventory of the efforts made by Russian scientific and educational community towards green chemistry. For this reason the collection
of articles titled *Green Chemistry in Russia* was gathered and published in
2004–2005 under the aegis and with financial help of IUPAC and the Italian
Consortium INCA (Prof. P. Tundo). The book includes contributions from
famous Russian scientists, pioneers of green chemistry investigations, such

as V. Charushin and O. Chupakhin, V. Lunin, K. Bogolitsyn, S. Yufit and many others. The articles encompass all fields of green chemistry, from biomass use for sustainable chemical materials production, catalysis, chlorine-free chemistry, to green organic synthesis. Two editions, one in Russian and another in English, were disseminated for free between libraries of universities, institutions in Russia and other countries, first of all, former Soviet republics.

Close collaboration with colleagues and organizations in these countries is the characteristic feature of the SEC SGC. It is organized not only through branches of MSU, situated in Baku (Azerbaijan), Tashkent (Uzbekistan), and Dushanbe (Tajikistan), but also through ecological organizations in Bishkek (Kyrgyzstan) and Donetsk (Ukraine), South China University of Technology (China) and many others. This collaboration was formally organized in the frames of IUPAC project *THE GREAT GREEN ROAD. Green Chemistry – Creation and Implementation of International Cooperation in Teaching and Investigations'* (2008-017-4-300, 2008–2010). This project aimed to provide the platform for chemists from developed and developing countries historically linked through the Great Silk Road to find collaborators for fruitful development of interdisciplinary green chemistry projects, both in science and in education, including public enlightenment, through establishing the network centred in Russia; to improve quality and to widen green chemistry collaboration from the West to the East. In the frames of project many educational initiatives were realized, for example, preparation and dissemination in Russia, China and Kyrgyzstan of short popular leaflet about green chemistry and questionnaire entitled 'Do you know about green chemistry?' to help everybody assess their own knowledge in the field and improve them by reading leaflet that follows.

The website of the SEC SGC has been working since 2006 and has attracted hundreds of subscribers. Educational programmes, reports about scientific activity of the centre, the books, issued by SEC, information about future events in the field of green chemistry in Russia and worldwide, as well as reports on completed events are presented on the website. The website is presented in Russian, which is very important, because in contrast to the abundance of web materials in foreign languages, the Russian web content on the topic is too scanty.

One of the major problems of GCE in Russia is the absence of textbooks in Russian. Even the book by P. Anastas and J. Warner, *Green Chemistry: Theory and Practice*, has so far not been translated into Russian. During the creation of educational programmes in the field of green chemistry for Master's and Bachelor's students the absence of available literature was the main obstacle for approval of the programme by authorities.

SEC SGC published the small collection of articles about their activity that includes material about the fundamentals of green chemistry: the description of the Twelve Principles, some examples of research works in this field and industrial applications. Surely that is not enough and the writing of the textbook is an urgent need for Russia.

The members of SEC GC collaborate with mass media. A special episode of Russian TV-5 channel's scientific-educational series *The History from the Future* hosted by the director of the Kurchatov Institute, Mikhail Kovalchuk, was organized as a discussion between the host, Valery Lunin and Ekaterina Lokteva.[10] Great interest of audience in the field was observed directly after filming the programme, when sound and video engineers and assistants surrounded the guests to clarify important questions about the safety of medicines to health, the possibility of producing new materials from biomass, and other green chemistry sub-topics.

Russian newspapers and journals, such as *Poisk* newspaper, *Chemistry and Life – XXI Century* and *Ecology and Life* magazines and many others regularly publish the materials about green chemistry.

Acknowledgements

The authors are grateful to the Russian Scientific Foundation for financial support (grant 14-33-0018). The authors thank all colleagues who provided information and illustrations for this chapter: Prof. K. Bogolytsin and N. Gorbova from North (Arctic) Federal University (Arkhangelsk); Prof. S. V. Meshcheryakov (Gubkin Russian State University of Oil and Gas), Prof. O. Martyanov (Novosybirsk State University); Prof. O. V. Burykina (South State University, Kursk); Prof. F. Gumerov (Kazan State University); Prof. B. N. Kuznetsov (Krasnoyarsk State University); Prof. E. Krasnokutskaya and Associate Professor N. Osipova from National Research Tomsk Polytechnic University; Prof. A. G. Tyrkov and Associate Professor S. B. Nosachev from Astrakhan State University; Prof. V. S. Petrosian and Prof. S. Vatsadze from Lomonosov Moscow State University; Prof. N. S. Popov and L. A. Mozerova (Tambov State Technical University); Assistant Professor A. E. Kurochkina (MUCTR).

References

1. *Opinion poll by Russian Public Opinion Research Center "Environmental situation in Russia: challenges and priorities" (Russian)*, http://www.mnr.gov.ru/regulatory/detail.php?ID=132085 (accessed February 2014).
2. N. P. Tarasova, A. S. Makarova, S. Yu. Vavilov, S. N. Varlamova and M. Yu. Shchukina, *Herald Russ. Acad. Sci.*, 2013, **83**, No. 6, 499.
3. *APEC Ministers issue Khabarovsk Statement on environment*, http://www.apec2012.ru/news/20120718/462768782.html (accessed December 2013).
4. *Prime Minister issues instructions following a meeting of the Presidium of Presidential Council on Economic Modernisation and Innovative Development*, http://government.ru/en/news/2171 (accessed December 2013).
5. http://www.oc-praktikum.de (accessed February 2014).

6. http://www.greenchemistry.ru (accessed February 2014).
7. http://ckp-nano.msu.ru/about/ (accessed February 2014).
8. http://www.chem.msu.ru/rus/supercritical-fluids/welcome.html (accessed February 2014).
9. http://www.festivalnauki.ru/ (accessed February 2014).
10. http://www.5-tv.ru/video/507075/ (accessed February 2014).

CHAPTER 13

Education in Green Chemistry: Incorporating Green Chemistry into Chemistry Teaching Methods Courses at the Universiti Sains Malaysia

MAGESWARY KARPUDEWAN*[a], WOLFF-MICHAEL ROTH[b], AND ZURIDA ISMAIL[a]

[a]School of Educational Studies, Universiti Sains Malaysia, 11800 USM, Penang, Malaysia; [b]MacLaurin Building A567, University of Victoria, Victoria, BC, V8P 5C2, Canada
*E-mail: kmageswary@usm.my

13.1   Introduction

Chemistry, as science generally, tends to be taught as a body of knowledge. Even when it is taught by means of experimentation and inquiry, teaching appeals to symbolic understandings that can be assessed. There is little concern for how certain phenomena, such as sensitivity to chemical substances in the environment, affect us physically and emotionally. But this is precisely where science really comes to matter. This was quite evident in the story of a well-known Canadian journalist, Wendy Mesley, to whom

Worldwide Trends in Green Chemistry Education
Edited by Vânia Gomes Zuin and Liliana Mammino
© The Royal Society of Chemistry 2015
Published by the Royal Society of Chemistry, www.rsc.org

chemical substances in environment and food became relevant only when she was diagnosed with breast cancer.[1] (See http://www.yorkregion.com/community-story/1446775-wendy-mesley-cancer-diagnosis-compelled-journalis-to-search-for answers/.) She found out about several studies that demonstrated the presence of 'varying levels of contamination from heavy metals, pesticides and other toxic chemicals (such as PCBs, mercury, lead)' among Canadians. Moreover, as the documentary *Last Call at the Oasis* shows, environmental exposure to chemicals also comes to the forefront in the minds of people when they and their loved ones are affected. There are therefore many contexts in chemistry which can become a preoccupying concern for students and their families.

In this chapter, we take a two-pronged approach. We first provide an argument why learning chemistry becomes relevant when the students are directly affected. We then describe an effort at the Universiti of Sains Malaysia to incorporate green chemistry into teacher education and, in so doing, respond to the UN declaration of a Decade of Education for Sustainable Development (UNDESD). It is a vision of education that seeks to empower people to assume responsibility for creating a sustainable future. Education is seen as an essential tool for achieving sustainable development. Education for Sustainable Development is centred on Man and the change to their behaviours. Education that succeeds in changing human's values will be able to change their behaviours. One approach that could be adopted to attain this goal is through greening the chemistry curriculum particularly the practical work. Green chemistry aims at preventing pollution and sustaining the earth; it is common practice in the production of industrial applications. While predominantly being applied in industrial applications it can be also adapted in education through laboratory-based experiments and classroom activities. It is also imperative to educate the future teachers (pre-service teachers) on green chemistry because these teachers have the power to access many generations of students in the near future. Incorporation of green chemistry experiments into the curriculum provides a platform for discussion of environmental issues in the classroom. In this chapter we discuss (1) how green chemistry can be adapted in chemistry teaching methods courses, (2) the feasibility of integrating green chemistry experiments, and (3) the effectiveness of green chemistry in bringing attitudinal, motivation and value change in solving environmental issues.

13.2 Background

13.2.1 Relevance of Chemistry

Chemistry is the study of matter, its properties and the changes. As a core science, chemistry has contributed to many remarkable technical advances that have increased the quality of human life.[1] However, manufacturing, processing, using, and disposing of chemical products, as exemplified in the now infamous Love Canal wastes (Niagara Falls, New York), led to many situations with tremendous negative impacts on human health and the environment.[2] For instance, the book *Silent Spring* has highlighted the harmful effects of chemical insecticides on the environment.[3] When the book was first published, there were many critics of the views articulated. Scientists considered pesticides as essential for crop protection. However, excessive use of fertilizers increasingly has been found to have several adverse affects on the environment, as shown, for example, by the worldwide effects of neonicotinoids on bee populations. There is therefore a need for a sustainable way for agriculture either by replacing the hazardous fertilizer with an environmentally friendly one or reducing the amount of fertilizer used. Additionally, many of those engaged in agricultural production gradually realize that pesticides may not be needed in large quantities to protect crops. As an alternative, biological pest control has been promoted and in some cases produced better results than using chemical pesticides.

Knowledge of chemistry is required to understand the environmental effects of pesticides and fertilizers. The role of chemistry in sustaining the environment was illustrated using a model on environmental chemistry.[4] In this model, environmental chemistry is relates to the anthropogenically disturbed environment and the undisturbed environment. As a prerequisite to understanding or controlling an anthropogenically disturbed environmental system, an in-depth understanding of the behaviour of undisturbed ecosystems is required. The whole ecosystem comprises two different environmental processes at two different levels: (1) micro-processes occur on the smaller scale; and (2) macro-processes occur at the larger scales. For example, chemical cycles are controlled by properties at molecular level and chemical reactions by properties at a more abstract level. Therefore, predictions of the behaviour of molecules at the global level will be impossible without good knowledge at molecular level. The border between micro-processes and macro-processes is very narrow, but the nature of processes, the techniques used to study them, and conceptual approaches are different. The teaching of environmental chemistry should cover all aspects proposed in the model of environmental chemistry to allow the students to see them in operation within the same phenomenon.[4] To understand chemical micro-processes, environmental chemistry has to correspond with analytical chemical method and mathematical modelling.[4]

13.2.2   Green Chemistry

Green chemistry, also known as sustainable chemistry, is a form of chemistry that aims at preventing pollution. Pollution prevention by means of green chemistry emphasizes the use of materials, processes, or practices that reduce or eliminate the creation of pollutants or wastes during chemical production. It includes practices that reduce the use of hazardous, non-hazardous materials, energy, water, or other resources as well as protect natural resources through efficient use.[2] The implementation of green chemistry experiments is guided by its twelve principles.[2] During instruction, one or more of these principles may be incorporated into each experiment that students conduct for the purpose of better understanding the working of green chemistry and its applications in the real world.

Green chemistry in teaching and learning process provides an opportunity or platform for the three pillars of sustainable development (economy, environment and society) to be discussed in the classroom. Such integration reflects the interdisciplinary nature of green chemistry and allows future chemists as well as the general population to have integrated knowledge of economical, environmental, and societal–political issues surrounding chemistry in the everyday world. The interdisciplinary nature also allows both students and teachers to consider ethical issues that arise from the potential impact of chemistry on the environment and which have repercussions for the local and global community.[5] In this way, those learning the principles of green chemistry do so in a holistic manner and connected to their everyday concerns.

One of the strengths of the green chemistry approach is that it connects chemistry concepts and students' everyday environment.[6] Since green chemistry is a student-centred, deep learning approach, the teaching and learning of green chemistry enables the development of higher-order cognitive skills such as communicative skills, problem solving skills and decision making ability.[7] Green chemistry bases learning in students' everyday experiences and helps them build appreciation, awareness, and a sense of shared responsibility for nature that students may carry throughout their lives.

13.3 Green Chemistry for Malaysian Pre-service and In-service Science Teachers

13.3.1 Introduction

The introduction of green chemistry experiments as laboratory-based pedagogy was an important step in our teacher education programme at the Universiti of Sains Malaysia. The implementation of green chemistry experiments turned out to be highly feasible as we could integrate these experiments into the existing curriculum without reforming its structure. The existing chemistry experiments as listed in the syllabus were modified or adapted into green chemistry experiments. For instance in order to teach rate of reaction, sodium thiosulfate is frequently used in the laboratory and to teach heating and cooling curve naphthalene is being used. Both these substances (sodium thiosulfate and naphthalene) are carcinogenic and are harmful to the environment and human health. In the green chemistry approach, these substances were replaced with lauric acid and vitamin C. Both these chemicals are safe to the environment and non-toxic. Introduction of biodiesel production, global warming, and biodegradable polymers was timely since these are knowledge issues relevant to our Malaysian context with a need for a sustainable lifestyle that can support future generations. The experiments we designed could be conducted in an environmentally safe manner while the appropriate chemistry concepts were taught at the same time. Furthermore, environmental issues were brought into the discussion of the results from the experiments as and when necessary. Hence, students could see the relevance of the content to their everyday activities and the impact of their behaviours on the environment.

13.3.2 Green Chemistry for Pre-service Science Teachers

In the School of Educational Studies of the Universiti Sains Malaysia green chemistry has been introduced to the pre-service teachers through integration of these experiments into chemistry teaching methods course.[8] This course is compulsory for those students majoring in science education. Prior to the commencement of the course a module with the details of green chemistry experiments was prepared. The duration of the course was 14 weeks with 2 hours of weekly lecture and another 2 hours of weekly tutorials. In total, 373 pre-service teachers enrolled in the chemistry teaching methods course. Three cohorts of future teachers thereby have been exposed to green chemistry as laboratory-based pedagogy. During the lecture, students were taught about green chemistry as an effective pedagogy. Mainly the lecture focused on assisting future teachers to understand (1) the differences between green chemistry and the traditional more polluting approach to chemical production and (2) how green chemistry actually reflects on addressing sustainability in the curriculum.

During tutorial sessions the pre-service teachers explored the application and concepts learned during the lecture. Peer teaching (also known as simulated teaching) is one of the strategies we use. Each week, two students assumed the role of a teacher and conducted a green chemistry lesson where their peers take the role of their future high school students. Teachers were required to prepare a complete lesson plan about the chosen topic for that particular week.[8] The instructor functioned as a facilitator during the lesson implementations. The lessons are student-centred and organized in the way that participants are intended to teach after becoming teachers themselves. A typical lesson plan for the teaching of polymers is presented in Table 13.1.

13.3.3 Green Chemistry for In-service Science Teachers

Following the successful implementation of green chemistry at the pre-service teachers' level, our efforts shifted to incorporate green chemistry into programmes designed for in-service teachers. A series of workshops have been conducted to familiarize practising teachers with green chemistry experiments, which allowed them to evaluate the relevance of the modified experiments. The workshops were attended by a total of 125 participants: six university lecturers, 116 secondary school chemistry teachers, and four inspectors from Departments of Education. The participants conducted three green chemistry experiments: rate of reaction with vitamin C, heating and cooling curve of lauric acid, and the production of biodiesel. Upon completion of the experiments the participants were asked to evaluate the feasibility

Table 13.1 Lesson plan: polymers.

Part of the plan	Lesson
Objective	Students should be able to differentiate between different type of polymers, identify the physical and chemical properties of the polymers; identify the usefulness and harmful effects of the polymers to human health and the surrounding
Phase one (engaging Phase)	The lesson begins with an introduction to polymer in which the teacher explains that polymer derived from combination of various monomers. The teacher showed example of polymers and the relevant monomers that forms the polymer. The teacher also explained the physical and chemical properties of the polymers.
Phase two (empowering phase)	
Step 1: Explaining	The teacher uses plastic bags, polystyrene boxes and plastic food wares as example of polymers and explaining the monomers that form these polymers, the chemical process involved and the benefits and harmful effects of these polymers to human health and the environment.
Step 2: Reinforcement	The teacher reinforces the non-biodegradable and biodegradable properties of the substances by allowing the students to conduct a laboratory experiment to identify which substance possesses faster degradation rate: paper; polystyrene or starch. For this purpose, two groups of separate beakers containing crushed paper, polystyrene and starch each were prepared and labelled as group A and group B. The content of the beakers were mixed with 10mL of water. Small amount of amylase powder (0.05g) was added to each sample in group A and stirred. The samples were allowed to sit for 10–15 minutes. Starch and glucose test was performed with the samples in all the six beakers. The teacher explains the purpose of performing starch and glucose test and how to design biodegradable polymers. The green principle: *use of renewable feedstock* is demonstrated here.
Step 3: Enhancing	The teacher divides the students into two groups and distributes worksheets to assist them in analysing the life cycle of these polymers. The two groups discuss the benefits and harmful effect of the polymers to the current and future generation.
Phase three: Closure	The teacher asks questions to assess the students' understanding of polymers and its properties. The teacher summarizes the lesson by reflecting on cognitive, economic, societal and environmental aspects of polymer production and asks students to review literature and prepare a folio on the environmental, economic and societal consequence of using polymer such as polystyrene.

of integrating green chemistry experiments into the integrated secondary school curriculum. A survey form consisting of five items was used for this purpose. The questions employed a 5-point Likert scale with 1 strongly disagree, 2 disagree, 3 not sure, 4 agree and 5 strongly agree. Question 1 of the survey form evaluates participants' views as to whether the green chemistry experiments are in accordance with the syllabus requirements. Questions 2, 3 and 4 focused on views regarding the experiments (effect of concentration on the rate of reaction, effect of temperature on the rate of reaction, effect of size on the rate of reaction and catalysis and rate of reaction). Question 5 asked for opinions regarding the implementation of green chemistry experiments.

The results show that large a number of participants strongly agreed (95%) and agreed (5%) that the green chemistry experiments are in accordance with the syllabus requirement and the objectives of the existing experiment can be achieved through the green chemistry technique. The majority of the participants strongly agreed (86%) and 13% agreed that the experiments facilitated students' understanding of chemistry concepts related to the experiments and the experiments were easy to conduct. Almost all the participants agreed that the implementation of green chemistry provided a platform for teachers and students to debate over environmental issues. These results are consistent with the calls for integrating environmental issues into conventional subjects to teach about sustainable development.[9]

A large number of the participants also agreed that the experiments reflect on real life beyond school, stimulate interest in chemistry/science, are easy to implement in schools and will increase the students' confidence in conducting the experiments. The existing Eurocentric curriculum (Western Science) is known to be mono-cultural and failed to reflect the traditional or indigenous life of local people.[10] In the traditional approach, students have and develop little interest of the subject matter and the learning of chemistry is considered irrelevant to everyday life. However, green chemistry is recognized by students as directly connecting them, affectively and cognitively, with the acquisition of new knowledge.

13.4   Green Chemistry Changes the Determinants of Learning

13.4.1   Introduction

In the preceding section, we describe how green chemistry was introduced in a Malaysian science teacher education programme and in programmes of in-service teacher enhancement efforts. We did not simply make the changes from traditional chemistry to green chemistry but simultaneously conducted studies to investigate whether green chemistry might be associated with changes in such achievement determinants as (1) learning motivation, (2) environmental awareness and concerns, and (3) attitudes, motivation towards the environment, and environmental values. In this section, we provide a general review of our findings.

13.4.2 Effectiveness of Green Chemistry in Enhancing Learning Motivation

One of the crucial aspects of learning is motivation, which is a key variable for students to perform well in examinations.[11] However, in the context of science learning, generally, and chemistry learning, specifically, low motivational levels are prevalent.[12] Self-efficacy belief, task value belief, goal orientation, and affect are important determinants of motivation.[13] The construct of self-efficacy belief refers to individuals' perceptions of their ability in particular situations or in accomplishing learning tasks.[14] The task value belief construct refers to students' beliefs about interest, utility, and importance of a course.[15] Goal orientations denote underlying purposes when approaching, engaging in, and responding to achievement situations.[16] Mastery and performance goal orientations are two commonly considered forms of such orientations.[17] Mastery goals refer to the intention to master a task to improve competence;[18] and performance goals refer to concerns for performing better than peers.[19] Affect orientation measures on the students' interest and anxiety in performing the task.[20]

In our studies we also conducted interviews to evaluate how pre-service teachers' motivation in learning chemistry changed after conducting a series of green chemistry experiments. A total of 20 pre-service teachers agreed to be interviewed. The changes in motivation to learn chemistry from the first to second interview were investigated. During the second set of interviews, we observed changes in the respondents' motivation to learn chemistry. The responses of the participants during the second interview reflected on higher level of motivation in learning chemistry. Relatively more student teachers were observed to exhibit high self-efficacy belief, high task value belief, mastery goal and interest during the second interviews than the first interview.

Out of the 20 pre-service teachers interviewed during the second interview, 14 exhibited high self-efficacy belief. Compared to the first interview there was an increase in the number of students with high self-efficacy belief. Only seven students noticed with high self-efficacy belief during the first interview. At the time of the second interview, 17 students demonstrated a high task value belief in discourses while only four students expressed high task value belief during the first interview. Similarly, for the goal orientation there was a notable increase in the number of students with mastery goal orientation following the green chemistry course. Out of 20 pre-service teachers interviewed, seven students demonstrated mastery goal orientation and 13 students exhibited performance goal orientation during the first interview, whereas during the second interview more students exhibited mastery goals (14 students) and fewer students expressed performance goals (six students). Our analyses showed that mastery goals were generally associated with intrinsic interest in learning activities,[21] positives attitudes toward learning[22] and increase amount of time spend on learning the task.[23]

Affect orientation refers to students' emotional reactions to learning tasks.[24] This includes students' interests and anxiety levels experienced while engaging in chemistry learning tasks. Student teachers exhibit interest when they are obviously engaged and involved in the learning task. Another component of affect orientations is anxiety. Students with anxiety orientations engage in the learning task because they worry about not being able to perform well in the examination. At the time of our first set of interviews, the discourse of the four students exhibited on affect orientation of interest and 16 reflected on the high levels of anxiety they experienced while studying chemistry. However during the second interview only three students commented in ways that exhibited anxiety orientations; the remainder (17 students) made apparent their affect orientation of interest.

The results of this study show that green chemistry contributes to a shift in pre-service teachers' perceived level of these constructs (self-efficacy belief, task value belief, goal orientation and affect orientation) after the change to green chemistry experiments associated with curricular changes that allowed them to make connections between school chemistry and their everyday lives. Thus, among those who initially expressed low self-efficacy beliefs subsequently tended to draw on high self-efficacy beliefs. Some individuals experienced a shift in emphasis from a low task value belief towards a high task value belief, performance goal orientation towards mastery goal orientation and some of the discourse draws on low level of anxiety and higher level of interest in learning chemistry.

13.4.3   Effectiveness of Green Chemistry in Enhancing Environmental Awareness and Concerns

A central presupposition for changing from the use of traditional substances in the chemistry curriculum to substances that are environmentally friendly is that there will be associated changes in students' environmental awareness and concerns. As part of our effort to implement environmental chemistry, we therefore conducted studies to investigate whether the anticipated changes would actually occur. Thus, for the purpose of measuring effects of the integration in enhancing environmental awareness and concern, 25 pre-service teachers were randomly selected and qualitatively interviewed before and after the integration of green chemistry experiments in the curriculum. During the pre-interviews, participants were asked to respond to the question 'What would you say are the most important environmental problems in present-day society?' All 25 respondents depicted air pollution and water pollution as most important environmental problem. During the post-interview the same question was asked. Diverse responses were obtained from the participants. A majority of the respondents (12) provided global warming as answer. The remainder talked about deforestation (two), climate change (five), natural resource depletion (four) and the extinction of certain species of flora and fauna (two) as the most important environmental problem faced by the present society. These answers are consistent with the hypothesis that, initially, students' awareness of environmental problems was very general. After completing the course their answers were apparently more sophisticated.

Our interview analyses revealed that students' beliefs about pollution and water pollution as environmental problems were rooted in what they previously had learned in high school and they have come to know from information gathered in their encounters with different types of media. Thus, for example, one of the respondents suggested:

> During secondary schooling, I learned about environment in geography. Lots of things were mentioned about environment though I can only remember some of it. Air pollution and water pollution were mainly talked about. Also, in newspapers I read about people getting sick from the polluted air. I think activities that contribute to air pollution and water pollution should be avoided.

Another participant explained:

> I learned about gases such carbon dioxide, nitrogen oxide, sulfur dioxides from chemistry lessons. I think these gases pollute the environment. I am not sure about greenhouses gases. I think the factories are the main reason for these gases in the atmosphere; and these factories should be banned from operating.

The responses obtained during the post-instruction interviews illustrate the participants understanding of the complexity of environmental issues. The responses reflect considerable integration of conceptual issues, including the interdependence of the living organisms and non-living materials. The respondents often began their answers by stating local circumstances and then expanded their explanations to include phenomena that occur at the global level. A number of respondents also included economic impacts in their responses. The post-instruction interview answers focused on issues such as over-consumption, poverty, population, crimes, policy-making and war. Such complex and differentiated views are expressed in the two following examples:

> I think deforestation is the main problem in my society. Trees are being cut down. We are slowly losing our pride. Malaysia is well known for tropical rain forest. Perhaps, in future there would be none. Animals living in the forest are gone. Deforestation also can cause flooding because there is less trees. Finally, the effect is on the community such as spreading of disease, losing job, poverty, destruction in food supply, hunger and you can just ad on the list.

> Global warming is what we are experiencing now. Greenhouse gases are accumulated in the atmosphere. These gases trap the heat from being released to the outer space. [There is an] over-consumption of fossil

fuels, in general consuming any material that functions on energy releases greenhouse gases. Deforestation for the purpose of economic development of the country contributes greenhouse gases [is bad] as well. Locally we are going through extreme weather. Sometimes it rains heavily and the next day it will be extremely hot. Globally, many places are affected by unusual heavy rain and unusual droughts and even sand storms like at Beijing, China.

In the first quotation, the individual refers to the well-known problem of illegal harvest of wood, which is rampant in present-day Malaysia. Because teachers have a potential impact on their students' attitudes and values towards the environment, we may anticipate that teachers understanding the impact of negative practices will influence attitudes and values of their own future high school students. The post-instruction interview results illustrate the interconnectedness of the entire environmental issues. The respondents begin the answer with one specific environmental problem and they tend to inter-relate with other environmental, economic, societal problems. The respondents' perception of environment has changed after taking up the course. Previously, each environmental problem was viewed discretely.

13.4.4 Effectiveness of Green Chemistry in Changing Attitudes, Motivations and Values

13.4.4.1 Introduction

Environmentally destructive activities such as the destruction of forest, land degradation, marine pollution and open burnings negatively affect the environment and its human inhabitants. These activities are rampant not only in Malaysia but in many other developing nations and emergent economies even though the ruling governments frequently impose stringent rules and heavy penalties on such activities. Various reasons were identified as the cause of the high incidence of such destructive activities. The environment is considered as a resource provider and people involved in such activities can or hope to raise their economic status. Other possible reasons cited include that the awareness of the effects of environmentally destructive activities tends to be minimal. It is suggested that environmental attitude is the primary determinant of environmental behaviour, *e.g.*, actions which contribute towards environment preservation or conservation.[25,26] Studies suggest that motivational change could lead to the change in the way people act toward the environment. For instance, persons with self-determined motivation tend to commit to self-determined behaviour as well; and this motivation is a significant determinant of pro-environmental behaviour.[26] In addition to attitude and motivation, environmental behaviours appear to be rooted in human values;[25] and environmental values are the basis for behavioural change.[27] Favourable environmental attitude, motivation and values are also to be desired in the Malaysian context and attitude, motivation and values are the reasons for citizens' involvement in environmentally destructive behaviours.[28]

We designed green chemistry with the intent to positively affect pre-service teachers' environmental attitude, motivation and environmental values. Therefore, green chemistry experiments have been integrated into the chemistry teaching methods course for pre-service chemistry teachers at Universiti Sains Malaysia. The aims of the course include bringing about changes in environmental attitudes, motivation and values towards being more environmental. It was hoped that these changes will promote the development of environmentally responsible behaviours among the pre-service teachers involved. A quantitative study was conducted to measure the effectiveness of green chemistry in changing attitude, motivation and values. In the following section, we provide details and outcomes of environmental attitude, motivation and value surveys obtained from the 2nd cohort of the student teachers enrolled in the course. The outcome obtained from the first cohort of the student teachers has been published previously.[29]

13.4.4.2   Environmental Attitudes

Environmental attitudes were measured using the New Ecological Paradigm.[26] The instrument contained 15 items and measured on a 5-point Likert-type scale ranging from 1 (strongly agree) to 5 (strongly disagree) (Cronbach $\alpha = 0.71$).[29] The study intended to investigate the effect of green chemistry on changing the attitude of 263 pre-service teachers. The New Ecological Paradigm was administered twice before the experimental treatment (Week 1) and after the treatment that was on the last week (Week 12). The experimental treatment lasted from Week 2 to Week 11.

Table 13.2 presents the pro-environmental attitude means, standard deviations and *t*-values for each item in the instrument. Significant differences in the mean values were obtained for all the items at 95% confidence level. This suggests that green chemistry experiments enhanced the students' pro-environmental attitudes. The outcomes of the pre-test on students' pro-environmental attitude were consistent with the results from an earlier study documented in the literature on Malaysian secondary school students.[30] The study gauged the level of environmental understanding, awareness and knowledge of secondary students; it suggested that the current practices of environmental education was ineffective in changing patterns of action and behaviours towards the environment. The higher post-test scores reported in the study with pre-service teacher is consistent with the hypothesis that green chemistry experiments had facilitated the development of pro-environmental attitudes. This could be probably due to the nature of green chemistry which is interdisciplinary.[7] The interdisciplinary nature of green chemistry provides an opportunity for students to ethically address the environmental issues. This further supported the development of pro-environmental attitudes.[5]

Table 13.2   Paired sample statistics for NEP.[a]

		Pre-test		Post-test		
Item		Mean	SD	Mean	SD	t-value
1	We are approaching the limit of the number of people the earth can support	3.34	1.13	1.61	0.83	17.38
2	Humans have the right to modify the natural environment to suit their needs	2.54	1.33	3.64	1.61	8.54
3	When humans interfere with nature it often produces disastrous consequences	3.43	1.38	2.25	1.43	9.62
4	Human ingenuity will ensure that we do NOT make the earth unlivable	2.81	1.30	3.54	1.43	6.13
5	Humans are severely abusing the environment	3.24	1.30	2.00	1.60	9.76
6	The earth has plenty of natural resources if we just learn how to develop them	2.36	1.46	3.43	1.37	8.67
7	Plants and animals have as much right as humans to exist	3.68	1.26	2.21	1.61	11.66
8	The balance of nature is strong enough to cope with the impacts of a modern industrial nation	2.43	1.20	3.74	1.60	10.62
9	Despite our special abilities humans are still subject to the law of nature	3.96	1.34	1.82	1.51	17.19
10	The so-called 'ecological crisis' facing humankind has been greatly exaggerated	2.04	1.20	3.13	1.41	9.54
11	The earth is like a spaceship with very limited room and resources	3.56	1.51	1.56	1.52	15.14
12	Humans were meant to rule over the rest of nature	2.20	1.42	3.26	1.76	7.61
13	The balance of nature is very delicate and easily upset	3.93	1.14	3.26	1.78	5.14
14	Humans will eventually learn enough about how nature works to be able to control it	2.65	1.31	3.51	1.55	6.87
15	If things continue on their present course, we will soon experience a major ecological catastrophe	3.82	1.16	2.17	1.59	13.59

[a]SD, standard deviation.

13.4.4.3   Motivation Towards the Environment

Motivation is an important construct in understanding how people orient themselves towards particular issues. In our work, we hoped that green chemistry would change the motivations that pre-service teachers exhibit towards the environment. Previous work showed that self-determined motivation is associated with greater interest, positive emotions, higher psychological well-being and stronger behavioural persistence.[31,32] In addition to constructs such as environmental satisfaction, environmental responsibility and self efficacy for environmental behaviours, self-determined motivation tends to be associated with environmental responsible behaviours. For instance, it was demonstrated that self-determined motivations were significant determinants of recycling.[33]

The Motivation Towards Environment Scale was used to measure the motivation of the participants towards the environment.[32] The scale is known to possess high level of validity and high level of internal consistency and satisfactory test-retest reliability.[34] The scale contains 24 items with six sub-scales: intrinsic motivation, integrated regulation, identified regulation, introjected regulation, external regulation and amotivation. These motivational types can be arranged in the continuum based on the level of self-determination. Intrinsic motivation represents the highest level of motivation with amotivation being the lowest and characterized by loss of personal control.[34] Behaviour is controlled when it is regulated either by external contingencies or introjected demands and autonomous when it is intrinsically motivated or regulated by identification. We studied the impact of a greener chemistry curriculum involving the same cohorts as before with a total of 263 pre-service teachers.

The results in Figure 13.1 demonstrate that a curriculum with green chemistry as a core feature significantly enhanced students' self-determined motivation (intrinsic motivation, integrated regulation and identified regulated motivation). Additionally, the higher mean value obtained for intrinsic motivation indicates that the attainment of self-determined intrinsic motivation is higher than the integrated and identified regulation. Lower mean values for external regulated motivation and amotivation indicate that, generally, the students' amotivation is low. Green chemistry has an impact on externally regulated motivation; and amotivation is minimal following the experimental intervention. Therefore, non-significant changes were observed for the non-self-determined extrinsic motivations of external regulation and amotivation. However, green chemistry has significantly influenced non-self determined introjection regulated motivation. This might have been due to peer

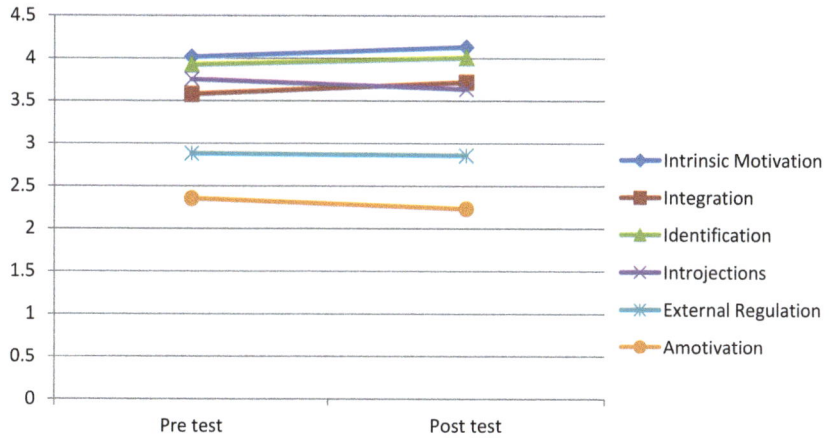

Figure 13.1 Motivation towards environment profile of the pre-service teachers.

pressure and societal demands. The demands and regulations thereby become part of a person's psychological make up but are not internalized in the person.

In general, the results of this study demonstrate that green chemistry experiments in the context of a curriculum that emphasizes connections with everyday life outside the university contribute to improving students' self-determined motivations. Self-determined motivation elevates an individual to perform behaviour that the individual freely chooses to carry out.[35,36] These behaviours are exhibited for reasons originating from within the individual and maintained without the need of external incentives or in the presence of barriers to action. Studies suggest that self-determined environmental behaviours can promote enduring behavioural changes because those behaviours are not controlled by external sources of motivation.[36] Furthermore, self-determined motivation is believed to advance positive behavioural consequences.[32] Our study shows that green chemistry may be employed as a tool to enhance the self-determined motivation.

13.4.4.4 Environmental Values

Values are important in the make-up of a person because they influence behaviours. Egocentric values centre on the individual. Individuals who articulate egocentric values generally emphasize activities that benefit themselves. They tend to engage in activities with a negative impact on the environment if the environment constitutes a threat to them. Homocentric values centre on human beings. Individuals with homocentric values judge the environment on the basis of costs or benefits to humanity generally and to concrete others specifically. Ecocentric values revolve around concerns for the whole ecosystem. Individuals with ecocentric value orientations attempt to protect the environment because of its intrinsic worth.

In our work, we used the Questionnaire on Environmental Values (Cronbach $\alpha = 0.89$) to measure the impact of green chemistry experiments on environmental value change.[36] The questionnaire contains 37 items: Items 1–11 are designed to measure egocentric value orientations, Items 12–26 measures of homocentric value orientations, and Items 27–37 measures ecocentric value orientations. For this purpose 110 pre-service teachers were involved. The questionnaire was administered three times: during the first week of the course prior to the treatment, after completing five green chemistry experiments in the manual (6th week), and finally after completing the remaining five experiments (13th week). The questionnaire was administered three times to increase the validity of the data obtained and to monitor possible changes in the construct measured.

Figure 13.2 shows the means for the overall value orientations obtained at three different times (pre-test, post-test 1 and post-test 2). In determining the overall mean value, the mean for all the questionnaire items representing egocentric, homocentric and ecocentric orientations (the entire 37 items) are taken into consideration. There is a decreasing trend in the overall score obtained from the pre-test to the second post-test. This decrease in the mean value suggests that the pre-service teachers' expressed value orientations have

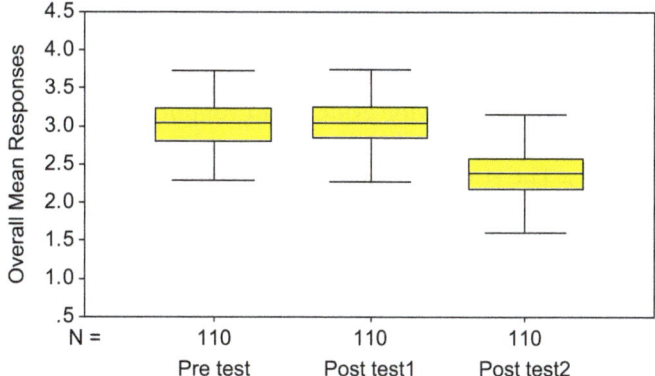

Figure 13.2 Environmental value profile of the pre-service teachers.

become more pro-environmental. The lower mean value due to the treatment also indicates that the pre-service teachers' value orientation have shifted from egocentric and homocentric towards ecocentric value orientation.[37,38]

Figure 13.2 depicts that, generally, there is lowering in mean values and interquartile range. The mean values of the pre-test and post-test 1 did not differ much. The largest lowering of mean value occurs in post-test 2. An overlap in interquartile range is observed between the pre-test and post-test 1. This indicates that the students' responses did not change much between pre-test and post-test 1. However, the responses changed significantly in post-test 2. Lowering of mean values and interquartile range shows that students' responses are being more environmental. Thus, the students are being more environmental after the post-test 2, suggesting that prolonged exposure to the treatment will develop a value change of being more environmental.

13.4.4.5 Summary

In sum, therefore, we can say that after completing the course in which green chemistry experiments play a central organizing role, students' environmental attitudes, motivations and environmental values changed towards being more pro-environmental. Participants expressed greater levels of intrinsic motivation and exhibited ecocentrism in their values after completing the course. Self-determined intrinsic motivations and ecocentric value are noted to be the antecedent of pro-environmental behaviours.[27,39]

13.5 Conclusion

In response to policies that oriented the community of Universiti Sains towards the environment, we implemented a curriculum change revolving around the use of chemicals that are less destructive to the environment than those normally used for certain chemical processes. Together with the change in chemicals were associated a greater integration of the curriculum with the students' experiences in their everyday world. The studies associated with these changes demonstrate a positive impact. For practising teachers, a change towards green chemistry is feasible because it fits within the already existing content structure. Taken together, our studies provide evidence in support of the positive effects of green chemistry experiments in improving students' motivation towards learning chemistry. The students' motivational beliefs changed towards high self-efficacy belief and high task value belief. Participants in the experimental studies also exhibited mastery and interest orientation after going through the green chemistry experiments. Additionally, the students' environmental attitudes, motivation and values improved. We observed more pro-environmental attitudes, improved self-determined intrinsic motivation and environmental values centred on ecocentric and homocentric.

Our findings constitute a significant contribution to the green chemistry literature. What is more important, though, is the potential impact the changes have for science education in Malaysia. Those hundreds of pre-service teachers experiencing green chemistry in their methods course will teach hundreds of students when they take up their positions following graduation and certification. If each teacher could bring about changes of a similar scope as we observed it in our studies, there would be a tremendous change in a nation where environmentally destructive behaviours are still the norm. We are confident that a change towards green chemistry experiments and the associated opportunities for making closer connections to students' everyday lives also would be effective in other developing nations and emerging economies. Green chemistry would thereby constitute an approach to teaching science supportive of the goal of sustainability of our global village in the new millennium and beyond.

References

1. J. J. Lagowski, *The Evolving Nature of Chemical Education: Current and Future Potential. Paper presented at the Singapore International Symposium on Chemical Education*, 2002. Retrieved March 2006 at http://www.chemistry.nus.edu.org.
2. P. Anastas and J. C. Warner, *Green Chemistry: Theory and Practice*, Oxford University Press, UK, 1988.
3. R. Carson, *Silent Spring*, Oxford University Press, UK, 1962.
4. D. J. Waddington, What is Environmental Chemistry? Global Environmental Change Science: Education and Training, Springer, UK, 1995.
5. J. Haack, J. E. Hutchison, M. M. Kirchhoff and I. J. Levy, *J. Chem. Educ.*, 2005, **82**, 974.
6. B. Braun, R. Charney, A. Clarens, J. Farrugia, C. Kitchens, C. Lisowski, D. Naistat and A. O'Neil, *J. Chem. Educ.*, 2006, **83**, 1126.
7. A. Parrish, *J. Chem. Educ.*, 2007, **84**, 245.
8. M. Karpudewan, Z. Ismail and N. Mohamed, *J. Educ. Sustainable Dev.*, 2011, **5**(2), 197.

9. *UNESCO, Report on Decade of Education for Sustainable Development*, www. unesco.org/educationon/desd, 2006.
10. G. S. Aikenhead, *Res. Sci. Educ.*, 2001, **31**, 337.
11. M. Boekaerts, S. Volet and S. Jarvela, *Motivation in learning context: Theoretical advances and methodological implications*, Pergamon, Amsterdam, 2001, pp. 17–31.
12. M. Jurisevic, S. Glazar, C. Pucko and I. Devetak, *Int. J. Sci. Educ.*, 2007, **30**, 87.
13. M. Bong, *J. Educ. Res.*, 2004, **97**.
14. F. Pajares, *Rev. Educ. Res.*, 1996, **66**, 543.
15. A. Wigfield and J. S. Eccles, *Dev. Rev.*, 1992, **12**, 265.
16. P. R. Pintrich, *Contemp. Educ. Psychol.*, 2003, **25**, 92.
17. C. S. Dweck and E. L. Leggett, *Psychol. Rev.*, 1988, **95**, 256.
18. C. Ames, *J. Educ. Psychol.*, 1992, **84**, 261.
19. C. S. Dweck, *Am. Psychol.*, 1986, **41**, 1040.
20. A. Zusho and P. R. Pintrich, *Int. J. Sci. Educ.*, 2003, **25**, 1081.
21. D. J. Stipek and P. S. Kowalski, *J. Educ. Psychol.*, 1989, **81**, 384.
22. C. Ames and J. Archer, *J. Educ. Psychol.*, 1988, **80**, 260.
23. R. Butler, *J. Educ. Psychol.*, 1987, **79**, 474.
24. N. Mousoulides and G. Philippou, *Students' motivational beliefs, self-regulation strategies and mathematics achievement, Proceedings of the 29th Conference of the International Group for the Psychology of Mathematics Education, Melbourne*, 2005, vol. 3, p. 321.
25. L. J. Axelrod and D. R. Lehman, *J. Environ. Psychol.*, 1993, **23**(1), 1.
26. R. Dunlap and K. D. Van Liere, *The J. Environ. Educ.*, 2008, **40**(1), 19.
27. R. Osbaldiston and K. M. Sheldon, *J. Environ. Psychol.*, 2003, **23**, 349.
28. M. Karpudewan, Z. Ismail and N. Mohamed, *Pre Service Teachers' Environmental Values an Indicator for Curriculum Reform to Address Education for Sustainable Development. Proceedings for World Conference on Science and Technology Education: Sustainable. Responsible. Global. Perth, Australia*, 2007.
29. M. Karpudewan, Z. Ismail and N. Mohamed, *J. Soc. Sci.*, 2011, **6**(3), 45.
30. N. Aini and Fakhrul, *Environ. Educ. Res.*, 2007, **13**(1), 17.
31. C. Seguin, L. G. Pelletier and J. Hunsley, *J. Appl. Soc. Psychol.*, 1999, **29**(8), 1582.
32. L. G. Pelletier, M. K. Tuson, G. I. Demers, K. Noels and M. A. Beaton, *J. Appl. Soc. Psychol.*, 1988, **28**(5), 437.
33. R. De Young, *J. Environ. Syst.*, 1989, **18**, 341.
34. C. Seguin, L. G. Pelletier and J. Hunsley, *Environ. Behav.*, 1998, **30**(5), 628.
35. A. Black and E. Deci, *Sci. Educ.*, 2000, **84**, 740.
36. L. G. Pelletier, R. I. Vallerand and P. Sarrazin, *Psychol. Sport Exerc.*, 2007, **8**, 615.
37. W. Kempton, J. S. Boster and J. A. Hartley, *Environmental Values in American Culture*. Cambridge: MIT Press, 1995.
38. M. Karpudewan, Z. Ismail and W. M. Roth, *Int. J. Sci. Math. Educ.*, 2012, **10**(3), 497.
39. P. C. Stern, Psychology and the science of human-environment interactions. *Am. Psychol.*, 2000, **55**(5), 523.

CHAPTER 14

Introducing Green Chemistry into Graduate Courses at the Brazilian Green Chemistry School

PETER R. SEIDL*[a], ESTEVÃO FREIRE[a], SUZANA BORSCHIVER[a], AND L. F. LEITE[a]

[a]School of Chemistry, Federal University of Rio de Janeiro, Brazil
*E-mail: pseidl@eq.ufrj.br

14.1 Introduction

Considering the many different aspects of chemical science and industry, Brazil is a relative newcomer to green chemistry. Although several isolated articles on the subject have appeared in journals since the basic concepts were proposed over 20 years ago, a comprehensive strategy involving industry, academia and government was only released in 2010.[1] It is the result of a study conducted by the Center for Management and Strategic Studies (CGEE), a think tank which works with the Brazilian Ministry of Science and Technology and is instrumental in designing new government initiatives involving technical innovation. Questions related to green chemistry got significant media coverage during the International Year of Chemistry,

Worldwide Trends in Green Chemistry Education
Edited by Vânia Gomes Zuin and Liliana Mammino
© The Royal Society of Chemistry 2015
Published by the Royal Society of Chemistry, www.rsc.org

celebrated in 2011, and were stimulated by the 4th International IUPAC Conference on Green Chemistry that was held in Brazil in 2012. World exposure to the issues related to green chemistry and corporate and institutional commitment to sustainability were reinforced by the United Nations Conference on Sustainable Development (also known as Rio+20) held in July of 2012.[2]

14.1.1 A Brief Historical Perspective

Green chemistry can play an important role in developing research and education in Brazil. In order to better understand this role, some background information on the country may be useful. Brazil corresponds to slightly less than half the surface area and population of South America. It has common borders with all but two other countries and shares most of their geology and ecology. There is one significant difference, however: as a colony it belonged to Portugal (and not Spain, as did most of its neighbours). This implies significant historical, political and cultural differences.[3] For example, before 1800 trade was restricted to the mother country while industrial and intellectual activities (including the press) were strongly suppressed whereas in other parts of the continent strong local cultures flourished. In fact, several universities in Latin America were founded at this time but not in Brazil since higher education was a privilege of the colonizers. When Spain was invaded by Napoleon, Spanish armies became involved in local conflicts and their presence overseas was significantly reduced. Ties to colonies grew weaker and, inspired by the ideals of the American and French revolutions, most of the Spanish speaking countries in the region gained their independence and adopted republican forms of government in the early 19th century. On the other hand, courtesy of his powerful ally (and Napoleon's enemy), Great Britain, the King of Portugal and important members of his court were escorted to Brazil by the Royal Navy. They immediately established the Empire of Portugal, Brazil and Overseas Colonies. Many of the restrictions on the exchange of ideas were quickly removed. The Empire opened its borders to trade and established important functions of government in its capital, Rio de Janeiro, which became the centre of intense cultural and economic activities. It was the emperor's son who declared the country's independence while his grandson remained as emperor of Brazil up to 1889 when he was dethroned and a republican form of government was adopted. This apparent curiosity had important implications for the economic and social development of the country. For example, significant economic activity remained almost entirely limited to agriculture and mining which, in turn, were largely dependent on slaves who were only freed in 1888.

Today the Brazilian economy is one of the tenth largest in the world as reflected by its gross national product. However, the country's economy remained dependent on extractive and agricultural products up to the 1930s (this was also when the first universities were established) and significant industrialization only took place in the 1950s and 1960s and mostly on the part of multi-national or state-owned companies. It is also worth noting that the Brazilian Society for the Progress of Science (SBPC) and agencies that offer fellowships and grants for research, such as the Councils for Scientific and Technological Development and for the Advancement of Graduate Education (better known by their abbreviations, CNPq and CAPES, respectively) and state agencies that support research, notably the Foundation established in the State of São Paulo (FAPESP) were also established in this period.

Although important contributions were reported before the 1900s chemistry in Brazil was taught at institutions that applied it to mining, agriculture or medicine (pharmacy, in particular). The first chemical society was established in 1922 and today there are societies in chemistry (ABQ and SBQ) and chemical engineering (ABEQ), as well as in specific areas such as polymers, catalysis and NMR. There are associations for the chemical industry (Abiquim) and the fine chemicals industry (Abifina) and related sectors such as pharmaceuticals, paint, chlorine, among others. Chemistry was recognized as a profession in 1956 and is regulated by a Federal Chemistry Council (CFQ), and by its Regional Councils (CRQs) that cover geographic regions where there is a significant activity in chemistry. Chemical engineers, industrial chemists, Bachelors in chemistry and chemical technicians are recognized as such professionals and their legal attributions depend on the respective curricula of the university or technical school. Presently, Brazil is making up for lost time. It has a significant chemical industry and a development bank and financial agency that promote its projects (BNDES and FINEP), large state-owned firms in chemical technology demanding areas, such as energy and fuels (Petrobras) and agriculture (Embrapa) and a legal and institutional framework for intellectual property (INPI) and standards (INMETRO). Science and technology are considered national priorities and have received significant financial support since the middle 1970s. The number of publications in chemistry and engineering and their citations in international journals well as the number of students enrolled in universities and graduate programs has increased accordingly.

14.1.2   The Chemical Industry

The first industrial activity in Brazil can be traced to the production of sugar. In fact, a sugar mill was installed shortly after discovery, and production during the colonial period was significant for the times. Dyes extracted from Brazilian wood along with other plants were exported by the first settlers. Soap and alkalis were later produced and, in the 1660s, table salt and gunpowder were also manufactured. Soaps, gunpowder and plant extracts were produced on a larger scale during the 19th century and the first sulfuric acid factory and other smaller factories were established as the 19 century was coming to an end. Two world wars exposed the deficiency of the nascent chemical industry in Brazil. Most of its raw materials were imported and their access was difficult during periods of conflict. There was an attempt to use locally available substitutes, but most of these initiatives were based on improvisation and only served to reveal the perils of dependence on sources of materials and manufacturing practices from abroad.[4]

The modernization and consolidation of the Brazilian chemical industry dates back to the 1960s. Petrochemical activity was expanding rapidly and the growing local automobile industry created a strong demand for fuels. Petrobras, the state-owned company that had a monopoly on much of the oil industry, built several refineries over a short period of time and had to overcome the difficulties involved in the operation of turnkey units. When its petrochemical subsidiary, Petroquisa, entered into joint ventures with other petrochemical companies it made sure that this know-how was included in the negotiations involving industrial projects. With strong incentives from the government three petrochemical complexes were built and several large firms in other fields (*e.g.*, banking, construction) took part in the projects that established the manufacture of polymers according to projected market demands. Nowadays, sales from the Brazilian chemical industry are significant and are increasing steadily.

However, historically, it has not been very innovative and the considerably negative balance of trade in chemicals is growing rapidly.

A National Pact for the Chemical Industry, based on an economic growth forecast for the coming years and involving potential investments of around US$ 170 billion, was proposed by Abiquim, the Brazilian Association of the Chemical Industry.[5] In the 2010–2020 period around a fifth of the total has to be invested in R&D aimed at correcting the deficit in the trade of chemical products, in expanding the renewable-based segment of the industry and on new opportunities that arise from exploration of the recently discovered petroleum reserves that lie far off-shore and deep under a salt layer. Its strategic goals are: position Brazil's chemical industry among the top five, generate a trade surplus in chemicals and take a leadership role in green chemistry. A top Brazilian petrochemical company already produces green plastics and another is introducing green alternatives to substances used in the formulation of consumer products.

14.1.3   A Strategy for Green Chemistry

Brazil is in a very favourable position in terms of the applications of biomass as a source of energy and raw materials for industry. It receives a constant and intense amount of solar radiation and has access to: the largest concentration of biodiversity on the planet, abundant sources of fresh water and considerable diversity in terms of regions with distinct microclimates and ecosystems. In a relatively short period of time the country has accumulated considerable experience in the large-scale production of fuel, in particular those related to ethanol from sugar cane.[1] A network to explore these comparative advantages was discussed in a workshop held in 2007, at Fortaleza, Ceará, a state in the northeast region of Brazil.[6] This is the basis for the establishment of a national strategy described in the text on: *Química Verde no Brasil (Green Chemistry in Brasil) 2010–2030*.[1] It proposes a network of centres, laboratories, pilot plants, and other specialized facilities that are involved in R,D&I on green chemistry. It works together with the Escola Brasileira de Química Verde (Brazilian Green Chemistry School) that is responsible for creating a knowledge base and preparing teams of qualified researchers who conduct the research, development and innovation activities of the network. The United Nations Conference on Sustainable Development (also known as Rio+20) provided an update of the strategies for green chemistry in Brazil and its contributions were discussed at one of the technical panels held by the Brazilian government.

14.2   The Brazilian Green Chemistry School

The Brazilian Green Chemistry School began its activities in November 2010 with a workshop on biorefineries. A series of subsequent meetings were held to draw up its agenda and engage its staff in the following tasks: (1) define the knowledge base for state-of-the-art R,D&I activities in green chemistry and form partnerships with industry to establish the priorities for these activities; (2) identify themes and academic units that are required in courses on green chemistry; and (3) organize outreach initiatives designed to promote public awareness of issues related to green chemistry.

The school is located at the School of Chemistry of the Federal University of Rio de Janeiro (UFRJ) which offers undergraduate courses in chemical engineering, industrial chemistry, food engineering and bioprocess engineering and a graduate programme 'Technology of Chemical and Biochemical Processes'. The programme grants a professional Master's degree in biofuel and petrochemical engineering and Master's and Doctoral degrees in chemical processes, biochemical processes, process engineering and technology management. Graduate students are mostly chemical engineers or chemists, but the multi-disciplinary nature of the programme also attracts biologists, microbiologists, physicists, pharmacists and engineers. The School of Chemistry is located on the university campus, close to important research centres, a technology park and an incubator for technology spin-offs.

The Green Chemistry School has a staff made up of professors from UFRJ and other universities, as well as researchers from technology centres and companies. It organizes courses, workshops and meetings on green chemistry and related topics. Present activities include: setting up a Professional Master of Sciences degree in green chemistry; development of educational materials, experiments and demonstrations for students at a high school level (that can also be used for outreach activities) and a survey of opportunities for biobased chemicals in Brazil.

14.2.1   Courses

The difficulties in introducing new courses and areas of concentration at the undergraduate level are readily apparent by an inspection of the rules and regulations that are applied by the Ministry of Education and the Federal

Chemistry Council to recognize degrees relative to the chemical professions. In the beginning of its activities, the Green Chemistry School only offered courses at the graduate level. They began shortly after the creation of the Green Chemistry School in Brazil was announced. As these courses were still not regularly held they took the form of 'special topics' which allow a certain degree of freedom in the definition of their content and the way that it is presented to students. Two of these courses were given in the second and third periods of 2010 (graduate courses are divided into four periods a year). These courses were partially based on classic texts and recent talks on the subject given at meetings in Brazil and abroad as well as the institutional arrangements and lines of research that are proposed in the green chemistry text.[1] It became immediately apparent that there were terms, like 'biorefineries' or 'green solvents' that are ambiguously (and often subjectively) defined and others such as 'design' or 'innovation' that require more thorough definitions and the use of examples. There are other questions, such as sustainability, climate change and chemical safety, which are probably familiar to students even before they are exposed to green chemistry (it is recommended that practitioners should have at least a basic understanding of the issues that are involved). Finally, since the potential applications of new products and processes are to be adopted by the chemical industry, it is important that students become familiar with local and international companies and their characteristics and business strategies.

A rather more complex situation became evident when students were asked to show their understanding of the material that was presented but discovered that processes based on one of the twelve principles did not necessarily represent the best combination of the other eleven, nor the best alternative in terms of the real world. This circumstance led to the interruption of classes and the organization of workshops. Here members of the staff and researchers from industry and other organizations got together to discuss topics of common interest and define tools and concepts that should be included in lectures, such as metrics, supply chains and life cycles. These discussions also served to identify researchers in other fields that work with these topics.

At this point international cooperation played an important role in structuring courses and selecting the material to be presented. Personal contacts at the Green Chemistry Centre of Excellence at the University of York, UK, provided a very good model of how green chemistry can be applied to projects related to industry. Certain adaptations in the lines of work were necessary since the UK has strong R&D activity in the pharmaceutical industry and green chemistry finds wide application in organic synthesis while in Brazil innovation is more closely related to fuels and consumer products and main industrial interest lies in developing processes based on renewable raw materials and sources of bioenergy. On the other hand the types of problems that arise in university relations with industry and their respective feedback mechanisms are quite similar. Besides, as outreach activities are being structured along with courses, this aspect of cooperation with the staff at York was also very fruitful.

The ACS Green Chemistry Institute (GCI) is very active in promoting activities related to education and cooperation with industry and had sent a representative to the workshop that outlined the Green Chemistry Network.[6] This was the starting point for a very useful exchange of ideas with GCI (including conversations by phone) which also provided materials used in courses and outreach activities. The text *Introduction to Green Chemistry*[7] was particularly useful but also had to be adapted to local situations.

14.2.2.1   Recent Courses

During the last three years graduate courses in green chemistry have been offered during the summer periods of 2012, 2013 and 2014. Green chemistry is also included in other courses, particularly those on the chemical industry, petrochemistry and organic processes in the oil, natural gas and biofuel sectors. A specific course on its impact on the chemical industry was offered this year.

Lectures in the courses are divided among the members of the staff of the graduate programme according to their experience in different aspects of green chemistry, usually among general topics covered by the authors and specific subjects given by professors in the respective fields and by invited lecturers. A typical outline of the course is given in Table 14.1. It is important to stress the origins of green chemistry concepts[8] and their applications and implications for the chemical industry.[9] Equally important is to point how the chemical industry has responded to the public on certain issues[10] and local practices to improve its safety, such as adoption of 'responsible care' commitments.[11]

Table 14.1   Outline of green chemistry courses.

Section	Content
Introduction	The development of green chemistry concepts: historical perspective and ethical guidelines
Sustainability	Common perceptions; growth and its limits; footprints; sustainable development
Climate change and the environment	Recent trends; indicators of change; negotiation and issues; potential consequences
Chemical safety	Public perception of chemistry; risk and contamination; regulation; goals
Criteria for evaluation	Value chains, metrics, life cycles
Green chemistry in Brazil	Recent activities; Fortaleza workshop on Network; studies by the CGEE; text on strategies for Brazil; the chemical industry and renewable raw materials
Greening processes and products	Case studies

Table 14.2 Green chemistry courses: invited talks.

Invited talk	Topic
Eduardo Falabella, Cenpes/Petrobras	Ethics, biorefineries, catalysis
Regina Lago, CTAA/Embrapa	Agricultural sources of raw materials
Gil Anderi Silva, Escola Politécnica/USP	Life cycles
Álvaro Schocair, Schocair Associates	Supply chains
Rogério Mesquita, Cenpes/Petrobras	Metrics
Andre Conde, Oxiteno	Greening consumer products
Mateus Lopes, Braskem	Synthetic biology
Lucia Appel, INT	Ethanol chemistry
Marcelo Kós Silveira Campos, Abiquim	Chemical safety, Rio +20
Jennifer Dodson, Instituto de Química/UFRJ	Outreach activities
Adão de Mattos Coelho, Oxiteno	Green surfactants

There are members of the staff working on catalysis, solvents, bio-processes, patents and technology evaluation and management. Their lectures are complemented by talks given by specialists on life cycles, metrics, biorefineries, ethics, synthetic biology, *etc.*, as well as examples of projects with specific industrial applications. A list of invited speakers is given in Table 14.2. Talks given by invited speakers are followed by a question and answer period and their implications are discussed in subsequent classes.

14.3 Students

Education is a challenging matter in Brazil. Although there has been considerable progress in recent years, the country consistently scores very low on international tests. Deficiencies in formal schooling are often pointed out as one of the main problems with technology development and international competitivity. By and large this is not the problem with the students that take the green chemistry courses. Most of them belong to the graduate programme at the Escola de Química, (School of Chemistry) which has reasonably high entrance requirements, so those who are admitted have shown good academic performance in their undergraduate courses and a certain degree of professional or research experience.

However, some of the students also teach at secondary schools and point out that there are certain aspects of secondary education (and even in some university courses) that still have not adopted modern approaches to teaching. They tend to emphasize the student's ability to reproduce the material that was presented in class or to solve problems selected from lists suggested as exercises rather than stimulate creativity and an ability to think independently. Unfortunately, some of the observations below, made by the physicist Richard Feynman, winner of a Nobel Prize, when he made his first trip to Brazil,[12] are still not completely out of date, even though he came before the 1970s (when modern graduate courses were introduced on a large scale and teaching at a federal university was still a part-time job).

14.3.1   Reflections by Feynman in Brazil

One of the first things to strike me when I came to Brazil was to see elementary school kids in bookstores, buying physics books. There are many kids learning physics in Brazil, beginning much earlier than kids do in the United States, that it's amazing you don't find many physicists in Brazil – Why is that? So many kids are working so hard, and nothing comes of it.

In regard to education in Brazil, I had a very interesting experience. I was teaching a group of students whom would ultimately become teachers, since that time there were not many opportunities in Brazil for a highly trained person in science. These students had already had many courscs, and this was to be their most advanced course in electricity and magnetism.

I discovered a very strange phenomenon: I could ask a question, which the students would answer immediately. But the next time I would ask the question – the same subject, and the same question, as far as I could tell – they couldn't answer it at all!

After a lot of investigation, I finally figured out that the students had memorized everything, but they didn't know what anything meant.

and

Everything was entirely memorized, yet nothing had been translated into meaningful words.

Finally, I said that I couldn't see how anyone could be educated by this self-propagating system in which people pass exams, and teach others to pass exams, but nobody knows anything. I said, 'I must be wrong. There were two students in my class who did very well, and one of the physicists I know was educated entirely in Brazil. Thus, it must be possible for some people to work their way through the system, bad as it is.'

In view of this situation it was quite clear that elements that stimulated the development of skills useful in the students' real-life situations should be included in the courses. Besides the knowledge received in formal lectures, a student's evaluation was based on specific assignments.

14.4   Assignments

In order to address such deficiencies the courses also included assignments that would force students to think and to communicate their findings. They are outlined below.

14.4.1   Literature Searches

The first assignment in the course is a literature search in Portuguese on Who's Who in Green Chemistry, carried out by individual students. This exercise reveals the disparity of entries on popular search machines and the shortcomings some students may recognize in carrying out their respective searches. For the next assignment they form groups that select one of the different aspects that are covered, such as: websites, courses, research groups, publications, meetings, *etc.* and rank them according to certain criteria. Results are presented in class by topic, each member of a group discussing his/her contribution. This exercise usually reveals synergies among members of the group and leads to tutoring by peers. It also serves to discuss some of the material covered by Appendix 1 in the book by Winterton,[9] notably the difference between the 'open' and peer reviewed sources of information, an important point that is not always familiar to students.

14.4.2   Panel Discussions

Panel discussions of current topics in the media that impact green chemistry are usually assigned to the groups that were formed for literature searches. Here the objective is to promote discussions on topics such as sustainability, climate change and chemical safety, which students were probably familiar with before their introduction to green chemistry. It is important to show how they are related to the material presented in the course (but cannot be covered in sufficient depth in the time allotted). Controversies on these subjects and arguments that arise on the part of groups that have particular interests in their discussion should also be stressed. Here, too, results are presented in class and important aspects such as the different perceptions of sustainability,[10] the role of the IPCC, issues discussed at specific conferences (such as the COPs or ICCMs.) and regulatory measures, such as REACH are usually brought up by students. If this is not the case, material found in the book by Johnson,[10] which covers the basic issues can be referred to but it is important to go over the sources of this type of information so that future practitioners of green chemistry can keep up to date on such questions.

14.4.3   Case Studies

Case studies of applications of green chemistry correspond to the final assignments in the courses. This work is clearly beyond the scope of the course but putting students through the required steps requires a basic understanding of the concepts involved and implies an understanding processes dedicated to 'greening' products or processes. It also provides an opportunity to verify how simple metrics can be employed for evaluating the impacts of green chemistry on the chemical industry. An important aspect is, for instance, the fact of using a renewable raw material does not mean the process is sustainable. There are many drawbacks in applying biochemical routes, as their kinetics are slower than conventional chemical processes, they require very dilute aqueous media and often undergo inhibition by the products that are formed. Moreover, to recover products such as alcohols and acids from dilute solutions demands a great deal of energy, and frequently difficulties occur due to the formation of azeotropic mixtures. Sometimes the E-factor of these roots is not favourable as higher amounts of effluents and waste are produced.

14.5   Conclusions

Introducing green chemistry courses into a graduate programme in Brazil is not a simple task. It must consider the main issues that are shaping the adoption of green chemistry on a worldwide scale while addressing its main applications to innovation on the part of local companies. Nevertheless, the large degree of interest that is generated by the students and the propensity of companies to cooperate provide strong incentives for investing the time and effort required. The proper course materials should cover concepts and metrics, include many activities in which students must search for outside information and interact with one another, and require oral communication in the form of reports and discussions. The Brazilian Green Chemistry School was the result of a study involving academics, industry and government followed by recommendations for its adoption over the appropriate length of time. One of the most important was to establish its initial activities at a university that has close ties with industry and does research on chemical and biochemical processes.

Acknowledgements

We thank the Graduate Program in Technology of Chemical and Biochemical Processes for the support of the Brazilian Green Chemistry School, our colleagues who took part in courses or discussions on their content who are mentioned in the text, and researchers in our respective groups who have gathered information or developed and tested materials, particularly Andrezza Lemos, Rafaela Nascimento, Gabriel Bassani, Yasmin Guimarães Pedro and Alexander Andrev. We are grateful to Mariana R. de P. A. Doria of Abiquim, for the data on the chemical industry. The interesting text containing Prof. Feynman's impressions of his visit to Brazil was a gift from Prof. Fabio H. Ribeiro of Purdue University.

References

1. CGEE, *Química Verde no Brasil: 2010–2030*, Centro de Gestão e Estudos Estratégicos, Brasília, 2010.
2. C. Hogue, *Chem. Eng. News*, 2012, **90**, 10.
3. W. P. Longo, *Brasil: 500 Anos*, Unama, Belem, PA, 2000.
4. P. Wongtschowski, in *Indústria Química – Riscos e Oportunidades*, ed. Edgard Blücher Ltda, 2ª Ed., São Paulo, S.P., 2002.
5. ABIQUIM, *Pacto Nacional da Indústria Química*, Associação Brasileira da Indústria Química, São Paulo, S.P., 2011.
6. J. O. B. Carioca, *Proceedings of the Workshop on the Brazilian Green Chemistry Network*, UFC, Fortaleza, 2007.
7. ACS, in *Introduction to Green Chemistry*, ed. M. A. Ryan and M. Tinnesand, American Chemical Society, Washington, D.C., 2002.
8. I. Amato, *Science*, 1993, **259**, 1538.

9. N. Winterton, *Chemistry for Sustainable Technologies A Foundation*, RSC Publishing, 1ª ed., Cambridge, U.K., 2011.

10. E. Johnson, *Sustainability in the Chemical Industry*, Springer Link, New York, N.Y., 1st edn, 2012.

11. ABIQUIM, *Programa Atuação Responsável – Requisitos do Sistema de Gestão*, Associação Brasileira da Indústria Química, São Paulo, S.P., 2011.

12. R. Feynman, *Surely You're Joking Mr. Feynman*, Bantam Books, New York, N.Y., 1985.

CHAPTER 15

Educational Efforts in Green and Sustainable Chemistry from the Spanish Network in Sustainable Chemistry

SANTIAGO V. LUIS*[a], BELÉN ALTAVA[a], M. ISABEL BURGUETE[a], AND EDUARDO GARCÍA-VERDUGO[a]

[a]Department of Inorganic and Organic Chemistry, University Jaume I, Av. SosBaynat s/n, E-12071 Castellón, Spain
*E-mail: luiss@uji.es

15.1 The Spanish Network of Sustainable Chemistry (REDQS)

No technological or social advance can be considered nowadays, after having accomplished the first decade of the 21st century, without taking into consideration its sustainability. The potential of such advances to make compatible an improvement in our current quality of life with a rational use of the limited resources of our planet is, necessarily, a key element for their evaluation and implementation. Those advances can never compromise or limit the quality of life of future generations.

This paradigm is of particular relevance in the case of chemistry. The understanding, in the second half of the 20th century, of the strong and complex relationships between the production of chemicals and their manipulation

Worldwide Trends in Green Chemistry Education
Edited by Vânia Gomes Zuin and Liliana Mammino
© The Royal Society of Chemistry 2015
Published by the Royal Society of Chemistry, www.rsc.org

and use and the environment,[1] led to a dramatic shift in the public perception of chemists, chemicals and chemistry. From being the leading scientific topic in the first half of the last century, the subject that was considered to have the highest potential to provide continuous and significant benefits to Mankind, chemistry was essentially banned from public life in advanced societies, becoming a synonym for anti-natural concepts, and of being a danger to Nature and to human beings.[2] However, we, as scientists, as chemists, are aware that there is no future without chemistry. No doubt, an educated society should also share this vision.

Professionals working in the areas of chemistry, materials, pharmaceutical drugs, agriculture, cosmetics, energy, *etc.* are the best prepared to understand the interactions between chemicals and the environment. Thus, it cannot be surprising that environmental chemistry has become a fundamental area of multi-disciplinary study and research that involves the collaborative work with a number of other disciplines such as statistics, physics, toxicology, biology or geology. In the same way, chemists were those who developed, in the last decade of the past century, the general concepts, principles and tools that could allow making compatible chemistry with the preservation of the natural resources and the quality of our environment. Since the initial enunciation of the green chemistry principles by Anastas and Warner in 1998,[3,4] multiple additional contributions have allowed the establishment of a solid conceptual body that is currently accessible to all those interested in the field.[5-14] As it has been enunciated many times, chemistry is not the problem. Chemistry is the solution that allows maintaining and improving the current welfare of our societies, without compromising their current sustainability or the resources for the expected quality of life and prosperity of future generations.

Within this context, the Spanish Network of Sustainable Chemistry (Red Española De Química Sostenible, REDQS) was formally created in 2003 by a group of university teachers and researchers from different institutions, including professionals and researchers from industry, sharing a common interest in developing and promoting the general principles of green chemistry. In this regard, the main task of the network was defined as:

> promoting the development of sustainable chemistry and the diffusion of its knowledge in the ambit of the university, research centers—public or private—industries and other centers of production and/or use of chemicals, scientific societies and in the society in general.[15]

For this purpose, carrying out different activities related to sustainable chemistry has been considered since then, including:

- Cooperation with universities, research and development centres, industrial companies and scientific societies, located in Spain and in other countries, in particular in developing countries
- Organization of courses and educational activities for teachers at different educative levels, including university teachers

- Promotion of educational and scientific exchanges between Spanish universities and scientific centres with centres in other countries
- General diffusion of scientific concepts, knowledge and achievements.

Thus, developing educational initiatives in the area of green and sustainable chemistry has been an essential concern of the RSEDQS from the beginning and this has been explicated through a diversity of activities as would be described below. Nevertheless, during this period of time, the design, promotion, contribution and support to a solid and fruitful inter-university postgraduate training programme in sustainable chemistry (Master's and PhD studies) has been a central motif for the efforts developed.

15.2   Education in Green and Sustainable Chemistry from the REDQS Perspective

Developing any efficient educational activity or strategy involves providing the correct answers to a number of essential questions. A preliminary question, of course, regards the opportunity of such educational activity: 'Is there a need for such activity?' A simplistic vision of the current situation of chemistry could suggest that green chemistry concepts have finally found a broad acceptance in the scientific, technological and educational communities in the last decade. This acceptance is, however, very superficial in some instances and does not involve a true understanding of the basic elements for the building up of chemical technologies for a sustainable society. The existence of a variety of external stimuli, some of them very relevant, explains this apparent contradiction. The appearance and success of journals and editorial series specifically devoted to green and sustainable chemistry, often associated with the main scientific societies and editorials in the area of chemistry, some of which have reached rapidly very high standards, as measured by the usual bibliometric parameters, is one of these stimuli.[16] In the same way, many national and international funding agencies have the principles of sustainable development as essential objectives for future research and development projects, including green and sustainable chemistry in the cases chemistry is involved. Of particular relevance in Europe has been the introduction of this general approach in the FP7 and Horizon2020 initiatives of the European Union, which has been immediately translated to national initiatives by the member states.[17] This has led to a general adoption of terms from green chemistry but also to a trivialization of their use. Unfortunately, it is not uncommon to now find scientific contributions in which a partial understanding of the green chemistry leads to products, systems or processes that, although they partially improve some specific aspects, provide, however, a significantly higher environmental footprint. Thus, there is a clear need for continuing the educational efforts towards all those involved in the field of chemistry. Moreover, the REDQS has always considered that there is an essential need to reach, with new and continued educational initiatives, a much broader audience as will be defined below.

The first traditional question to be addressed involves the definition of the subject or subjects of the activity: 'Which is the audience to whom it is directed?' Of course, every specific activity has its own specific audience, but the overall activities of the REDQS in this field have tried to cover the broader possible spectrum of audiences. This is illustrated in Figure 15.1, which identifies the different levels taken into consideration.

The first level involves activities addressed to society as a whole as the consumers and end users of most of the chemical systems and technologies, but also as passive subjects of their side effects. The main outreach actions of the REDQS should be included here. No specific definition for this audience is required, but when designing the corresponding events, understanding that most of the audience will lack explicit technical knowledge or training in the field is essential. A very particular effort, at this level, needs to be taken for actions and activities focused towards people and organizations having any kind of social leadership, those that are involved, to any extent, in the definition of policies or in the generation of the public opinion. Trades unions, political parties and associations or mass media, for instance, cannot be forgotten when defining the outreach activities.

Professionals having a direct or indirect connection to chemistry represent the second level presented in Figure 15.1. In this case, our experience shows that it is very important to specifically distinguish a very particular profile in the case of teachers. Most efforts here should be devoted to the training in the area of green and sustainable chemistry of teachers involved in pre-university levels. There is no doubt that 'teaching the teachers', in this case, can provide

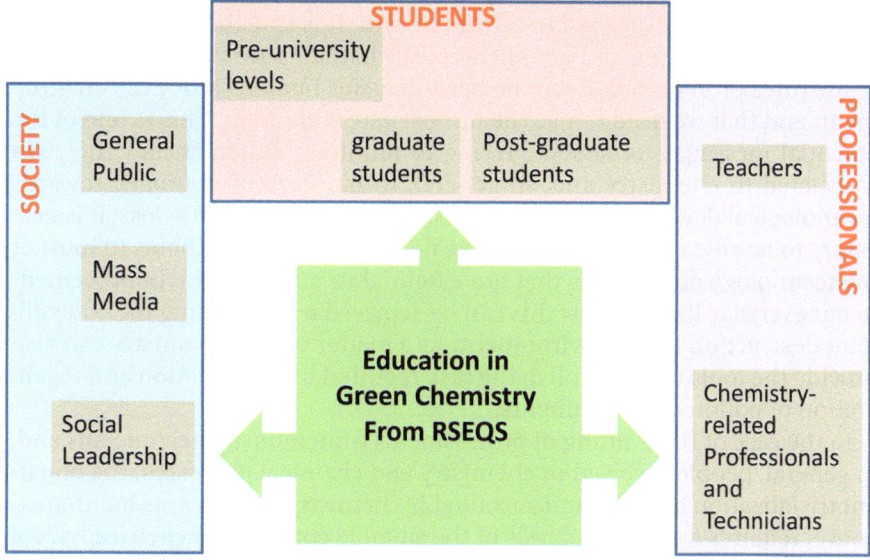

Figure 15.1   General audiences considered for education in green chemistry by the Spanish Network in Sustainable Chemistry.

higher returns for the future. We must bear in mind that news and simplified concepts and ideas associated to the effects of chemical pollution, chemical hazards and chemical risks arrive very easily nowadays to school and high school students. Thus, it is absolutely vital to be able to reach this audience on a regular base, and particularly through their teachers, to instil in them the most elemental concepts of green and sustainable chemistry and the understanding that only scientific and technological achievements have the answers to solve environmental problems and to provide a viable future.

Finally, the third kind of audience is that of those registered as students at the different stages. A clear differentiation is possible here between pre-university students and graduate and postgraduate students. Pre-university students can be reached directly through specific outreach activities or indirectly, as mentioned above, through educational activities for their teachers. Graduate and postgraduate students can also be addressed through different events, but the most important task here should be the incorporation of green chemistry concepts in academic curricula both at the graduate and postgraduate levels.

The second essential question involves a definition of the objectives and contents of the educational activities considered: 'What is to be taught and which skills and competences are to be provided?' Apparently, this seems to be an easy question, although a correct answer has been for us one of our main concerns for the last years. A clear difference must be assumed, here, between the objectives to be fulfilled and the skills and competences to be reached in the training of professionals and workers in the area of chemistry and those intended in educational activities for the society in general. In the second case, we have approached the different educational initiatives based on a single essential concept: chemistry has the answers for the current challenges our society is confronted with. It is true that the production, storage, manipulation and use of chemicals in their pure state or in the form of mixtures or in materials can be hazardous for humans or for the environment, and that we need to use the utmost care with them. This is one of the essential messages to be sent. Trying to minimize systematically the risks associated to chemistry and, in general, to our current approach towards technological development is not the correct attitude. Nevertheless, it is necessary to be able to convincingly show how chemistry contributes to most of the technological advances that are essential to maintain the improvement in our everyday life and how this can be achieved without being the concomitant destruction of the environment. As a matter of fact, chemistry can also provide the tools to accomplish the task required for remediation and regeneration of polluted environments.

In the case of the training of professionals from industry, technicians and, in general, people involved in chemistry and chemical processes, the apparent trivialization of green and sustainable chemistry concepts, as mentioned above, requires a careful analysis of the suitable contents for each individual activity. In line with the discussion in the former paragraph, from the RSEQS we have considered that the most important objective is to *develop a critical thinking* in those to be trained. Any human activity, including there chemistry, has a cost, an environmental footprint. We intend to train professionals being able to properly assess this multifaceted balance between costs and benefits. This has to be done with a complementary training in the concepts, techniques and tools that make up green chemistry. Nevertheless, it is worth mentioning here that in many instances this second aspect has been the priority. Education in green chemistry has focused, many times, in the training on individual tools that have been developed and implemented associated to this field. On the contrary, our starting point is the full integration of those elements as is illustrated in Figure 15.2. Only this approach can allow obtaining a true *integrated vision of the field*, achieving in those professionals the corresponding critical thinking skills that are needed to solve current and future challenges. The need for a given level of training in environmental chemistry is also considered in Figure 15.2. There has been a long-lasting debate regarding the interactions between green chemistry and environmental chemistry, and, in many cases, they have been considered as divergent fields. Our option, however, has been to integrate environmental chemistry as one of the subjects essential for an appropriate professional education in green chemistry. This can be achieved by including a specific course(s) in this subject, but preferably through its introduction in a transversal way in the different courses dealing with the tools and methodologies of green chemistry. In this regard, environmental chemistry can represent an integration motif for them as well as an essential instrument to properly assess the benefits and risks of each individual tools. It is important to bear in mind that, historically, the main negative impacts of chemistry in our environment took

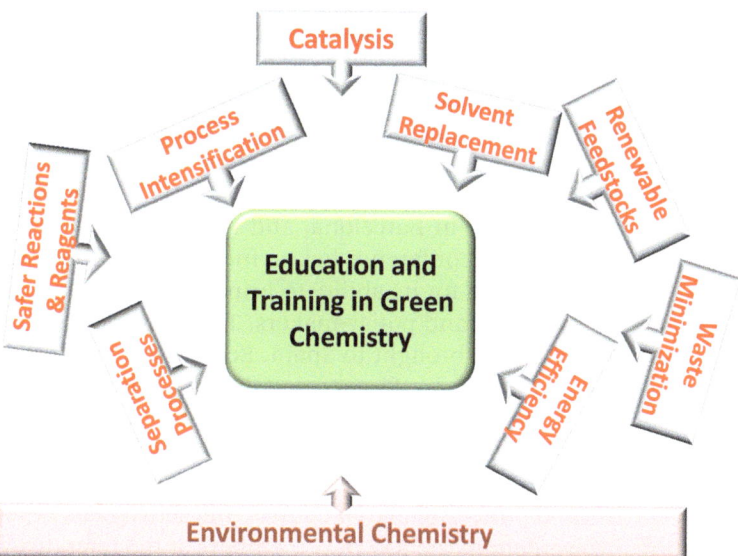

Figure 15.2 Integration of concepts for education in green chemistry.

place because of our lack of knowledge of the interactions between chemicals being created and used by men and the ecosystems and living beings, interactions that sometimes are only observable on a long-term basis. Not including environmental chemistry elements in the teaching of green chemistry can undoubtedly lead to repeating those historical mistakes. This is sometimes detected when a detailed analysis of some individual contributions in the field is carried out. In some instances, attempted solutions to the negative impacts of some traditional chemical processes through individual specific changes provide processes that can be much more harmful, as much as (eco) toxicological, biodistribution, transformation and other related aspects are not considered in detail.

Finally, a last essential question should define the way in which the activity is to be developed: 'Which is going to be the format?' In this regard, any format is valid as much as we are able to fine tune the requirements of the format with the contents expected and the potential audience. As we will discuss below, we have used a large variety of formats, taking advantage of the specific features they individually present. In a simplified way, we could signal that the RSEDQS has never rejected participating in any activity that could potentially have an educational output in green chemistry.

15.3 Educational Initiatives from the REDQS

15.3.1 General Initiatives

As mentioned above, development of educational initiatives has been deemed essential since the first stages of the Spanish Network in Sustainable Chemistry. A variety of different actions have been carried out by the members of the Network either at an individual level or collectively. It is not easy to provide a reasonable summary of such activities, although probably the best option is to classify them according to their expected audiences. Regarding diffusion of green chemistry concepts towards scientists and chemistry-related professionals, with special emphasis on those working in industry, the REDQS has maintained the Green Chemistry Conference as its main initiative.[18] Usually held in Barcelona, and reaching in 2013 its 10th edition, the Green Chemistry Conference has an international character and has gained a sound reputation for its ability to bring together high-level scientists, industrials, educators and policy-makers. This has allowed many of the pioneers in the field to be brought to Spain. Education has always been an important component of the conference and different presentations and round tables on this issue have been held at the different editions. Associated with the conference it has been usual to add one-day sessions devoted to facilitate the interaction between the members of the RSEDQS and other interested scientists and professionals. Of course, this activity has never precluded a strong involvement of the RSEDQS members at the reference international events in the field,[19] such as the International IUPAC Conference on Green Chemistry, the Annual Green Chemistry & Engineering Conference from the ACS and the Green Chemistry Institute, the Gordon Research Conference on Green Chemistry or the International Conference on Green and Sustainable Chemistry (GSC) or the International Conference on Chemistry and the environment (ICCE) promoted by the Division of Chemistry and the Environment (DCE) of the EUCHEMS. The ICCE series of conferences have included, in several editions, symposia directly related to green and sustainable chemistry under the influence of members of the RSEDQS, considering the importance and the need of keeping close contact between the specific fields of green chemistry and environmental chemistry. The participation in those events has not been limited to the presentation of research activities of the different groups, but the participation at educational forums and symposia has also been very active. In this regard, we should mention the presentation of the RSEDQS initiatives at the Green Chemistry Education Schedule in Eugene, Oregon, in 2001, at the Post-Graduate Summer Schools on Green Chemistry in 2000 and 2001, at the ACS Annual Green Chemistry & Engineering Conference in Washington in 2005, at the 4th IUPAC Conference on Green Chemistry in Foz de Iguaçu, Brazil, in 2012,[20] at several green chemistry conferences in Barcelona, the 6th ANQUE International Congress of Chemistry (Puerto de la Cruz, Tenerife, 2006) or through the participation at the corresponding round tables at the 6th Green Chemistry Conference (Barcelona, 2004) or at the RSC/SUSCHEM forum (London, 2006).

Diffusion of green chemistry concepts for general public has been approached through talks in cultural associations and a constant presence in local media, with interviews and taking advantage of the presentation of specific activities. A second category of activities correspond to the organization of summer courses or related workshops. Thus, the RSEDQS and its members have been at the core for the organization of summer courses and related workshops in different places in Spain:

- *Green Chemistry* (University Jaume I-Bancaixa, Castellón, 2002)
- *Workshop on Sustainable Chemistry* (University Menendez Pelayo, Formigal, Huesca, 2004)
- *Climate Change and the Protocol of Kyoto* (University Complutense of Madrid, El Escorial, Madrid, 2004)
- *Sustainable Chemistry. Towards a Cleaner Production* (International University of Andalusia, Seville, 2005)
- *Environmental Education, the Way Towards Sustainable Development* (Suances, Cantabria, 2006)
- *Workshop on Hazardous Wastes* (Huelva, 2006)
- *Environmental Values in Education* (University of Alcalá de Henares, Cáceres, 2013).

Our experience is that, in general, the audience attending summer courses in Spain is very heterogeneous and includes university students, professionals, teachers and general public. Often some of the participants only have a very limited knowledge of chemistry and attend the courses with a high level of curiosity but have great difficulty in following the technical discussions. This determines that the selection of topics and the level of the presentations are very delicate and sometimes the educational efficiency of these activities is relatively restricted. However, they have played, in particular at the initial steps of our work, an important role for improving the general visibility of the field. This type of course is usually presented in detail in local media, allowing the appropriate messages of our capacity to provide the correct solutions for sustainable development to be sent. On the other hand, these courses and workshops need to be complemented by more specific courses oriented towards more restricted audiences, even if the general content is considered in many cases. Thus, for instance, different workshops and seminars have been developed with a clear orientation towards managers and professionals with responsibilities in companies with a direct or indirect connection with chemistry. Some illustrative examples of the first actions in this line are the following:

- *Waste Recycling* (Chamber of Commerce of Granada, Granada, 1997 and 1998)
- *Economy and Environment* (EOI, Seville, 1999)
- *Workshop in Green Chemistry* (University Institute of Science and Technology – IUCT, Mollet, Barcelona, 2003)
- *Is it Possible to have a Non-polluting Chemical Industry?* (Chamber of Commerce of Barcelona, Barcelona, 2004)
- *Is it Viable to have Non-polluting Chemistry?* (Chamber of Commerce of Tarragona, Tarragona, 2004)
- *Green Chemistry Thematic Worskshop* (SUSCHEM Spain, EXPOQUIMIA, Barcelona, 2005).

Alternatively, the same effort has been made to involve trades unions in educational activities in this field:

- *Workshop on Green Chemistry* (Department of Environment – CCOO-Aragón, Zaragoza, 2005)
- *Workshop on Green Chemistry* (FITEQA-CCOO, Sant Celoni, Barcelona, 2005)
- *The Challenge of Sustainability for Chemistry in Andalusia* (FITEQA-CCOO, Huelva, 2006)
- *Workshop on Green Chemistry* (FITEQA-CCOO, Valencia, 2006).

Some initiatives have been carried out involving both the Association of Spanish Chemical Industries (FEIQUE) and the two main trades unions of Spain (CCOO and UGT): the *Workshop on Sustainable chemistry* (FEIQUE, FIA-UGT, FITEQA-CCOO, Madrid, 2005) is an example.

In some instances, the cooperation with the Professional Associations of Chemists has also been fruitful in the development of courses seminars and workshops as those held in Valencia in 2006 (*Chemistry and Sustainable Development*) and 2007 (*Climate Change: a Scientific Debate*).

The very particular situation of many agricultural and agriculture-based companies requires considering separately activities such as the participation at the workshop for agricultural companies in the region of Girona: *Biorefineries as a Reality: Application to Specific Cases* (Girona, 2007).

In most cases, the former activities were focused to make available the most general and basic concepts of green and sustainable chemistry to owners, managers, technicians and workers involved in industrial activities related to chemistry. This, of course, needs to be complemented by other actions devoted to a more specific training in the field. In this regard, it is worth mentioning here the efforts carried out for more than 30 years by Dr F. Velázquez de Castro in Granada, but also in other locations in Andalusia, Madrid and other regions of Spain, for the training of technicians and specialists in different aspects of environmental management, in particular sustainable environmental management (Madrid, 2011 and 2012). A second important contribution is that of IUCT (Drs J. Castell and C. Estevez) a private company located in the heart of the area concentrating most of the fine chemicals and pharmaceutical industries in Spain. IUCT has been one of the pioneers in green chemistry in Spain and, besides, has defined training of technicians of this industry in different specific aspects, always considering the contribution of sustainability issues, as one of its main goals.

As mentioned above, teaching the teachers is, without a doubt, one of the fastest approaches to spread the ideas of sustainability and, in particular, sustainable chemistry. Again here we must mention the important contributions of Dr F. Velázquez de Castro with his important participation in courses *Experts in Environmental Education* (Madrid, 2006–2013; Oviedo, 2010; Granada, 2005–2013), and a strong involvement in the activities of the Centers for Teachers (CEPs) intended to implement the continuous training of high school teachers, and different institutes and departments of education. Other groups of the RSEDQS have also participated actively in the events organized by the CEPs, as is the case of the group of the University of Castellón (*Chemistry at the Microscale. Approaches to Sustainability*, Castellón, 2010). Besides, school and high school students have been approached through talks in their centres and through a continuous participation in science weeks and open days. It is clear that the concepts behind green chemistry, if properly selected, can easily capture the attention of young students. For this purpose, we have observed that a combination of concepts of green and environmental chemistry always provide the best outputs.

In the case of early stage university students, they have been approached through similar strategies. Interestingly, we have found that, in general, the presentation of the concepts and tools of green and sustainable chemistry by professors and researchers not belonging to the same university or centre than the students is the best option. Of course, university teachers and researchers in general also need to be educated in the principles of green chemistry. It is our experience that, most likely, this was one of the most difficult tasks at the beginning of the field. Besides personal and departmental discussions, the organization of talks, presentations, conferences and workshops at the different universities and research centres have played an important role to highlight the current importance in this, the field. A second level of activities has been the publication of scientific literature regarding this field. A strong and excellent activity has been carried out, in this regard, by Dr R. Mestres, from the University of Valencia and the former President of the RSEDQS, with the publication of a continuous series of works—articles and books—in Spanish, Catalan and English in journals of very broad scope, but also in specialized journals.[21-27] However, the capacity of the members of the RSEDQS to introduce green chemistry concepts in official chemistry curricula has been rather limited, although, at a personal level, different teachers have been able to introduce some specific elements in their classes, this area represents an important target to be achieved in the next future.

Educational cooperation in this field is another important question to be considered. The initiatives and activities of the RSEDQS have been presented and debated at different European universities in France (Dr P. Cintas, 2002) Poland (Dr R. Mestres, 2005) Portugal (Dr S.V. Luis, 2005) Slovakia (Dr P. Cintas, 2008) or Italy (Dr A. Alcántara, 2010, 2012). However, the most important cooperation actions in this regard have taken place with third countries in the two natural areas for Spain, corresponding to its geographical location (North-African countries) and its historical and linguistic location (Ibero-American countries). Thus, for instance, the groups at the University of Extremadura set up different initiatives and workshops to favour the gathering of Ibero-American researchers and teachers interested in the areas of sustainable chemistry. The group of biotransformations at the University Complutense have maintained a lasting and fruitful educational collaboration with different universities of Peru, in particular Trujillo, Tingo María and San Marcos in Lima, for more than one decade, but have also participated in academic activities related to green chemistry in Argentina (U.N. Quilmes, 2002 and 2012) and in Chile (U.C. Valparaiso, 2012). The cooperation with Peru (Piura and San Marcos) has also been continued in the case of Dr R. Mestre from the University of València. In the same way, the groups at the University of Zaragoza have maintained collaborations with Morocco (Marrakesh), Tunisia (U. El Manar) and Mexico (UNAM). Collaboration with Mexico has also been important for the group of the University of Castilla–La Mancha. In the case of the University Jaume I of Castellón, important collaborations have been established with Argelia (Sidi-Bel-Abbés), Morocco (Fès) and Peru (Trujillo) but particularly with Cuba (University of Oriente) where there have been contributions to the creation of a symposium on green and sustainable chemistry at their biannual Cuban Conference of Chemistry, and set up the bases for the creation of common curricula for education in sustainable chemistry. According to their sound tradition, extended collaborations, at very different levels, have been also maintained by the group involved with supercritical fluids at the University Complutense and the groups at the Institute of Chemical Technology (ITQ, CSIC–Technological University of Valencia) with a variety of Ibero-American countries.

There are three specific audiences we have not considered here until now. The first corresponds to postgraduate students. Education at this level has been one of the main goals of educational activities of the RSEDQS and this will be reflected in the next section. The second specific target is politicians. They are the responsible for decision-making at many levels. Accordingly, they need to understand the main basic concepts behind the words of environmental, green and sustainable chemistry. They need to understand that chemistry has the power to solve many of the challenges our societies are confronted with and be aware they need to get the proper advice and the proper balance between the benefits and risks that political decisions involving technological and scientific issues can provide. Sometimes politicians know they need this training: Very recently, EUCHEMS carried out several presentations in this field at the European Parliament under the demand of the own parliament. We have approached this target mainly at a local level, through contacts and discussions with political representatives at local governments, in particular in Cataluña and at the Valencian Community. Interestingly, in the context of cooperation activities, Dr Mestres gave a presentation on pollution by organic chemicals at the Commission of Environment and Ecology of the Congress of the Republic of Peru.

The third target is journalists. We must acknowledge that, up to now, we have not been very efficient in this regard and this is one on the pending actions of RSEDQS. We have been regularly in contact with journalists for presenting our activities in newspapers, local radio and so on. These individual contacts have been, in general, very positive but we are still lacking the definition of activities devoted to the training of journalist in issues related to the fields we are involved in. This need to be designed covering two different alternatives: scientific journalists on one hand but also journalists covering local and national general activities whose articles play often a key role in developing the public opinion.

As has been shown in this section, a group of different but complementary educational activities has been developed by the RSEDQS in the last few years in an attempt have broad coverage of possible audiences. There is still much work to be done in each of the fields mentioned as, besides, education is never a completed task. It needs to be renewed and restarted continuously. A graphical summary of those activities can be found in Figure 15.3.

15.3.2 The Spanish Inter-University Master and PhD Programmes in Sustainable Chemistry

As has been mentioned before, the setting up—for more than a decade—of the Spanish Inter-university Master and PhD Programmes in Sustainable Chemistry is one of the main achievements of the REDQS. The network has contributed to postgraduate education through the involvement in different Master's programmes in the areas of environmental chemistry or business and environment held at different academic institutions in Spain, showing

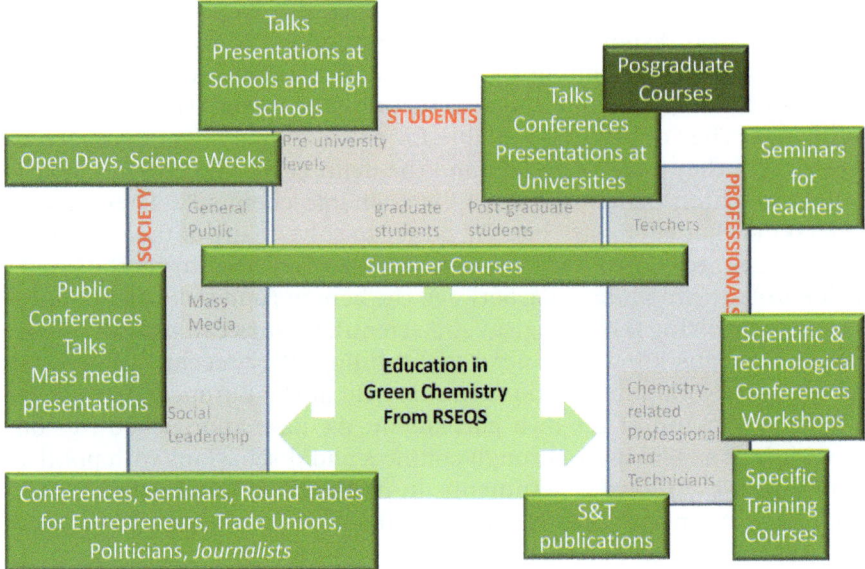

Figure 15.3   Main educational activities developed by the RSEDQS.

again the important role we have given to the integration of aspects of environmental chemistry into green chemistry studies. Additionally, several members of the REDQS were involved in the development of the degree Master in Sustainable Chemistry that was available for some years at the University of Zaragoza.

The design, development and, particularly, the maintenance in a truly operative status of an Inter-university Programme is not a simple task. Most likely it is even more difficult in the context of Spanish education. An inter-university programme in Spain needs to face a number of important challenges, as coordinating the administrative bodies of the different academic institutions involved is complex. Each university has its own philosophy and internal regulations and has a high degree of autonomy with regard to its academic decisions. Additionally, the legal framework can be different. Most educational issues have been transferred, in Spain, to the local governments (autonomic communities) and, thus, an inter-university degree needs to be approved simultaneously by the corresponding organism of several different local governments (often after the participation of their local evaluation agencies). After this is achieved, the degree needs the final approval of the central government through an evaluation and accreditation process assigned to a specific national agency (ANECA: National Agency for the Evaluation and Accreditation).[28] This multiple and endless process of evaluation and approval, associated with the changing legislation as we discuss below, requires complex and continuous administrative labour and it is difficult to constantly provide the programme with the updates required both in terms of legislative issues and in terms of educational needs.[29,30]

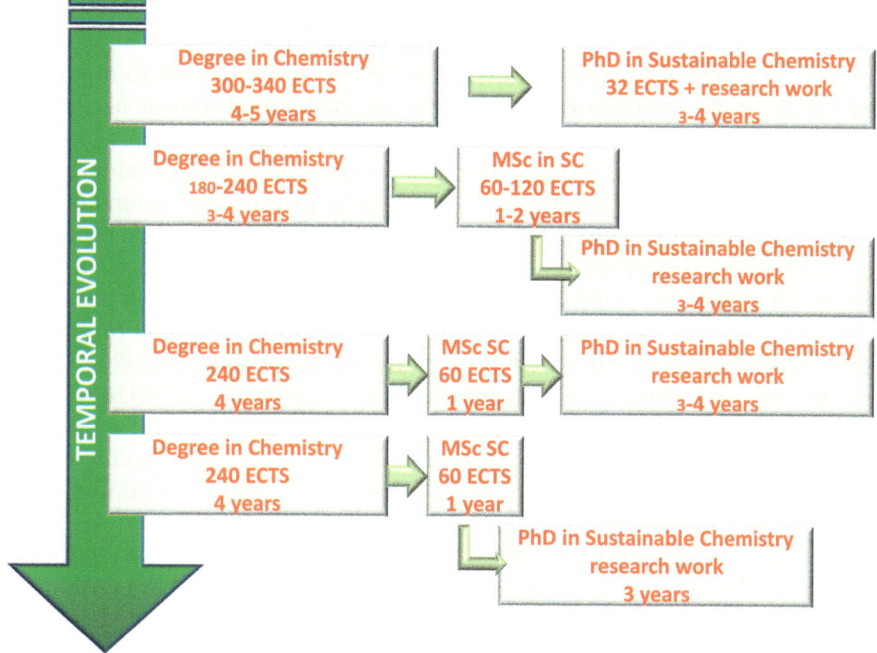

Figure 15.4   Historical evolution of postgraduate studies in Spain.

The second element of complexity has been the continuous changes in the applicable national educative legislation than then needs to be incorporated to the autonomic legislation. This has been made in the framework of the completion of the Bologna process.[31,32] This is clearly illustrated by the graph shown in Figure 15.4 displaying the variable and sometimes contradictory configurations for the studies of Master's and PhD degrees in Spain since the RSEDQS started with its efforts towards an integrated education, at the postgraduate level, in the field of sustainable chemistry.

According to the legislation, more than 10 years ago, our first efforts were directed towards the creation of the Spanish Inter-university Program in Sustainable chemistry. At this time the university education system was based on a degree with a variable length of 300–340 ECTS (European Credit Transfer System) distributed in 3–4 years, followed directly by PhD studies in which the student has to follow 32 ECTS in training and complete a research project.[33] Most often, the period required for achieving the PhD was 4 years. In general, the courses to complete 32 ECTS were followed by the students in the first year. It was in this context that the first Spanish Inter-university PhD programme in sustainable chemistry, which was one of the first to obtain the Quality Award by the Spanish Department of Education in 2003, was created. Although this scheme was simple, several important variations took place, in the next years, although most of them affected the grades. Thus, for instance, in the case of the University Jaume I of Castellón, we started the degree in chemistry with 340 ECTS, to accommodate about 40 ECTS of transversal subjects that were a specific feature of the newly created university, organized in four academic years. This situation changed soon to organize the courses in five academic years, which provided a total of nine academic years for those intending to obtain the PhD degree. In this academic organization, Master's studies, which played a secondary role, were considered as non-official studies and usually were handled by private institutions in the form of specialization courses. This situation changed in 2005 when trying to adapt the Spanish system to the European Higher Education Area (EHEA) in the so-called Bologna process.[34] At this time, a clear differentiation between graduate and postgraduate studies took place, with the introduction of the Master's studies as an intermediate step between graduate and PhD studies. The corresponding Spanish legislation offered a flexible model in which, for instance, degrees were defined ranging from 160 to 240 ECTS, while the Master's could be offered with an extent of 60–120 ECTS. After having completed a total of 300 ECTS, the student could enter the PhD programme (not needing, then, the MSc degree) for which only the research project was considered, although the standard extent continued being 4 years.[35] At the time, most Spanish universities selected the option of grades with 240 ECTS, distributed in 4 years. Interestingly, students could enter MSc studies after having followed most of the credits of their degree, but not needing to complete the degree in full. We adapted immediately to this change with the creation of the postgraduate programme involving Master's and PhD degrees in sustainable chemistry. Taking into account that most of the expected students for the initial editions of the MSc in Sustainable Chemistry would correspond to students still following the old scheme with degrees of 300–340 ECTS (5 years), we selected a mixed system for the Master's degree, with an initial stage of 60 ECTS, expected for these initial students, that should change to 90 ECTS when the students following this new scheme eventually arrived at MSc level. This regulation did not last very long, as in 2007 new legislation was approved for education at university level,[36] which was modified again in 2010.[37] The new system defined the Master's degree as a compulsory prerequisite for entering PhD studies. Although it was not legally defined in that way, there was strong pressure to keep the MSc studies in 60 ECTS during an academic year. During this period, the average extension of the PhD was still 4 years. Finally, a full legal redefinition of the PhD studies was dictated in 2011.[38] The complexity of this legal normative required a long period of adaptation so that universities could develop all the associated regulation allowing its proper development. In the case of the University Jaume I of Castellón, those regulations were completed in 2012.[39] This opened again the process of evaluation and accreditation of the degrees according to the new normative. Only the course for 2013–2014 started the process of registration of students in the new programmes: in the pioneering universities—as is the case of the University Jaume I of Castellón—in many cases with some steps of the accreditation process still pending. The main changes of the PhD system affected the length of the corresponding studies, which are now considered to take no longer than 3 years, except under special circumstances. A second important change is the introduction of the so-called complementary training activities. The exact definition and quantification of those activities was led to the universities, through the creation of a PhD school, and to each specific PhD degree. A final change was that, as in former regulations, the Master's degree was not compulsory for entering PhD studies. Thus, the student can now enter the PhD programme after having completed a validated Master's degree or after having completed 300 ECTS of university studies, with 60 of them corresponding to disciplines of one or several Master's programmes.

Unfortunately, we cannot be confident we are now working with a stable legislative framework. Thus, developing an educational inter-university postgraduate programme in an over-regulated and changing legal system is not a simple matter, in particular in a context, for the last years, of economic crisis with an important shortage of education funding. In spite of this the programme developed by the RSEDQS has been successful during this period and has survived, in a healthy way, all those vicissitudes. This has been possible because of a very sound design of the programme from the very beginning. The main features associated with this design are summarized in Figure 15.5.[40]

As mentioned before, the starting point of the programme was the proper definition of an integrated design for the training needs of students in green/sustainable chemistry at the postgraduate level. This initial design was independent of the legal framework, in particular the existence or not of the studies at Master's level. Instead, training at the highest possible level (PhD) was the essential goal, but defining, simultaneously, alternative intermediate exits for those not completing the PhD studies. These alternative exits could then take the form of specialization courses or, as was later on the case, Master's degrees. This was carried out after a period of open discussion and contributions from a large number of members of the RSEDQS,

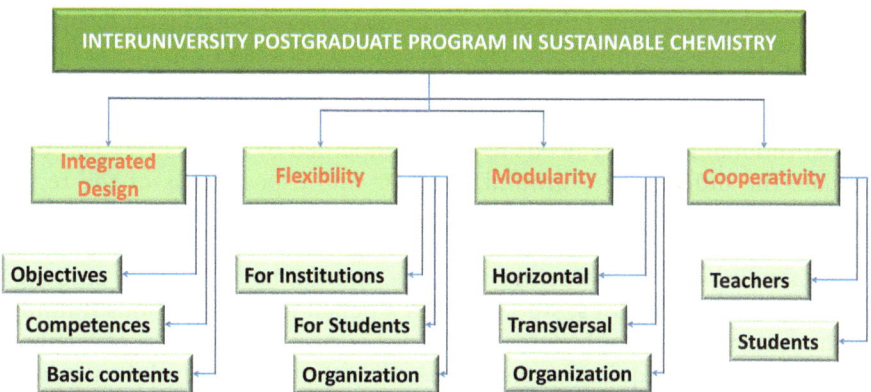

Figure 15.5 Main design features for the Inter-university Postgraduate Programme in Sustainable Chemistry.

leading to the final clear definition of the three basic elements of the programme: the training objectives, the expected competences and finally the basic contents that should conform the core training of the programme, independently of any contextual change or individual needs of students or institutions. Currently, the adoption of competences-based educational programmes is a hot topic still subject to much controversy, in particular because of the often diffuse definition and lack of appropriate quantitative measurements.[41] It is true that this term has been very often adopted without any criticism and without a proper understanding of its meaning and, consequently, over-expressed in many educational debates and programmes. However, if we consider for competence the generic description of mastering the ability to efficiently deal with complex situations characterized by 'ill-defined problems, contradictory information, informal collaboration and abstract, dynamic and highly integrated processes',[41] this can be very helpful in defining the essential goals of our design. In this regard, the Spanish Inter-university Postgraduate Programme intends to develop in our students the required competences to allow them to properly evaluate the benefits and drawbacks of any specific chemical process or the use and application of any specific chemical or material not only in terms of classical parameters like productivity, efficiency or cost, but also taking into account environmental aspects like the environmental footprint (including, for instance, the carbon and carbon dioxide footprint) and (eco)toxicological aspects. This includes the competences required to properly select a given chemical, material and chemical process from different competing alternatives under this complex context, exploring both short term and long term aspects.

The second element shown in Figure 15.5 is flexibility. The former core elements of the programme needed to be integrated and expressed in one (or several) programmes that could be flexible enough to accommodate the different needs expressed by changing legislations at the central government level, but also the variances in the local legislations developing the general normative issued by the different autonomic governments involved. Finally, it is necessary to consider the different regulation and working philosophies at each participating university. The combination of shared courses at one or several common sites with courses at each participating university favours the introduction of specific elements for each contributing centre. This level of institutional flexibility needs to be accompanied by the facility to accommodate the different background, contexts and expectations of the registered students. This is not always easy to define, from the beginning, in the legal documents describing the educational programme. In many cases the corresponding needs (part-time students, multi-lingual context, students with economic difficulties, students with a deficit in their previous training in chemistry, *etc.*) are only detected when the specific individual cases need to be solved. This requires a continuous adaptation of the legal documents, when revised, to incorporate those issues, but, most importantly, the clear willingness of searching for immediate solutions to each individual situation, without the borderlines of the core elements of the programme. Finally, the third flexibility element is the practical organization of the programme. Aspects such as teaching schedules, location of the classes, assigned teachers and so on need to be handled very carefully and with a high degree of flexibility in order to adapt to the changing conditions (number of students, origin, budgetary aspects...) that have to be expected for each academic course.

In part, these flexibility requirements are associated to a modular design of the courses and other training and educational activities. In the system developed, the concepts and subjects have been grouped in modules associated to the main training areas defined. Of course, these modules are very permeable, as some of the subjects contain elements that could allow their classification, at least partially, in several modules. As can be seen in Figure 15.6, seven different thematic modules have been considered. On the other hand, the gradation of these individual contents allows their classification in terms of basic and advanced subjects. To the first category correspond the subjects under the description of complementary training in chemistry and those that are marked as '(B)' in the other modules. In the case of advanced subjects, they have been further classified in terms of applied, professionally oriented ('A') and research-oriented ('R'). This last classification can look somehow arbitrary in some instances, but it is of interest to favour the organization of the full-curriculum and the decision-making of students.

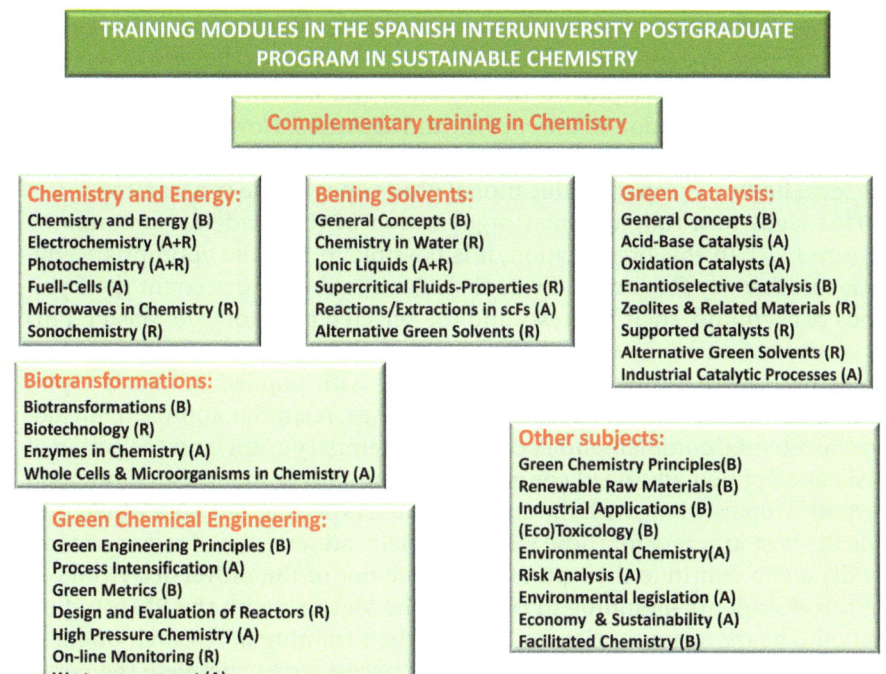

Figure 15.6   Thematic modules considered in the Spanish Inter-university Post-graduate Programme in Sustainable Chemistry.

The first modulus includes any content that could be required to achieve an appropriate homogenization of the background in chemistry of the students entering the programme. Although throughout this text we have, in general, simplified the origin of the students by considering the Postgraduate Programme in Sustainable Chemistry as an advanced step in the studies of chemistry, this is not always the case. Students registering in the Programme can have degrees in chemistry, pharmacy, chemical engineering and other. Moreover, students from different geographical origins can have clear differences in their backgrounds even if their original degree is in chemistry. The five main thematic modules include energy aspects, benign solvents, green catalysis, biotransformations and green chemical engineering. The interaction between energy and chemistry is analysed in a double way: the potential contribution of chemistry to new more efficient and sustainable energies, and the development of new, alternative sources of energy for efficiently carrying out chemical reactions. In a final modulus we have included a series of complementary concepts and subjects, some of the critical for an appropriate training in this field.

Those contents can be grouped or split, if appropriate, to generate the corresponding specific courses that give place to the yearly academic organization of the programme. In our initial design of the programme, this was expected to take place with the maximum level of flexibility, which would allow us to immediately adapt to the changes and to properly fulfil any detected need. The normative system, however, and unfortunately according to our experience, has been continuously evolving towards a very rigid administrative system that strongly difficult introducing changes in the academic programme. The process of evaluation and accreditation is based on a very detailed description of the courses and activities. However, we have tried to maintain the maximum level of modularity and flexibility, sometimes at the legal limit, to keep this educational philosophy in the programme.

This leads to a fully modular organization of the academic programme (Figure 15.7). In this organization, it is possible to find the appropriate educational itinerary for each individual student, taking into account their previous background, their interests and the selected point of exit: Master, PhD or specialization courses. Thus, for instance, the selection of some courses involving basic concepts and others dealing with applied advances topics should be appropriate for specialization courses. A similar approach, including the required complementary courses in chemistry and a larger selection of basic and applied advanced courses, and eventually some research-oriented advanced topics, along with the corresponding experimental work would provide the best itinerary for those students solely interested in obtaining a Master degree to continue their professional life out of the university. However, for those students intending to continue the Master's with the PhD studies, it should be more appropriate to broaden their training in research-oriented topics. Of course, in this case it should be necessary to complete the corresponding research period required for a PhD.

The last essential feature displayed in Figure 15.5 is cooperation. The success of a educational programme such as the one here discussed is linked to the cooperative efforts of the members of the RSEDQS, in particular those involved as teachers in the programme. Nevertheless, the role of the students must also be highlighted. The cooperative contribution of both teachers deeply involved and compromised with the programme and the students is, besides, significantly implemented as a consequence of the teaching periods at the common site(s). This approach has important advantages:

- All subjects are covered by high level specialists in the field.
- The programme is set up by a combination of different original ideas. Each participating centre contributes with its own expertise area and its own approach.
- Students from different universities, interests and backgrounds are forced to interact very closely.
- Researchers and teachers from different universities, interests and backgrounds interact between them and with the students, in particular during intensive sessions at common sites.
- Development of cooperation in experimental work and development of researches combining different subjects and expertise are favoured.
- Scale economy is gained: it is easier to reach a critical mass and the cost for individual universities is very much reduced.
- The involvement of high level experts from industry and at an international level is greatly facilitated.
- Combines a common core of training with some degree of specialization based on the expertise at the home institution.

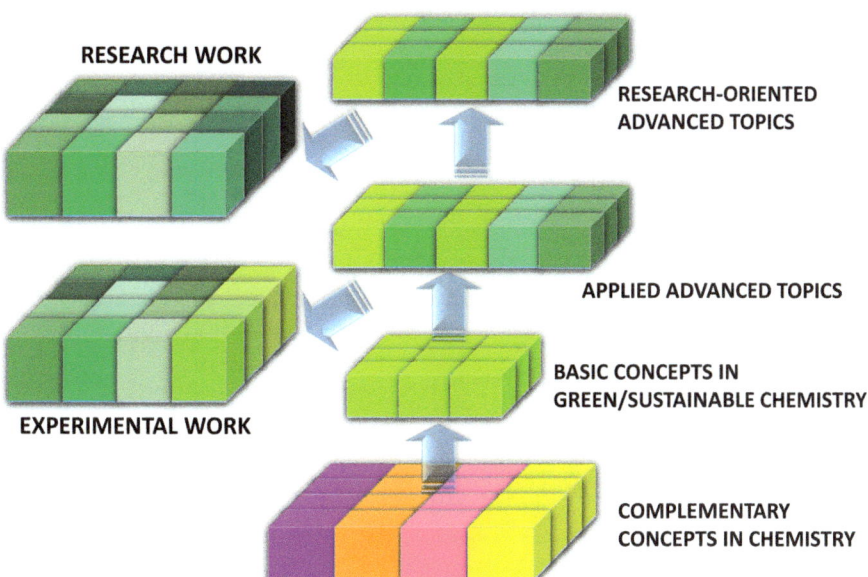

Figure 15.7 Modular design for the organization of the programme.

Overall, the practical organization of the Spanish Inter-university Programme in Sustainable Chemistry follows the general scheme displayed in Figure 15.8. As can be seen, this scheme is flexible enough to accommodate the different normative changes that were described above. The flexible combination of courses at the common sites with those at the home institutions and with research activities and additional training events (workshops, seminars, research exchanges and stays abroad, scientific and technological publications, participation in conferences, *etc.*) is compatible with a scheme in which the PhD degree is the main expected outcome (Figure 15.8) with a system combining a Master's and a PhD degree, but also with the simultaneous development of specialization courses of limited length.

In the current structure, the organization of the Master's studies is based in a programme of 60 ECTS, from which 18 are allocated to the corresponding

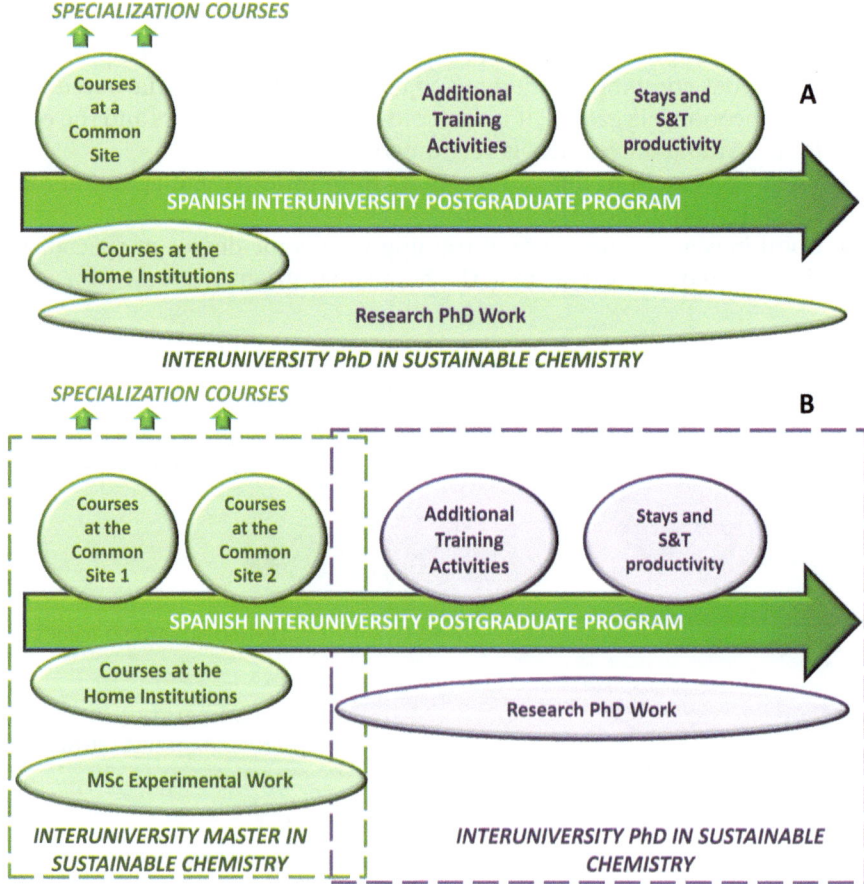

Figure 15.8   General organization of the Inter-university Postgraduate Programme. A: PhD as the main outcome. B: Considering both the degrees of Master and PhD.

experimental work. It is guaranteed that the yearly offer will always be higher than 78 ECTS in order to ensure that students have a minimum number of options. Students can register for any of the courses offered by the participant universities. The teaching periods at the common sites are organized in an intensive way with an extent of 3 weeks, every period involving six to seven courses of three ESCT each. Although the sessions at the common site(s) could allow the completion of the credits required (along with the experimental work), usually the students select one or several courses at their home institution (or at other participant institutions), which allows them to design their specific training profile. At least 18 ECTS must be followed at the common sessions, thus ensuring a sound training in the core subjects of the programme. Figure 15.9 displays the courses currently offered at the common sites. As an example, the specific courses that have been offered at the University Jaume I of Castellón include: *Biomimetic and Supramolecular Chemistry*, *Sustainable Engineering*, *Fine Chemistry*, and *Safety and Risk Analysis in Chemistry*. The participation of part-time students has also been considered. Essentially, in our experience, they belong to two different categories: those sharing the studies with regular work activities and those experiencing economic difficulties to complete the Master's degree in one academic year. In both cases the programme can be completed in two academic years and besides, when properly justified, some levels of on-line tutoring and self-study can be authorized. In the same way, the experimental work can be based on the regular work activities of the student, through the analysis and/or implementation of green elements in their work place. Evaluation of the courses involves small tests either during the classes or on-line and the study and critical analysis of selected cases. The evaluation of the experimental work requires its public defence, with an evaluation committee of three researchers or teachers participating in the Master's degree.

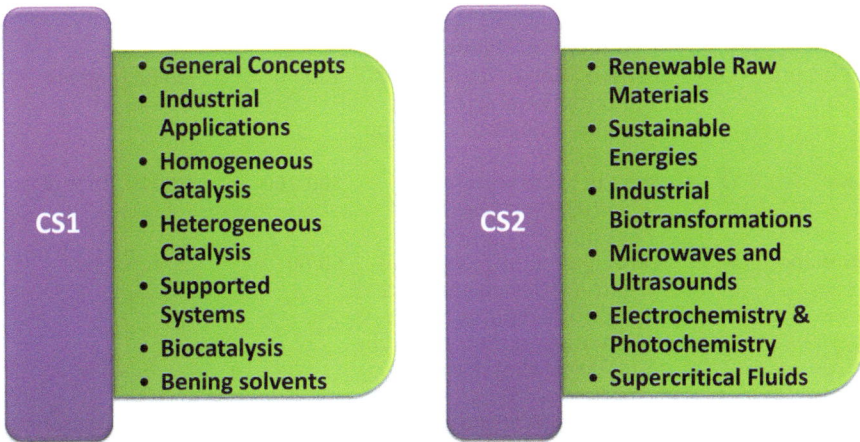

Figure 15.9 Courses offered at the common site(s) (CS) with an extent of three ECTS each.

In order to favour the introduction of complementary training activities and also in order to provide a smooth entrance into the new PhD system, in the last few years, the University Jaume I of Castellón has organized yearly a workshop in green chemistry, with the participation of external and international experts in one or two selected fields. It is intended that the experts and the subject vary each year. The Master's and PhD students are invited to actively participate in these workshops to complete their training during the 2–3 days of presentations and discussions. Of course, this workshop is of particular relevance for the students of the new PhD system, as this should be a core element regarding the additional training activities required to achieve their PhD degree. The regulation of the PhD establishes the need for a student to complete an equivalent to 600 hours of a series of different complementary activities, which exact nature was to be defined through the regulations of each university and for each individual PhD programme. In the case of the Spanish Inter-university PhD Program in Sustainable Chemistry, the definition of these activities is provided in Table 15.1. Seven different categories are considered, but only two of them are mandatory, the other being considered as optional.

The first one evaluates the contribution to training of the contribution of the student to publications and patents (and if appropriate other elements of S&T productivity) coming from their research experimental work. The

Table 15.1 Training activities considered for the PhD Spanish Inter-university Programme in sustainable chemistry.[a]

Activity	Type	Nature	Max/Min (h)	Other
Publications	M	Publications or patents	500/200	200 h/publication or patent
Courses	O	Workshops, summer courses or other courses related to sustainable chemistry. Transversal training	200/30	30 h/course
S&T events	O	Participation in national or international scientific events (congresses, symposia *etc.*)	150/50	50 h/congress
Stays	O	Stays in other research centres for research and training	500/160	480 h/3 months
Seminars	M	Participation in group seminars: (1) design and write scientific reports, (2) presentation of their research; (3) discussion of their research	60/60	20 h per year
Projects	O	Participation in competitively funded research projects	60/5	Full dedication: 30 h per year
Others	O	Other training activities	50	—

[a]M, mandatory; O, optional.

accreditation of these activities provides 200 hours-equivalent per publication or patent. Any student is expected to obtain at least one publication/patent during the PhD period, and the maximum of hours that can be accredited is 500. The second mandatory activity is the regular participation at the group seminars. In this case, the activity values the ability of the student to design and write scientific reports, to present their research in public and to be able to defend and discuss their research results. At least 20 hours per year must be devoted to these activities, providing the accreditation of 60 hours-equivalent at the end of the PhD period. Of course, the participation in courses, workshops and related activities, in particular those in green and sustainable chemistry or related fields, is an important aspect to complement the training of the students. In this context, the organization of a workshop on green chemistry every year, within the programme, whenever the budgetary factors allow it, plays a key role. Under this section, we also consider the courses to be introduced by the schools of doctorate at each university mainly focused on training in complementary skills (communication, writing, entrepreneurship, *etc.*). Only courses with enough entity are expected to be considered in this section and will be accredited, by the academic committee of the PhD studies, by 30 hours per course, with a maximum of 300 h allowed to be obtained from these activities. Participation in scientific and technological conferences and congresses is also considered an element of the training in the PhD programme and can be accredited by 50 hours per congress up to a maximum of 150 hours at the end of the PhD. Stays abroad have been considered traditionally an important element of the training of a PhD student, and this is acknowledged in the present system. The equivalence in hours is given here in terms of the period required to obtain the international award for the PhD (480 hours per 3 months). As it is considered that only sufficiently long stays provide a significant contribution to the student, only stays longer than 160 hours are accredited, up to a maximum of 500 hours. Finally, the participation as a researcher in competitively funded research projects, along with other alternative activities will also be evaluated by the academic committee that is the responsible for the accreditation of the different activities.

15.4   Lessons Learnt after a Decade

Educational activities in the field of green and sustainable chemistry can be very rewarding. However, they need to be continuously adapted and modified in order to properly reach the desired audiences. The presentations, talks, activities, *etc.* designed for a specific public cannot be used directly in a different context. This is one of the main elements defining the success of our efforts. In this regard, approaching the global training needs from a cooperative perspective, through the joint efforts of the RSEDQS members, has been demonstrated to be very appropriate. The different backgrounds and professional and personal skills of the RSDQS members provide a unique cluster of resources from which the best suited combination of activity and people in charge of its accomplishment can be always found.

A simple analysis of the data gathered regarding the different activities carried out provides some interesting trends. First of all, a serious attempt has been made to reach the different audiences considered. However, there are still some areas in which the impact has been very limited. We have mentioned before the lack of activities intended for the training of journalists and the limited access to people with political responsibilities. An even more important element is the limited access achieved to students at any level (school, high school or graduate students) and the difficulty with which green chemistry concepts can be incorporated in the context of environmental chemistry. This is an issue that needs clearly to be carefully analysed and implemented in the future years with the design of the appropriate strategies. A second and very important trend is the observed decrease in the number and intensity of the activities carried out in the last years. This is particularly visible when some major activities such as summer courses or specific training courses are analysed. There are two main reasons associated to this. One of them is indicative of the fact that green chemistry is not considered any more a novelty and that, as a matter of fact, green is a word that has been over-exploited very often. From this perspective, the participant members of the REDQS can feel less prone to organize this kind of events that, on the other hand, can be considered less attractive by the responsible institutions and organizations. This decrease in the individual activity of REDSQ members is natural, but this trend needs to be reversed. There is a need to find new ways in which green and sustainable chemistry issues can be presented in an attractive and efficient way and maintain a strong involvement of the people who, from the beginning, have considered these training tasks as an essential component of the network. The second element explaining the decrease in major activities is both simpler and more complex. During the last few years in Spain we have been suffering from a significant economic crisis. This has led, with an absence of logic and of vision of the future, to a sharp decrease in the budgets for education and research. Although an important effort has been made to reduce the costs for any individual activity, in particular in terms of the expenses associated with the members of the REDS, there is always a need for a minimum budget to carry out many of the programmed activities.

When considering the Spanish Inter-university Programme in Sustainable Chemistry, its development over this decade has been a clear success. About 250 students from more than 10 different nationalities have attended the programme. Almost 30% the students came from abroad, mainly from Latin America, but also from North Africa and Europe. Some of the former students have now obtained permanent positions at different Latin American universities, where they continue to transfer the acquired knowledge in the field of green chemistry to new generations of students. Many others continue their careers in Europe where they are developing their research careers in Spain and other countries through completion of PhD or postdoctoral studies. Some others have obtained positions at local or international chemical companies.

The main features and advantages of the programme have been highlighted above. However, it is also important to be aware of its limitations, difficulties and drawbacks from a practical point of view. Some of them can be summarized as follows:

- *An important effort of mobility for students and teachers is required.* This is directly reflected in the need of an applicable budget to appropriately cover the resulting expenses, in terms of a direct budget and/or in terms of student grants.
- *Management aspects become more complex.* The participation of different universities in the programme multiplies the administrative processes (registration, assessment records, *etc.*).
- *Legal aspects need to be considered very carefully.* This includes the generation of the corresponding agreements between the participating institutions and an academic organization of the programme compatible with the different individual regulations at each centre and with the legislation provided by each autonomic government.
- *Participation of working students is not easy*, considering the presence of intensive sessions and the existence of courses at different geographical locations.
- *A strong and continuous coordination effort is required.* This is needed to fully integrate the management aspects but also to properly accommodate the individual needs generated by teachers and students from different origins, expectative and backgrounds.
- *Active participation of all the contributors to the programme is essential.* This is fundamental not only to guarantee an efficient educational system every year, but also to identify, well in advance if possible, additional needs and new improvements to be made.
- *A permanent improvement of conceptual and practical aspects of the programme is needed.* This is most likely a common aspect to most educational programmes, but it is made more complex in the multi-faceted Spanish Inter-university Master and PhD Programmes in Sustainable Chemistry analysed here.

As discussed above, the process of implementation of the programme has been significantly affected by the continuous legislative changes affecting the university system in Spain and, particularly, education at postgraduate level. This has overlapped with the application of the Bologna Process in Spain. In our experience no clear benefits have been obtained after all this confusing period in terms of an enhancement of the educational system. In general, all those changes and the accompanying evaluation and accreditation processes have been carried out in a very formal and purely administrative way, without an actual analysis of the resulting educational outputs. The emphasis has been put into the administrative aspects but not on the quality of the training processes themselves. The transmission of knowledge, skills (and competences) is not any more the centre of the educational system as this has been shifted towards the administrative processes behind the so-called 'quality systems' driven by several different administrative bodies and a variety of alternative and overlapped instances at each university. The whole system is now burdened with the need of completing intractable amounts of administrative forms. Teachers as well as people responsible of the academic and educational processes need to devote more time to those administrative tasks, in many cases, than to the improvement of the actual educational elements. In spite of this, the Spanish Inter-university Programme in Sustainable Chemistry has healthily survived during this period. This has been possible for its high level of flexibility and its capacity to adapt to changing situations, as formerly described. Besides, the strong commitment of the RSEDQS members with the programme has allowed running the programme with the more flexible and broadest interpretation of the normative.

The above process has been accompanied, besides, as mentioned, with a dramatic shortage of the financial resources assigned to education, and particularly to postgraduate training. This aspect is becoming dramatic year after year. This should never be the way to confront an economic crisis, as this not only becomes an additional element of socio-economic crisis by itself, but significantly compromises the potential of our society to be able to efficiently tackle the crisis. Although members of the REDQS participate in the programme on a voluntary basis and do not require any stipend, there are a number of mobility expenses that require to be covered. They can be slightly reduced through small continuous changes in the organization of the programme, but still there is a minimum budget required for these purposes. In the case of the students, the high increase in registration fees for Master's students that has taken place in the last years has been accompanied by a similar significant decrease in the number of registered students, which immediately affects the budget available for the programme. This decrease in students has affected both to the registration of Spanish students and to the registration of Latin American students.

15.5 Future Perspectives

The participation of the RSEDQS in a broad series of educational initiatives has not only demonstrated a very positive contribution to education in green chemistry at the different levels considered, but also a need whose continuity and consolidation is required, fulfilling the new demands of our society and the gaps detected from previous activities. The Spanish Inter-university Postgraduate Programme in Green Chemistry has been, no doubt, one of the main contributions of the RSEDQS in this regard. However, the continuity of those educational initiatives and in particular the postgraduate programme, is currently confronting important challenges. First of all, it is necessary to convince academics and politicians in charge of educational issues of the need for education in green chemistry. This is an essential activity as it is the only way to guarantee the possibility of technological sustainable development and also to reverse the current trend in which most societies are strongly critical of technological advances. Additionally, young generations are tending to abandon science and technology, which could seriously compromise their future development. For this, the appropriate funding is required, even understanding that in the current context the maximum efforts are to be made to reduce the associated expenses. It is not possible to achieve educational objectives without having the corresponding financial backing. On the other hand, nothing is cheaper for a society than education. No other activity produces greater rewards and improvements in its future perspectives.

The Spanish Inter-university Programme in Sustainable Chemistry would not be able to survive in a scenario of continuously decreasing educational budgets and particularly if there were mobility issues of teachers and students. On the other hand, it is necessary to find the way to again focus education onto its essential objectives, and counterbalance the absurd predominance of formal and administrative issues that, in the end, do not provide any improvement in the transmission of knowledge, skills and competences. Taking advantage of our former experiences, different initiatives have been launched in order to simplify as much as possible some of those administrative issues. Besides, the programme needs to be able to attract the best students from the different universities involved, but also from other Spanish universities and from abroad. For this purpose, a very important communication effort is essential and this has to be achieved in the very next future. Some of the aspects of the programme, in particular the original PhD programme, have been pioneers at an international level in training in green chemistry. Some of them have inspired other related initiatives at different countries. It is essential to maintain this pioneering spirit and to continue developing new initiatives, approaches and strategies, but, at the same time it is important to take advantage of other initiatives in this field to establish the corresponding links and cooperation activities. Although some exchanges and interactions have been established with different European (UK, Portugal, Poland, Germany and Finland) and non-European countries, the internationalization of the programme is an important aspect to be executed in the future.

Acknowledgements

As mentioned in different sections, the educational initiatives carried out by the RSEDQS would not have been possible without the help and cooperation of many people and institutions and their contribution is warmly acknowledged. Special mention is deserved by the members of the RSEDQS and particularly the teachers contributing to the postgraduate programme. Some of them or their groups have been mentioned in this chapter, but many others have provided an essential contribution to this initiative. They are responsible for its success.

References

1. R. Carson, *Silent Spring*, Houghton Mifflin Harcourt, New York, 1st Mariner Books Edition, 2002, First published by Houghton Mifflin in 1962.
2. *The Chemical Element. Chemistry's Contribution to Our Global Future*, ed. J. Garcia-Martinez and E. Serrano-Torregrosa, Wiley-VCH, Mannheim, 2011.
3. *Green Chemistry, Designing Chemistry for the Environment*, ed. P. T. Anastas and T. C. Williamson, ACS, Washington DC., 1996, vol. 626, ACS symposium series.
4. P. T. Anastas and J. C. Warner, *Green Chemistry: Theory and Practice*, Oxford University Press, Oxford 1998.
5. P. T. Anastas and M. M. Kirchhoff, *Acc. Chem. Res.*, 2002, **35**, 686.
6. M. Poliakoff, J. M. Fitzpatrick, T. R. Farren and P. T. Anastas, *Science*, 2002, **297**, 807.
7. R. Mestres, *Environ. Sci. Pollut. Res.*, 2005, **12**, 128.
8. I. Horváth and P. T. Anastas, *Chem. Rev.*, 2007, **107**, 2169.
9. P. T. Anastas and N. Eghbali, *Chem. Soc. Rev.*, 2010, **39**, 301.
10. N. Winterton, *Green Chem.*, 2001, **3**, 73.
11. P. T. Anastas, *Chem. Eng. News*, 2011, **89**, 62.
12. J. Andraos, *Pure Appl. Chem.*, 2011, **83**, 1361.
13. *Green organic chemistry in lecture and laboratory*, ed. A. P. Dicks, CRC Press, Boca Raton, FL, 2012.
14. P. J. Dunn, *Chem. Soc. Rev.*, 2012, **41**, 1452.
15. http://redqs.s43.eatj.com/redqs/.
16. Some of the leading journals in the field include *Green Chemistry*, the pioneering journal edited since 1999 by the RSC (http://www.rsc.org/publishing/journals/gc/about.asp) *ACS Sustainable Chemistry & Engineering* from the ACS (http://pubs.acs.org/journal/ascecg) *ChemSusChem* from Wiley (http://onlinelibrary.wiley.com/journal/10.1002/(ISSN)1864-564X) *Green Chemistry Letters and Reviews* from Taylor and Francis (http://www.tandfonline.com/toc/tgcl20/current#.UvIByE2PKUl) or *Current Green Chemistry* from Bentham (http://benthamscience.com/journal/index.php?journalID=cgc).
17. http://ec.europa.eu/programmes/horizon2020/en.
18. http://www.iuct.net/.
19. https://www.acs.org/content/acs/en/meetings/greenchemistryconferences.html.
20. http://www.congresscentral.com.br/sbq/ufscar/icgc4/.
21. R. Mestres, *Anal. Quím.*, 2003, **99**, 58; (in Spanish).
22. R. Mestres, *Rev. R. Acad. Farm. Cataluña*, 2003, **24**, 19 (in Catalan).
23. R. Mestres, *Green Chem.*, 2004, **6**, 583.
24. R. Mestres, *Química Sostenible*, Ed. Síntesis, Madrid, 2011, (in Spanish).
25. R. Mestres, *Quím. Ind.*, 2008, **576**, 30 (in Spanish).
26. R. Mestres, in. *About the nature and meaning of Green Chemistry*, ed., B. Nelson, Oxford University Press, in press.
27. R. Mestres, *Educ. Quim.*, 2013, **24**, 103; (in Spanish).

28. http://www.aneca.es/eng.
29. B. Altava, M. I. Burguete and S. V. Luis, *Educ. Quím.*, 2013, **24**, 132 (in Spanish).
30. B. Altava, M. I. Burguete, E. García-Verdugo and S. V. Luis, *SmartQuimic*, 2014, **2**, 68–69 (in Spanish).
31. *Recognition in the Bologna Process: policy development and the road to good practice*, ed.S. Bergan and A. Rauhvargers, Council of Europe Higher Education Series, No 4, 2006.
32. *For a complete information on the Bologna process*: http://www.coe.int/t/dg4/highereducation/EHEA2010/BolognaPedestrians_en.asp#P132_13851.
33. *For the applicable legislation, see*: http://www.boe.es/boe/dias/1998/05/01/pdfs/A14688-14696.pdf.
34. http://www.ehea.info/.
35. http://www.boe.es/boe/dias/2005/01/25/pdfs/A02846-02851.pdf.
36. http://www.boe.es/boe/dias/2007/10/30/pdfs/A44037-44048.pdf.
37. http://www.boe.es/boe/dias/2010/07/03/pdfs/BOE-A-2010-10542.pdf.
38. http://www.boe.es/boe/dias/2011/02/10/pdfs/BOE-A-2011-2541.pdf.
39. *Regulation of the PhD studies at the University Jaume I (approved January 26th, 2012)*, according to the new legislation (ref. 36): http://www.uji.es/bin/infoest/estudis/doctorat/norma.pdf; for the corresponding regulation of the Master studies (approved June 25th, 2012): http://www.uji.es/bin/uji/norm/est/nage1213-e.pdf.
40. For additional information the following web pages are of interest (most in Spanish and/or Catalan, some with English versions): http://www.miqs.uji.es/; http://www.uji.es/ES/infoest/estudis/postgrau/oficial/e@/22891/?pTitulacionId=42108; http://e-ujier.uji.es/pls/www/!gri_ass.lleu_ficha_d?p_titulacion=14025; http://www.quimicasostenible.uji.es/index.php.
41. See for instance W. Westera, *J. Curric. Stud.*, 2001, **33**, 75.

Subject Index